EMERGING
POSTHARVEST TREATMENT
OF FRUITS AND VEGETABLES

Postharvest Biology and Technology

EMERGING POSTHARVEST TREATMENT OF FRUITS AND VEGETABLES

Edited by
Kalyan Barman, PhD
Swati Sharma, PhD
Mohammed Wasim Siddiqui, PhD

AAP | APPLE ACADEMIC PRESS

Apple Academic Press Inc.   Apple Academic Press Inc.
3333 Mistwell Crescent   9 Spinnaker Way
Oakville, ON L6L 0A2 Canada   Waretown, NJ 08758 USA

Exclusive worldwide distribution by CRC Press, a member of Taylor & Francis Group

ISBN 13: 978-1-77463-399-1 (pbk)
ISBN 13: 978-1-77188-700-7 (hbk)

Library and Archives Canada Cataloguing in Publication

Emerging postharvest treatment of fruits and vegetables / edited by Kalyan Barman, PhD, Swati Sharma, PhD, Mohammed Wasim Siddiqui, PhD.

(Postharvest biology and technology book series)
Includes bibliographical references and index.
Issued in print and electronic formats.
ISBN 978-1-77188-700-7 (hardcover).--ISBN 978-1-351-04631-2 (PDF)

1. Fruit--Postharvest technology. 2. Vegetables--Postharvest technology. I. Siddiqui, Mohammed Wasim, editor II. Barman, Kalyan, editor III. Series: Postharvest biology and technology book series

SB130.E44 2018   634   C2018-903729-6   C2018-903730-X

Library of Congress Cataloging-in-Publication Data

Names: Barman, Kalyan, editor.
Title: Emerging postharvest treatment of fruits and vegetables / editors: Kalyan Barman, Swati Sharma, Mohammed Wasim Siddiqui.
Description: Waretown, NJ : Apple Academic Press, 2019. | Includes bibliographical references and index.
Identifiers: LCCN 2018030073 (print) | LCCN 2018034676 (ebook) | ISBN 9781351046312 (ebook) | ISBN 9781771887007 (hardcover : alk. paper)
Subjects: LCSH: Horticultural crops--Postharvest technology.
Classification: LCC SB319.7 (ebook) | LCC SB319.7 .E44 2019 (print) | DDC 635--dc23
LC record available at https://lccn.loc.gov/2018030073

ABOUT THE EDITORS

Kalyan Barman, PhD

Kalyan Barman, PhD, is Assistant Professor in the Department of Horticulture, Institute of Agricultural Sciences, Banaras Hindu University, Varanasi, Uttar Pradesh, India. Formerly, he served as Assistant Professor in the Department of Postharvest Technology, K.R.C. College of Horticulture, University of Horticultural Sciences, Bagalkot, Karnataka, and

Department of Horticulture (Fruit & Fruit Technology), Bihar Agricultural University, Sabour, Bihar, India. He is the author or co-author of 23 peer-reviewed research articles published in various international and national journals, 12 popular articles, and 14 book chapters. He is an active reviewer of 25 international journals, such as *Postharvest Biology and Technology* (Elsevier), *Scientia Horticulturae* (Elsevier), *Food Chemistry* (Elsevier), *Acta Physiologia Plantarum* (Springer), *Food Science and Nutrition* (Wiley), *Journal of Plant Growth Regulation* (Springer), *International Journal of Food Properties* (Taylor & Francis Online), and many more. He has been invited as a guest editor for special issue of *Journal of Food Quality* (Wiley Hindawi). He has been guiding 7 students for MSc and PhD programs. He is a dynamic teacher, researcher, and guide in the area of postharvest technology of horticultural crops.

He acquired his BSc degree (Horticulture) from Uttar Banga Krishi Vishwavidyalaya, Cooch Behar, West Bengal, India and completed both his MSc and PhD from Indian Agricultural Research Institute, New Delhi, India, specializing in postharvest technology of horticultural crops. He was awarded ICAR Junior Research Fellowship (JRF) and INSPIRE fellowship from the Department of Science and Technology, Government of India for pursuing his MSc and PhD, respectively. He has worked on alleviating chilling injury of mango and pomegranate during their low-temperature storage, reducing pericarp browning of litchi and reducing sapburn in mango. He has also identified a new physiological disorder in mango named 'Stem-end blackening' and associated with development of one variety each of litchi and bael.

Swati Sharma, PhD

Swati Sharma, PhD, is currently working as a Scientist at the ICAR-Indian Institute of Vegetable Research, Varanasi, Uttar Pradesh, India. Her research work has been published in various international and national journals, and she has been author or co-author of more than 17 peer-reviewed research articles, 12 popular articles, 20 conference papers, a technical manual, and five book chapters. She is also an active reviewer for several national and international research journals. She has worked on maintaining the fruit quality and enhancing marketability of plums using treatments, such as 1-methylcyclopropene, salicylic acid and nitric oxide. She has also worked on modified atmosphere packaging and ethylene absorbents packaging of plum.

Dr. Sharma completed her BSc (Agriculture) degree at the Institute of Agricultural Sciences, Banaras Hindu University, Varanasi, Uttar Pradesh, India. She completed both her MSc and PhD degrees at the Indian Agricultural Research Institute, New Delhi, India, specializing in postharvest technology of horticultural crops. She has been recipient of ICAR Junior Research Fellowship (JRF) and INSPIRE Fellowship by the Department of Science and Technology, Government of India. She has also qualified for the ICAR Senior Research Fellowship (SRF), acquiring first rank in the horticulture discipline and got second All-India rank in the ICAR Agricultural Research Scientist (ARS) Examination in Fruit Science discipline.

Mohammed Wasim Siddiqui, PhD

Mohammed Wasim Siddiqui, PhD, is an Assistant Professor and Scientist in the Department of Food Science and Post-Harvest Technology, Bihar Agricultural University, Sabour, India. His contribution as an author and editor in the field of postharvest biotechnology has been well recognized. He is an author or co-author of many peer-reviewed research articles, over 30 book chapters, two manuals, and more conference papers. He has over 20 edited books and an authored book to his credit. Dr. Siddiqui has established an international peer-reviewed journal, the *Journal of Postharvest Technology*. Dr. Siddiqui is a Senior Acquisitions Editor in Apple Academic Press for Horticultural Science and is editor of the book series "Postharvest

Biology and Technology." He has been serving as an editorial board member and active reviewer of several international journals, such as *PLoS ONE*, (PLOS), *LWT- Food Science and Technology* (Elsevier), *Food Science and Nutrition* (Wiley), *Acta Physiologiae Plantarum* (Springer), *Journal of Food Science and Technology* (Springer), *Indian Journal of Agricultural Science* (ICAR), etc.

Recently, Dr. Siddiqui was conferred with several awards, including the Best Citizen of India Award (2016); Bharat Jyoti Award (2016); Best Young Researcher Award (2015) by GRABS Educational Trust, Chennai, India; and the Young Scientist Award (2015) by the Venus International Foundation, Chennai, India. He was also a recipient of the Young Achiever Award (2014) for the outstanding research work by the Society for Advancement of Human and Nature (SADHNA), Nauni, Himachal Pradesh, India, where he is an Honorary Board Member and Life Time Author. He has been an active member of the organizing committees of several national and international seminars, conferences, and summits. He is a key member in establishing the World Food Preservation Center (WFPC), LLC, USA. Presently, he is an active associate and supporter of WFPC, LLC, USA. Considering his outstanding contribution in science and technology, his biography has been published in *Asia Pacific Who's Who* and *The Honored Best Citizens of India.*

CONTENTS

LIST OF CONTRIBUTORS

Ram Asrey
Division of Food Science & Post Harvest Technology, ICAR-Indian Agricultural Research Institute, New Delhi, India

Kalyan Barman
Department of Horticulture, Institute of Agricultural Sciences, Banaras Hindu University, Varanasi, India

Salvador Castillo
Department of Food Technology, Miguel Hernández University, Ctra Beniel km 3.2, 03312, Orihuela, Alicante, Spain

Changbao Chen
Laboratory of Chemical Biology, College of Chemistry and Material Science, Shandong Agricultural University, Taian 271018, Shandong, China

M. B. Darshan
ICAR-AICRP on PHET, University of Agricultural Sciences, GKVK, Bengaluru, Karnataka, India

J. E. Dávila-Aviña
Universidad Autónoma de Nuevo León, Facultad de Ciencias Biológicas, Apdo. Postal 124-F, Ciudad Universitaria, San Nicolás de los Garza, Nuevo León 66451, México

S. García
Universidad Autónoma de Nuevo León, Facultad de Ciencias Biológicas, Apdo. Postal 124-F, Ciudad Universitaria, San Nicolás de los Garza, Nuevo León 66451, México

L. E. Garcia-Amezquita
Centro de Investigación en Alimentación y Desarrollo, Coordinación de Fisiología y Tecnología de Alimentos de la Zona Templada, Av. Río Conchos S/N Parque Industrial, C.P. 31570 Cd. Cuauhtémoc, Chihuahua, México

A. García Heredia
Universidad Autónoma de Nuevo León, Facultad de Ciencias Biológicas, Apdo. Postal 124-F, Ciudad Universitaria, San Nicolás de los Garza, Nuevo León 66451, México
Current address: University of Massachusetts Amherst, Amherst, MA 01003, USA

Fabián Guillén
Department of Food Technology, Miguel Hernández University, Ctra Beniel km 3.2, 03312, Orihuela, Alicante, Spain

Alok Kumar Gupta
ICAR-National Research Centre on Litchi, Muzaffarpur, Bihar, India

N. Heredia
Universidad Autónoma de Nuevo León, Facultad de Ciencias Biológicas, Apdo. Postal 124-F, Ciudad Universitaria, San Nicolás de los Garza, Nuevo León 66451, México

Sunil Kumar
Department of Horticulture, North-Eastern Hill University, Tura, Meghalaya, India

Susan Lurie
Department of Postharvest Science, Agricultural Research Organization, Volcani Center, Bet Dagan, Israel

B. V. C. Mahajan
Punjab Horticultural Postharvest Technology Centre, Punjab Agricultural University, Ludhiana, Punjab, India

Evening Stone Marboh
ICAR-National Research Centre on Litchi, Muzaffarpur, Bihar, India

Alejandra Martínez-Esplá
Department of Food Technology, Miguel Hernández University, Ctra Beniel km 3.2, 03312, Orihuela, Alicante, Spain

Domingo Martínez-Romero
Department of Food Technology, Miguel Hernández University, Ctra Beniel km 3.2, 03312, Orihuela, Alicante, Spain

Vishal Nath
ICAR-National Research Centre on Litchi, Mushahari, Muzaffarpur, Bihar, India

Serdar Öztekin
Farm Machinery and Technologies Engineering Department, Agriculture Faculty, Cukurova University, Adana, Turkey

Alemwati Pongener
ICAR-National Research Centre on Litchi, Muzaffarpur, Bihar, India

Ovais Shafiq Qadri
Department of Bioengineering, Integral University, Lucknow, Uttar Pradesh, India
Department of Postharvest Engineering and Technology, Aligarh Muslim University, Aligarh, Uttar Pradesh, India

S. Vijay Rakesh Reddy
ICAR-Central Institute of Arid Horticulture, Beechwal, Bikaner, Rajasthan, India

María Serrano
Department of Applied Biology, Miguel Hernández University, Ctra Beniel km 3.2, 03312, Orihuela, Alicante, Spain

R. R. Sharma
Division of Food Science & Postharvest Technology, ICAR-Indian Agricultural Research Institute, New Delhi 110012, India

Swati Sharma
Division of Crop Production, ICAR-Indian Institute of Vegetable Research, Varanasi, Uttar Pradesh, India

Anil Kumar Singh
Department of Horticulture, Institute of Agricultural Sciences, Banaras Hindu University, Varanasi, Uttar Pradesh, India

Sanjay Kumar Singh
ICAR-National Research Centre on Litchi, Mushahari, Muzaffarpur, Bihar, India

Abhaya Kumar Srivastava
Department of Postharvest Engineering and Technology, Aligarh Muslim University, Aligarh, Uttar Pradesh, India

Gaurav Srivastava
Department of Bioengineering, Integral University, Lucknow, Uttar Pradesh, India

Daniel Valero
Department of Food Technology, Miguel Hernández University, Ctra Beniel km 3.2, 03312, Orihuela, Alicante, Spain

Juan M. Valverde
Department of Food Technology, Miguel Hernández University, Ctra Beniel km 3.2, 03312, Orihuela, Alicante, Spain

Vijay Yadav Tokala
Curtin Horticulture Research Laboratory, School of Molecular and Life Sciences, Curtin University, Perth, WA, Australia

Aabon W. Yanthan
ICAR Research Complex for North Eastern Hill Region, Nagaland Centre, Jharnapani, Nagaland, India

Kaiser Younis
Department of Bioengineering, Integral University, Lucknow, Uttar Pradesh, India

Pedro J. Zapata
Department of Food Technology, Miguel Hernández University, Ctra Beniel km 3.2, 03312, Orihuela, Alicante, Spain

Lili Zhang
Laboratory of Chemical Biology, College of Chemistry and Material Science, Shandong Agricultural University, Taian 271018, Shandong, China

Jie Zhou
Laboratory of Chemical Biology, College of Chemistry and Material Science, Shandong Agricultural University, Taian 271018, Shandong, China

Shuhua Zhu
Laboratory of Chemical Biology, College of Chemistry and Material Science, Shandong Agricultural University, Taian 271018, Shandong, China

C. Zoellner
Department of Food Science, Cornell University, 306 Stocking Hall, Ithaca, NY 14850, USA

LIST OF ABBREVIATIONS

1-MCP	1-methylcyclopropene
6-BAP	6-benzylaminopurine
ACC	1-aminocyclopropane-1-carboxylic
ACO	ACC oxidase
ACS	ACC synthase
AEW	alkaline electrolyzed water
AOX	alternative oxidase
APX	ascorbate peroxidase
ASC	ascorbate
ATP	adenosine triphosphate
BI	browning index
BRs	brassinosteroids
BS	black splendor
CA	controlled atmosphere
CAT	catalase
CCO	cytochrome C oxidase
CHI	chitinase
CI	chilling injury
CRM	confocal Raman microspectrometry
CYS	cysteine
DETANO	diethylenetriamine/nitric oxide
DHAR	dehydroascorbate reductase
EOs	essential oils
ETDA	ethylenediamine tetraacetic acid
ETR	ethylene receptors
EW	electrolyzed water
FAD	fatty acid desaturase
FDA	Food and Drug Administration
FT	Fourier transform
FW	fresh weight
GABA	γ-aminobutyric acid
GLU	β-1,3-glucanase
GPAT	glycerol-3-phosphate acyltransferase
GPX	glutathione peroxidase

GR	glutathione reductase
GRAS	generally recognized as safe
GSH	glutathione
GSSG	oxidized glutathione
GST	glutathione-S-transferase
HAT	hot air treatment
HPP	high-pressure processing
HSPs	heat shock proteins
HWB	hot-water brushing
HWRB	hot water rinsing and brushing
HWT	hot water treatment
JA	jasmonic acid
LOX	lipoxygenase
MA	modified atmosphere
MAP	modified atmosphere packaging
MC	methylcellulose
MDHAR	monodehydroascorbate reductase
MeJA	methyl jasmonate
MMT	montmorillonite
MNV	murine norovirus
NAI	negative air ions
NEW	neutral electrolyzed water
NO	nitric oxide
ODC	ornithine decarboxylase
OP	oxygen permeability
ORP	oxidation–reduction potential
OTR	oxygen transmission rate
PAL	phenylalanine ammonia lyase
PAs	polyamines
PBN	N-$tert$-butyl-a-phenylnitrone
PDH	proline dehydrogenase
PDJ	n-propyl dihydrojasmonate
PG	polygalacturonase
pI	isoelectric point
PO	oxygen permeance
POD	peroxidase
PPO	polyphenol oxidase
PR	pathogenesis related
RF	radiofrequency

RH	relative humidity
ROS	reactive oxygen species
SA	salicylic acid
SAM	*S*-adenosyl methionine
SAR	systemic acquired resistance
SDH	succinic dehydrogenase
SERS	surface-enhanced Raman scattering
SiOx	silicon oxide
SNAP	*S*-nitroso-*N*-acetylpenicillamine
SNP	sodium nitroprusside
SOD	superoxide dismutase
SPIs	soy protein isolates
SS	soluble solids
SSC	soluble solid concentration
TA	titratable acidity
TBZ	thiobendazole
TIL	temperature-induced lipocalins
TS	tensile strength
TSD	type II Sk2 dehydrin
TSS	total soluble solids
UV	ultraviolet
WP	whey protein
WVP	water vapor permeability
WVTR	water vapor transmission rate

ABOUT THE BOOK SERIES: POSTHARVEST BIOLOGY AND TECHNOLOGY

As we know, preserving the quality of fresh produce has long been a challenging task. In the past, several approaches were in use for the postharvest management of fresh produce, but due to continuous advancement in technology, the increased health consciousness of consumers, and environmental concerns, these approaches have been modified and enhanced to address these issues and concerns.

The Postharvest Biology and Technology series presents edited books that address many important aspects related to postharvest technology of fresh produce. The series presents existing and novel management systems that are in use today or that have great potential to maintain the postharvest quality of fresh produce in terms of microbiological safety, nutrition, and sensory quality.

The books are aimed at professionals, postharvest scientists, academicians researching postharvest problems, and graduate-level students. This series is a comprehensive venture that provides up-to-date scientific and technical information focusing on postharvest management for fresh produce.

Books in the series address the following themes:

- Nutritional composition and antioxidant properties of fresh produce
- Postharvest physiology and biochemistry
- Biotic and abiotic factors affecting maturity and quality
- Preharvest treatments affecting postharvest quality
- Maturity and harvesting issues
- Nondestructive quality assessment
- Physiological and biochemical changes during ripening
- Postharvest treatments and their effects on shelf life and quality
- Postharvest operations such as sorting, grading, ripening, de-greening, curing, etc.
- Storage and shelf-life studies
- Packaging, transportation, and marketing
- Vase life improvement of flowers and foliage

- Postharvest management of spices, medicinal, and plantation crops
- Fruit and vegetable processing waste/byproducts: management and utilization
- Postharvest diseases and physiological disorders
- Minimal processing of fruits and vegetables
- Quarantine and phytosanitory treatments for fresh produce
- Conventional and modern breeding approaches to improve the post-harvest quality
- Biotechnological approaches to improve postharvest quality of horti-cultural crops

We are seeking editors to edit volumes in different postharvest areas for the series. Interested editors may also propose other relevant subjects within their field of expertise, which may not be mentioned in the list above. We can only publish a limited number of volumes each year, so if you are interested, please email your proposal to wasim@appleacademicpress.com at your earliest convenience.

We look forward to hearing from you soon.

Editor-in-Chief:
Mohammed Wasim Siddiqui, PhD
Scientist-cum-Assistant Professor | Bihar Agricultural University
Department of Food Science and Technology | Sabour | Bhagalpur | Bihar | INDIA
AAP Sr. Acquisitions Editor, Horticultural Science
Founding/Managing Editor, *Journal of Postharvest Technology*
Email: wasim@appleacademicpress.com
wasim_serene@yahoo.com

BOOKS IN THE POSTHARVEST BIOLOGY AND TECHNOLOGY SERIES

Postharvest Biology and Technology of Horticultural Crops: Principles and Practices for Quality Maintenance
Editor: Mohammed Wasim Siddiqui, PhD

Postharvest Management of Horticultural Crops: Practices for Quality Preservation
Editor: Mohammed Wasim Siddiqui, PhD, Asgar Ali, PhD

Insect Pests of Stored Grain: Biology, Behavior, and Management Strategies
Editor: Ranjeet Kumar, PhD

Innovative Packaging of Fruits and Vegetables: Strategies for Safety and Quality Maintenance
Editors: Mohammed Wasim Siddiqui, PhD, Mohammad Shafiur Rahman, PhD, and Ali Abas Wani, PhD

Advances in Postharvest Technologies of Vegetable Crops
Editors: Bijendra Singh, PhD, Sudhir Singh, PhD, and Tanmay K. Koley, PhD

Plant Food By-Products: Industrial Relevance for Food Additives and Nutraceuticals
Editors: J. Fernando Ayala-Zavala, PhD, Gustavo González-Aguilar, PhD, and Mohammed Wasim Siddiqui, PhD

Emerging Postharvest Treatment of Fruits and Vegetables
Editors: Kalyan Barman, PhD, Swati Sharma, PhD, and Mohammed Wasim Siddiqui, PhD

PREFACE

The ultimate aim of science is to achieve better understanding of the world we live in and to improve the living standards for human beings. Agricultural science, in particular, has evolved enormously with the advancement in human civilization. The postharvest preservation and storage practices for utilization of agricultural produce date back to the time of initialization of human settlements and horticultural produce and crop cultivation.

Postharvest technology of horticultural crops has undergone many changes with the changing times. The discipline has emerged from simple and important technologies like drying and fermentation to the modern-day innovative and highly effective postharvest treatments encompassing irradiation, biological control using microbial treatments, application of biosensors to evaluate the different stages, and applications that address diseases to the issues of residues of harmful agrochemicals, plant genetic resources, and other pesticides.

Postharvest science is evolving continuously. The research impact of safe and simple postharvest grassroot technology has the potential to save agricultural produce and provide food for millions of hungry people worldwide. Endeavors are being made by scientists involved in postharvest research for maintenance of the quality and safety of fresh horticultural produce to enhance the postharvest life and to extend the availability of the produce in both time and space. There is demand for the development and application of adequate technologies for preservation of the perishable food products, particularly fresh fruits and vegetables.

It has been proven that the consumption of fruits and vegetables is very important for the prevention of many life-threatening diseases, like cancer, diabetes, cardiovascular diseases, etc., and to maintain good health. However, the rapid loss in quality of the harvested fruits and vegetables and high percentage of spoilage drastically lower their availability and result in appreciable increase in costs. The postharvest loss in developing countries reaches ever-escalating heights of 25–40% in different fresh produce at different stages of handling. Consequently, the burden of huge nutritional and economical losses falls upon all the growers, traders, as well as consumers.

This present book has been designed with the main consideration to serve as a consortium of information on new postharvest treatments that are used for postharvest quality maintenance and extension of storage life of fresh and minimally processed produce. It has been framed to act as a reference guide as well as to formulate and provide further new thrust areas for research to find new, easy to apply, and effective techniques for reduction in postharvest losses and enhancing nutritional security.

The present book is divided into 12 chapters, all of which cover exhaustively the most significant postharvest treatments employed for maintenance of the quality, safety, visual acceptability, and availability of the fruit and vegetables and their minimally processed products for a longer duration after harvest.

The postharvest storage of tropical and subtropical fruits at low temperature is limited due to the occurrence of chilling injury. This results in serious reduction of the quality as well as the shelf-life of such produce. This issue is addressed in Chapter 1 in the form of the various treatments to alleviate chilling injury during postharvest storage at low temperatures.

Heat and calcium treatments for enhancing storability of fruits and vegetables are detailed in Chapters 2 and 3. Chapters 4 and 5 elaborate on the significance and mechanisms of action of methyl jasmonate and nitric oxide as postharvest treatments for preservation of the quality of fresh produce. The potential applications of nanotechnology in postharvest field are discussed in Chapter 6.

The basic concepts as well as applications of biologically safe, effective, and promising technologies like biological control are discussed in Chapter 7. Chapter 8 discusses in detail the influences of ozone treatments on the postharvest quality of horticultural produce. Chapter 9 describes comprehensively the advances in edible coatings and films for fresh fruits and vegetables. The present times of busy schedules of the family members and increasing number of working women demand convenience and ready-to-eat horticultural food products. Hence, Chapter 10 has been framed to deal elaborately with the various postharvest treatments to reduce browning in minimally processed products. Chapter 11 focuses on natural antimicrobial agents of plant origin, essential oils and plant extracts, their source, and antimicrobial activity so as to adjudge their probable postharvest application to the fruit and vegetable industry. Chapter 12 gives insight into the role of polyamines in delaying ripening and senescence of fruits and vegetables.

We hope that this book will contribute immensely as essential reference reading for the students, teachers, professors, scientists, and entrepreneurs engaged in fresh horticultural produce handling related to this field. It will also lead to further progress in various postharvest technologies for maintenance of nutritional, visual, and sensory acceptance of the fresh horti-cultural produce. The compilation of the significant research studies associated with each chapter of this book will result in giving a focused direction of the present status of research activities and thrust areas for future studies.

The editors will appreciate suggestions and constructive comments for improvement for future works from the readers.

CHAPTER 1

POSTHARVEST TREATMENTS TO ALLEVIATE CHILLING INJURY IN FRUITS AND VEGETABLES

KALYAN BARMAN[1], SWATI SHARMA[2*], and RAM ASREY[3]

[1]*Department of Horticulture, Institute of Agricultural Sciences, Banaras Hindu University, Varanasi, Uttar Pradesh, India*

[2]*Division of Crop Production, ICAR-Indian Institute of Vegetable Research, Varanasi, Uttar Pradesh, India*

[3]*Division of Food Science & Post Harvest Technology, ICAR-Indian Agricultural Research Institute, New Delhi, India*

Corresponding author. E-mail: swtsharma92@gmail.com

ABSTRACT

Fresh fruits and vegetables are stored at low temperature to delay ripening, senescence, thereby extend their shelf life. But, most tropical and subtropical origin horticultural crops develop chilling injury during storage at low temperature. In response to chilling stress, different physiological and biochemical alterations leading to cellular dysfunctions take place in chilling sensitive produce. These alterations and dysfunctions cause development of a variety of internal and external chilling injury symptoms. In this chapter, different postharvest technologies using physical and chemical treatments such as prestorage temperature conditioning, intermittent warming, ultraviolet radiation and salicylic acid, nitric oxide, methyl jasmonate, polyamines, brassinosteroid, etc. in alleviating chilling injury of fruits and vegetables and their mechanism of action have been discussed.

1.1   INTRODUCTION

Low-temperature storage has been widely used as main strategy to prolong shelf life and preserve quality of fresh fruits and vegetables. Storage at low temperature reduces rate of metabolic activity, delay ripening and senescence, and incidence of decay-causing microorganisms (McGlasson et al., 1979). However, fruits and vegetables of tropical and subtropical origin are sensitive to low temperature during postharvest storage, negatively affecting their quality. The occurrence of chilling injury (CI) is manifested at temperatures above 0°C, depending upon the origin and nature of crop. On the basis of incidence and severity of CI in produce, fruits and vegetables can be classified into three categories, namely, (1) chilling resistant, (2) chilling sensitive, and (3) slightly chilling sensitive (Wang, 1994a). The postharvest life of fruits and vegetables under group 1 is inversely proportional to the storage temperature, the lower being the temperature the higher will be the shelf life provided the temperature is not below the freezing point. Shelf life of fruits and vegetables under group 2 increases with decrease in storage temperature up to a certain limit called critical temperature thereafter decreases with lowering the temperature. The critical temperature of commodities under group 2 in which most of the crops of tropical and subtropical origin belong generally ranged between 10°C and 13°C. The critical temperature of commodities belonging to the group 3 is lower than group 2 and often ranged between 3°C and 4°C. The occurrence of CI in harvested horticultural produce varies among species, variety of a single species, and even the climatic condition in which the crop is grown (Lyons, 1973).

1.2   SYMPTOMS OF CHILLING INJURY

The symptoms of CI vary with commodities. The most common symptoms of CI in tropical and subtropical horticultural produce are surface pitting, discoloration of peel, sunken lesions, internal breakdown, water-soaked appearance, lenticels spotting, shriveling, incomplete or impaired ripening, poor color development, off-flavor development, and increased susceptibility to microbial attack (Hardenburg et al., 1986). The development of symptoms depends not only on species or cultivars but also on maturity stage, type of tissue, and storage environment such as relative humidity. Among the symptoms, pitting on the produce surface is the most common symptom which takes place during CI in many tropical and subtropical fruits and vegetables. This is found in crops such as papaya, pomegranate, citrus, cucumber, sweet pepper, okra, melon, eggplant, and sweet potato. Due to the onset of CI, some produce such as mango, banana, papaya, avocado, melon, sapodilla, and tomato do not ripen properly or completely fail to ripen. Some symptoms of CI are very specific to certain commodities such as brown streaking of vascular tissues in banana, membranous staining in lemon, and mahogany browning of potato (Wang, 1994a). Likewise, internal discoloration occurs in pineapple, avocado, taro, and sweet potato. The symptoms at microscopic level include swelling and disorganization of the mitochondria and chloroplast in which dilation of thylakoid and unstacking of grana occurs. Lipid droplets accumulate inside the chloroplast, nuclear chromatin condenses, and reduction in size and number of starch granules takes place in the cell (Sevillano et al., 2009).

1.3 FACTORS AFFECTING PRODUCE SUSCEPTIBILITY TO CHILLING INJURY

The development of CI in a specific commodity depends upon several factors such as produce origin, genetic makeup, maturity stage, chemical composition of the produce tissue, and the storage environment such as temperature, relative humidity, atmospheric composition, and light. The origin and genetic makeup of a commodity determines the resistance or sensitivity of the product toward CI (Patterson and Reid, 1990). For example, produce originated in the temperate region are resistant to CI while those from tropical and subtropical region are sensitive to it. However, the critical threshold temperature below which CI takes place in commodities varies from crop-to-crop and with stage of maturity. Fruits and vegetables such as mango, papaya, avocado, melon, and tomato are more sensitive to CI at immature stage compared to maturity (Paull, 1990). Likewise, CI is also affected by the chemical composition of the produce tissue exposed to chilling temperature. It has been reported that the commodities resistant to CI tend to have higher degree of unsaturations in the fatty acids of membrane lipids than those of the chilling sensitive tissues (Tabacchi et al., 1979). Higher level of reducing sugars and proline content in the produce tissue also induce resistance toward CI (Purvis, 1981; Purvis and Grierson, 1982). Apart from these, several other mechanisms induce chilling tolerance in fruits and vegetables (Fig. 1.1). Among the environmental factors, temperature plays the predominant role in inducing CI. The prevalence of low relative humidity in the storage environment also promotes CI while its severity reduces at high relative humidity.

1.4   RESPONSE OF COMMODITIES TO CHILLING TEMPERATURE

Fruits and vegetables of tropical and subtropical origin when exposed to low temperature below their critical temperature initially cause primary responses including alteration in cell membrane conformation and structure, thereby affecting its membrane permeability with an increase in electrolyte leakage and lipid phase transition (Raison and Orr, 1990; Lyons and Raison, 1970). The changes in composition of membrane lipid including lipid peroxidation, increase in fatty acid saturation index, sterol:phospholipid ratio, and degradation of phospholipids and galactolipids takes place during storage at chilling temperature (Parkin and Kuo, 1989; Whitaker, 1992; Matsuo et al., 1992). As a result, decrease in fluidity and functionality of cell membrane and its associated proteins occur. It has been reported that higher adaptation of a commodity to chilling temperature is mainly attributed to increase in proportion of unsaturated fatty acids by inducing activity of fatty acid desaturase (FAD, EC 1.14.19.1–3) and some specific isoforms of glycerol-3-phosphate acyltransferase (GPAT, EC 2.3.1.15) thereby reducing the fluidity of membranes (Vigh et al., 1998; Murata et al., 1992). Storage of produce below the critical temperature thus leads to reorganization of membrane lipids from liquid crystalline into a rigid solid gel state. Exposure at this temperature for a sufficiently long time leads to loss of membrane elasticity which causes alternation in functionality of membrane proteins and membrane rupture that leads to leakage of water, electrolytes, and other metabolites from the cell (Sevillano et al., 2009). As a consequence of above primary responses, a cascade of secondary responses develops such as leakage of electrolytes, loss of turgidity and metabolic energy, and finally death of cell (Lyons, 1973).

In addition to the above direct effect of chilling temperature on membrane lipids, low temperature also causes a secondary response causing an increase in the level of reactive oxygen species (ROS) leading to oxidative stress in the commodity. ROS such as singlet oxygen, superoxide radical, hydrogen peroxide, hydroxyl radical, etc. oxidize different cellular components and thereby damage the cells (Sevillano et al., 2009). Due to oxidative stress death of cell occurs due to disintegration of cell membrane by lipid peroxidation, protein oxidation, inhibition of enzyme activity, and damage to DNA and RNA (Scandalios, 1993; Mittler, 2002). Plants protect themselves from this oxidative stress by two ways: first is by activation of ROS avoidance genes such as alternative oxidase and the second is by inducing the activity genes for ROS scavengers such as superoxide dismutase (SOD), catalase (CAT), ascorbate peroxidase (APX), glutathione peroxidase, glutathione-S-transferase, monodehydroascorbate reductase, dehydroascorbate reductase (DHAR), and glutathione reductase (GR) (Möller, 2001).

1.5 POSTHARVEST TREATMENTS TO ALLEVIATE CHILLING INJURY

1.5.1 TEMPERATURE CONDITIONING

Prestorage temperature of the commodity can significantly affect the susceptibility of tropical and subtropical produce toward CI (Hatton, 1990; Paull, 1990). Exposure of chilling sensitive produce before storage to a temperature slightly above the critical threshold temperature have been found to increase tolerance of produce to CI during low-temperature storage. Such low-temperature conditioning delays the development of CI symptoms by inducing some physiological and biochemical modifications. These includes increase in sugar, starch, and proline content, maintaining higher content of membrane phospholipids, polyamines (PAs), squalene and long-chain aldehydes, and an increase in ratio of unsaturated to saturated fatty acids. The susceptible produce is exposed to the conditioning temperature either directly in a single step or gradually by multiple steps while, the later process is more effective in alleviating CI. For example, in eggplant, temperature conditioning at 15°C for 2 days followed by 10°C for 1 day was more effective in reducing CI than at 15°C for 2 days alone, during storage at 6.5°C (Nakamura et al., 1985). Alleviation of CI by low-temperature conditioning have been found effective in fruits such as mango, papaya, lime, lemon, and grapefruit and vegetables such as tomato, cucumber, eggplant, sweet pepper, sweet potato, watermelon, and zucchini squash (Hatton, 1990) (Table 1.1).

Unlike low-temperature conditioning, exposure of produce to a high temperature above 35°C for a short duration also reduce incidence of CI. The beneficial effect of prestorage high-temperature conditioning against CI was first reported in grapefruit by Brooks and McColloch in 1936. They reported that conditioning at 38°C for 17–22 h reduces skin pitting of grapefruit during storage at 4.5°C. High-temperature conditioning can be done either by hot water dipping, hot water rinsing and brushing, hot forced air or water vapor. High temperature induces expression of stress genes that encode heat-shock proteins (HSPs) and other stress proteins, reduces chromatin condensation and DNA breakdown, and suppresses oxidative activity. These HSPs exhibit protective role against CI by molecular chaperone activity that constitutes of (1) identifying and binding with unfolded proteins so as to correctly complete their folding, (2) preventing aggregation of proteins, and (3) facilitating renaturation of aggregated proteins (Aghdam and Bodbodak, 2014). The temperature conditioning induced the expression of various genes that encode enzymes which modify the membrane lipid composition (Sapitnitskaya et al., 2006). Exposure of commodity to high temperature has also been reported to increase desaturation index of membrane lipids thereby membranes become more fluid and leakage of electrolytes become lower during storage at low temperature (Lurie, 1998). Besides this, high temperature also causes an increase in antioxidant activity and PA levels in produce during low-temperature storage (Ghasemnezhad et al., 2008; Mirdehghan et al., 2007a). Heat treatment of peach fruit before storage at low temperature reduced CI by increasing activities of antioxidant enzymes (SOD, CAT, APX, and GR) and decreasing lipoxygenase (LOX) activity, which led to reduction of oxidative stress by decreasing accumulation of H_2O_2 and O_2^- contents (Cao et al., 2010). Reduced accumulation of H_2O_2 lowers membrane unsaturated fatty acids peroxidation, maintains higher unsaturated fatty acids:saturated fatty acids ratio and reduces lignin synthesis thereby maintains higher membrane integrity (Hodges et al., 2004; Shao and Tu, 2013). Alteration of phenylalanine ammonialyase (PAL) and polyphenol oxidase enzymes in response to high temperature treatment has also been associated with mitigating the effect of CI in banana (Chen et al., 2008) and loquat (Shao and Tu, 2013). In addition to this, enhanced sugar metabolism in response to high-temperature conditioning also plays pivotal role mitigating CI in fruits and vegetables. Hot air treatment of loquat fruit increased reducing sugars (glucose and fructose) content and decreased sucrose content during cold storage by enhancing activities of acid invertase, neutral invertase, sucrose synthase, and sucrose-phosphate synthase enzymes (Shao et al., 2013; Li et al., 2011). The higher content of glucose in fruit led to increase in ascorbic acid and glutathione content, increased activities of APX and GR, activation of AA/GSH cycle, thereby mitigating oxidative stress (Aghdam and Bodbodak, 2014). In tomato, Zhang et al. (2013a) reported that hot air treatment alleviate CI by increasing arginase activity. Arginase catalyzes conversion of arginine to ornithine which causes synthesis of PAs (putrescine, spermidine, and spermine) and proline that provide chilling tolerance to the chilling sensitive fruits and vegetables (Zhang et al., 2011; Shang et al., 2011). High-temperature conditioning has been found beneficial in alleviating CI in mango, citrus, pomegranate, grape, avocado, persimmon, tomato, sweet pepper, cucumber, zucchini squash, etc. (Table 1.2). However, in some cases the beneficial effect of high-temperature conditioning to alleviate CI was found to be cultivar specific. For example, high-temperature conditioning of tomato cv. Rutgers prior to cold storage did not show beneficial effect (Whitaker, 1994). Several factors such as time and temperature of exposure of commodity to high temperature, size, and shape affect the effectiveness of the treatment to alleviate CI.

TABLE 1.1 Effect of Low-temperature Conditioning in Alleviating CI.

Crop	Conditioning temperature	Storage temperature	Reported results	References
Loquat	5°C for 6 days	0°C for 54 days	Alleviated CI by maintaining higher phenol content	Cai et al. (2006)
Mango	0°C for 4 h, and then 20°C for 20 h	2°C 85–95% RH for 12 days	Increased chilling tolerance by enhancing activities of CAT, APX, and glutathione and phenolic content	Zhao et al. (2006)
Zucchini squash	15°C for 2 days	5°C for 14 days	Reduced CI by increasing APX, AFR, DHAR enzyme activities	Wang (1996, 1994c)

AFR, ascorbate free radical; APX, ascorbate peroxidase; CAT, catalase; CI, chilling injury; DHAR, dehydroascorbate reductase; RH, relative humidity.

TABLE 1.2 Effect of High-temperature Conditioning in Alleviating Chilling Injury.

Crop	Treatment	Storage condition	References
Alleviate chilling injury by increasing antioxidant enzyme activity			
Banana	Hot water (52°C for 3 min)	7°C for 10 days	Wang et al. (2012a)
Orange	Hot water (41°C for 20 min)	1°C for 20 days	Bassal and El-Hamahmy (2011)
Lemon	Hot water (53°C for 3 min)	1.5°C for 56 days	Safizadeh et al. (2007)
Mandarin	Hot water (50°C for 2 min)	2°C for 56 days	Ghasemnezhad et al. (2008)
Peach	Hot air (38°C for 12 h)	0°C for 35 days	Cao et al. (2010)
Loquat	Hot air (38°C for 36 h)	4°C for 28 days	Shao and Tu (2013)
	Hot air (45 °C for 3 h)	5°C for 35 days	Shao et al. (2013)
Alleviate chilling injury by accumulation of HSPs			
Grapefruit	Hot water (62°C for 20 s)	2°C for 56 days	Rozenzvieg et al. (2004)

TABLE 1.2 *(Continued)*

Crop	Treatment	Storage condition	References
Banana	Hot air (38°C for 3 days)	8°C for 12 days	He et al. (2012)
Cherimoya	Hot air (55°C for 5 h)	20°C for 5 days	Sevillano et al. (2010)
Grape	Hot air (38°C for 10 h)	−2°C for 3 days	Zhang et al. (2005)
Tomato	Hot air (38°C for 3 days)	2°C for 21 days	Lurie et al. (1996), Sabehat et al. (1998a, 1998b)
Reduce chilling injury by reducing electrolyte leakage and increasing membrane integrity			
Pome-granate	Hot water (45°C for 4 min)	2°C for 90 days	Mirdehghan et al. (2007a)
Loquat	Hot air (38°C for 5 h)	1°C for 35 days	Rui et al. (2010)
	Hot air (38°C for 36 h)	4°C for 28 days	Shao and Tu (2013)
	Hot air (45°C for 3 h)	5°C for 35 days	Shao et al. (2013)
Cucumber	Hot air (37°C for 24 h)	2°C for 9 days	Mao et al. (2007a, 2007b)
Improve chilling tolerance by increasing arginase activity, polyamine, and proline content			
Pome-granate	Hot water (45°C for 4 min)	2°C for 90 days	Mirdehghan et al. (2007a)
Tomato	Hot air (38°C for 16 h) Hot air (38°C for 12 h)	2°C for 28 days	Zhang et al. (2013b, 2013c)
Pepper	Hot water (53°C for 4 min)	8°C for 28 days	González-Aguilar et al. (2000)
Reduce chilling injury by altering PAL and PPO activity			
Banana	Hot air (38°C for 2 days)	8°C for 12 days	Chen et al. (2008)
Loquat	Hot air (38°C for 36 h)	4°C for 28 days	Shao and Tu (2013)
Alleviate chilling injury by enhancing sugar metabolism			
Pome-granate	Hot water (45°C for 4 min)	2°C for 90 days	Mirdehghan et al. (2006)
Peach	Hot air (39°C for 3 days)	20°C for 7 days	Lara et al. (2009)
Loquat	Hot air (45°C for 3 h)	5°C for 35 days	Shao et al. (2013)

HSPs, heat-shock proteins; PAL, phenylalanine ammonialyase; PPO, polyphenol oxidase.

1.5.2 INTERMITTENT WARMING

In this method, intermittent warming of produce for a short time is performed for one or more periods during storage at low temperature. This interruption in low temperature facilitates to alleviate CI and extend storage life of chilling-sensitive produce, provided the treatment is applied before the CI becomes irreversible. In this method, the time of treatment application and early detection of CI is very important which otherwise proceeds beyond recovery and warming of produce will accelerate degradation. In Israel, intermittent warming is commercially practiced to reduce CI in lemon fruit (Cohen, 1988). "Eureka" and "Villa Franca" lemons can be stored successfully for up to 6 months at 2°C by warming the fruit at 13°C for 7 days at every 21 days interval. Likewise, application of intermittent warming was also found beneficial in reducing CI in sweet pepper, cucumber, and zucchini squash (Kramer and Wang, 1989; Wang and Baker, 1979). During intermittent warming when the produce is exposed to high temperature, it induces higher metabolic activities which cause the tissue to metabolize excess intermediate products, thereby replenish deficiencies that developed during chilling. Exposure of produce to high temperature for a short period causes warming of chilled tissues which helps to repair damage to cell membranes or organelles and increases the synthesis of polyunsaturated fatty acids (Lyons and Breidenbach, 1987). Intermittent warming have been found effective in alleviating CI in several fruits and vegetables such as peach, nectarine, lemon, grapefruit, tomato, cucumber, sweet pepper, zucchini squash, etc. (Forney and Lipton, 1990).

FIGURE 1.1 Mechanism of chilling tolerance in fruits and vegetables.

1.5.3 SALICYLIC ACID

Salicylic acid (SA) is an endogenous signaling molecule that belongs to the group of phenolic compounds. It is present ubiquitously throughout the plant kingdom and is involved in regulating several plant developmental processes such as photosynthesis, respiration, transpiration, stomatal closure, cell growth, ion uptake and transport, senescence-associated genes expression, etc. (Klessig and Malamy, 1994; Clarke et al., 2004; Morris et al., 2000; Rajjou et al., 2006; Harper and Balke, 1981; Khan et al., 2003). SA has been reported to play a significant role in modulating the response of plants to various biotic and abiotic stresses (Asghari and Aghdam, 2010). It is involved in activating local and systemic disease resistance in response to the pathogen attack (Alverez, 2000) and also modulates plant response to various abiotic stresses such as drought, salinity, heat shock, chilling stress, and UV light (Ding and Wang, 2003; Ding et al., 2001).

Postharvest treatment of fruits and vegetables with SA has proved its effectiveness in alleviating CI during their low-temperature storage (Table 1.3). One of the important mechanisms of SA in reducing CI is induced expression of ROS avoidance genes and ROS scavenging genes such as SOD, CAT, APX, etc. (Asghari and Aghdam, 2010). Likewise, it also induces synthesis and further accumulation of HSPs conferring protection against CI (Tian et al., 2007). In peach fruit, SA treatment alleviated CI during storage at 0°C by enhancing activities of APX, GR and increasing reduced-to-oxidized ascorbate ratio (AsA/DHAsA) and reduced-to-oxidized glutathione ratio (GSH/GSSG). This also increased accumulation of heat-shock proteins (HSP101 and HSP73) in the cells (Wang et al., 2006). Luo et al. (2011) reported that SA alleviates CI in plum by increasing accumulation of PAs and reducing electrolyte leakage and MDA accumulation. This was also found by Aghdam et al. (2012a) and Cao et al. (2009) in tomato and cucumber, respectively. Postharvest immersion treatment of pomegranate fruit in 2.0 mM SA solution is found to be highly effective in reducing CI and electrolyte leakage during low-temperature storage (Sayyari et al., 2009). Barman and Asrey (2014) also found it effective in alleviating CI in mango during storage at 8°C.

TABLE 1.3 Effect of Salicylic Acid in Inducing Chilling Tolerance.

Crop	SA treatment	Storage conditions	Reported results	References
Lemon	SA 2.0 mM	−0.5, 2, or 4.5°C for up to 28 days + 7 days at 23°C	Enhanced chilling tolerance by increasing total phenolics and PAL activity and decreasing activity of POD	Siboza et al. (2014)
Loquat	Acetyl SA 1.0 mM L⁻¹	5°C for up to 39 days + 5 days at 20°C	Reduced CI by impairing accumulation of superoxide free radicals	Cai et al. (2006)
Mango	SA 2.0 mM L⁻¹	5°C for up to 30 days + 5 days at 25°C	Alleviated CI by inhibiting O_2^- accumulation, delayed decrease of H_2O_2, and higher reducing status of ascorbate and glutathione.	Ding et al. (2007)
Mango	SA 2.0 mM	8°C for up to 30 days + 3 days at 25°C	Reduced CI by inducing expression of ROS avoidance and scavenging genes and accumulation of HSPs	Barman and Asrey (2014)
Peach	SA 1.0 mM	0°C for up to 28 days + 3 days at 20°C	Suppressed CI by inducing antioxidant systems and HSPs	Wang et al. (2006)
Plum	SA 1.5 mM	1°C for up to 60 days + 3 days at 20°C	Reduced CI symptom by delayed activities of PPO, POD, and increased accumulation of polyamines	Luo et al. (2011)
Pomegranate	Acetyl SA 1.0 mM	2°C for up to 84 days + 4 days at 20°C	Alleviated CI by increasing antioxidant capacity	Sayyari et al. (2011a)
Pomegranate	SA 2.0 mM	2°C for up to 3 months + 3 days at 20°C	Reduced CI symptoms during storage	Sayyari et al. (2009)
Tomato	SA 2.0 mM	1°C for up to 3 weeks + 3 days at 20°C	Reduced activity of phospholipase and lipoxygenase thereby reduced CI	Aghdam et al. (2012a, 2014)
Cucumber	Chitosan-g-SA conjugate (0.57% w/v)	2°C for up to 12 days + 2 days at 20°C	Alleviated CI by increased endogenous SA concentrations and antioxidant enzyme activities including SOD, CAT, APX, and GR	Zhang et al. (2015)

APX, ascorbate peroxidase; CAT, catalase; CI, chilling injury; GR, glutathione reductase; HSPs, heat-shock proteins; PAL, phenylalanine ammonialyase; POD, peroxidase; ROS, reactive oxygen species; SA, salicylic acid; SOD, superoxide dismutase.

1.5.4 NITRIC OXIDE

Nitric oxide (NO) is a highly reactive, free-radical gas that acts as a multifunctional signaling molecule in regulating several physiological processes in plants (Wendehenne et al., 2001). It is involved in plant growth and development starting from germination, root development, stomatal closure, flowering, fruit ripening, reproduction, and senescence (Beligni and Lamattina, 2001; Correa-Aragunde et al., 2004; Neil et al., 2003; He et al., 2004; Leshem, 2000; Prado et al., 2004; Leshem et al., 1998). NO also reported to induce resistance against various biotic and abiotic stresses (Manjunatha et al., 2010). Nitric oxide can be applied either by fumigation or by immersing the fruits and vegetables in sodium nitroprusside (SNP) solution, a donor of NO.

CI in fruits and vegetables during low-temperature storage is associated with generation of ROS, which induces oxidative stress. NO acts as a reaction cascade breaker and prevents damages due to CI (Farias-Eisner et al., 1996; Durzan and Pedroso, 2002). In addition to this, NO reverses the effect of ROS directly, either by suppressing the activities of ROS enzymes or relevant signaling cascade in a tightly coordinated manner (Clark et al., 2000). It is found that application of NO upregulates the activities of SOD, CAT, and peroxidase (POD) enzymes, which are involved in removal of ROS (Flores et al., 2008; Zhu et al., 2008). Fumigation of banana (Wu et al., 2014) and cucumber (Yang et al., 2011a) with NO has been reported to induce chilling resistance of these fruits by increasing the activities of antioxidant enzymes such as SOD, CAT, POD, and APX thereby decreasing the accumulation of ROS. Likewise, application of NO has been found to alleviate CI in many fruit crops such as peach, mango, plum, and loquat during their low-temperature storage (Singh et al., 2009; Zhu et al., 2010; Xu et al., 2012; Aghdam and Bodbodak, 2013) (Table 1.4).

TABLE 1.4   Effect of Nitric Oxide in Alleviating Chilling Injury.

Crop	NO treatment	Storage temperature (°C)	Reported results	References
Banana	60 µL L^{-1} NO fumigation	7	Improved chilling-resistance by increasing antioxidant enzyme activities (SOD, CAT, POD, and APX) and related gene expression, and decreasing accumulation of ROS	Wu et al. (2014)

TABLE 1.4 *(Continued)*

Crop	NO treatment	Storage temperature (°C)	Reported results	References
Peach	15 µL L^{-1} NO fumigation	5	Alleviated chilling injury by protecting cell membrane and cellular integrity	Zhu et al. (2010)
Loquat	NO-specific scavengers and NO synthase inhibitors	1	Chilling tolerance can be enhanced by elevating endogenous accumulation of NO	Xu et al. (2012)
Mango	1.5 mM SNP immersion	8	Reduced chilling injury and electrolyte leakage	Barman et al. (2014)
Mango	10, 20, and 40 µL L^{-1} NO fumigation	5	Alleviated chilling injury by reducing oxidative stress	Zaharah and Singh (2011)
Plum	10 µL L^{-1} NO fumigation	0	Reduced chilling injury by inhibition of ethylene production	Singh et al. (2009)
Tomato	0.02 mM SNP immersion	2	Reduced chilling injury by inducing NO accumulation and expression of LeCBF1	Zhao et al. (2011)
Cucumber	25 µL L^{-1} NO fumigation	2	Improved chilling-resistance by increasing activities of SOD, CAT, POD, and APX enzymes and decreasing accumulation of ROS	Yang et al. (2011a)

APX, ascorbate peroxidase; CAT, catalase; NO, nitric oxide; POD, peroxidase; ROS, reactive oxygen species; SNP, sodium nitroprusside; SOD, superoxide dismutase.

1.5.5 POLYAMINES

PAs are low molecular weight small aliphatic amines that are ubiquitous in all living organisms. The most common PAs found in every plant cells are putrescine (Put), spermidine (Spd) and spermine (Spm), the concentration of which depends greatly on environmental conditions, especially stress (Galston and Sawhney, 1990). In plants, PAs have been implicated in a wide range of biological processes including growth, development, and responses to abiotic stresses (Evans and Malmberg, 1989; Flores et al., 1989; Galston and Sawhney, 1990). They are mainly localized in vacuoles, mitochondria, and chloroplast of the cells (Slocum, 1991).

Fruits and vegetables when exposed to chilling temperature, changes in cell membrane lipid take place from liquid-crystalline to solid-gel state that lead to increase in membrane permeability and leakage of ions (Stanley, 1991; Gómez-Galindo et al., 2004). PAs play a very crucial role in alleviating CI of fruits and vegetables (Table 1.5). Exogenous application of PAs is reported to induce cold acclimation thereby protecting the produce from CI by increasing the level of endogenous PAs that lead to maintenance of membrane fluidity and reducing electrolyte leakage during storage at low temperature (Mirdehghan et al., 2007a). The alleviation of CI by PAs is attributed to protecting the membrane lipids from being conversion in physical state and its antioxidant property, which prevents lipid peroxidation of cell membrane (Mirdehghan et al., 2007a; Barman et al., 2011). Due to polycationic nature at physiological pH, PAs bind with negatively-charged macromolecules such as phospholipids, proteins, and nucleic acid thereby stabilizing the cell membranes under chilling stress conditions (Smith, 1985). Moreover, PAs have also been reported to have antioxidant properties which are helpful in removal of ROS generated during CI. The antioxidant property of exogenously applied PAs is reported due to its ability to reduce level of hydrogen peroxide, malondialdehyde content, and increase antioxidant enzyme activity (SOD, POD, and CAT) in plant (Nayyar and Chander, 2004). Further, it has been reported that ethylene biosynthesis in plants sensitive to low temperature increases along with development of CI due to increase in precursors of ethylene biosynthesis and enzyme activities involved in ethylene biosynthesis (Concellón et al., 2005; Lederman et al., 1997). The PAs and ethylene use the same precursor SAM (S-adenosyl methionine) for their biosynthesis. Therefore, an increase in PA levels affects the level of ethylene production. However, increase in ethylene production with the onset of CI cannot be generalized in all the chilling sensitive plants. For example, in some commodity beneficial effect of ethylene against CI has also been found. In banana, postharvest application of propylene, an ethylene analogue induced resistance toward development of CI.

TABLE 1.5   Effects of Polyamines in Inducing Chilling Tolerance.

Crop	PA treatment	Storage temperature (°C)	Reported results	References
Apricot	Put and Spd (1.0 mM)	1	Alleviated CI by increasing antioxidant enzyme activity (SOD, POD, and CAT)	Saba et al. (2012)
Pome-granate	Put and Spd (1.0 mM) by pressure-infiltration and immersion	2	Reduced CI by inducing cold acclimation	Mirdeh-ghan et al. (2007b, 2007c)
Pome-granate	Put (2.0 mM) by immersion	3	Reduced CI due to antisenescence property of Put	Barman et al. (2011)
Zucchini squash	Put, Spd, and Spm (0.1, 0.25, 0.5, 2.0, and 4.0 mM) by infiltration	2	Alleviated CI by stabilizing and protecting cell membranes	Martínez-Téllez et al. (2002)

CAT, catalase; CI, chilling injury; PA, polyamine; POD, peroxidase; Put, putrescine; SOD, superoxide dismutase; Spd, spermidine; Spm, spermine.

1.5.6 METHYL JASMONATE

Jasmonates are the class of endogenous plant growth regulators, play an important role in plant growth, development, fruit ripening and responses to biotic and abiotic stresses (Creelman and Mullet, 1997). Among the jasmonates, methyl jasmonate (MeJA) has received much attention owing to its ability to enhance chilling tolerance to harvested fruits and vegetables (Li et al., 2012a). Exogenous application of MeJA has been found to alleviate CI by the following mechanisms: (1) increasing membrane integrity, (2) increasing expression and accumulation of HSPs, (3) enhancing antioxidant system, (4) enhancing arginine pathway, and (5) altering PPO and PAL activities (Table 1.6). In general, higher levels of unsaturated fatty acids (linolenic acid and linoleic acid), increased fatty acid desaturase enzymes activity causing increase in membrane saturation degree and its fluidity are responsible for increased chilling tolerance of fruits and vegetables (Hernández et al., 2011). Increase in membrane fluidity decreases the affinity of membrane toward phase change of membrane lipids from liquid crystalline to solid gel state (Los and Murata, 2004). Membrane lipid peroxidation is also carried out by the oxidation of unSFA by the enzyme LOX or ROS (Aghdam and

Bodbodak, 2014). Exogenous application of MeJA has been found to alleviate CI by maintaining higher unSFA/SFA ratio, reducing LOX activity thereby maintaining higher membrane integrity. In peach, treatment with MeJA reduced excess accumulation of ROS (O_2^-, singlet oxygen, and H_2O_2) by increasing activities of antioxidant enzymes such as SOD, CAT, APX thereby reduced membrane lipid peroxidation, and CI (Cao et al., 2009). Ding et al. (2002) reported that MeJA-treated tomatoes when stored at low temperature, it caused accumulation of HSPs which enhanced GSH level, activities of antioxidant enzymes gene expression, thereby increases their chilling tolerance. Likewise in cherry tomato, Zhang et al. (2012) found that MeJA induced activities of arginase, arginine decarboxylase, and ornithine decarboxylase enzymes during cold storage. Arginase and arginine decarboxylase are the crucial enzymes that catalyses the conversion of arginine to ornithine, which serves as the precursor for the biosynthesis of PAs and proline. Therefore, production of PAs and proline lead to increased tolerance toward CI (Zhang et al., 2011; Shang et al., 2011). In addition, MeJA treatment also increased activity of glutamate decarboxylase which caused increased production and accumulation of γ-aminobutyric acid (GABA) that enhanced chilling tolerance (Cao et al., 2012). Further, reduced CI incidence in response to MeJA treatment is also reported to be associated with reduced activity of PPO and POD enzymes (Jin et al., 2009).

TABLE 1.6　Effect of Methyl Jasmonate on Alleviating Chilling Injury of Fruits and Vegetables.

Crop	MeJA treatments	Storage temperature	Reported results	References
Loquat	MeJA (10 µM L⁻¹) for 24 h MeJA (16 µM L⁻¹) for 6 h	1°C and 95% RH for 35 days	Reduced CI by enhancing antioxidant enzyme activity, higher unsaturated:saturated fatty acid ratio, reduced lignin accumulation, increased cell-wall polysaccharide solubilization, and increasing proline and γ-aminobutyric acid content	Cao et al. (2009, 2010, 2012), Cai et al. (2011), Jin et al. (2014a)

TABLE 1.6   *(Continued)*

Crop	MeJA treatments	Storage temperature	Reported results	References
Peach	MeJA vapor (1 μM L^{-1})	0°C for 5 weeks + 20°C for 3 days	Reduced CI by increasing activities of PAL, SOD and lowering PPO and POD activities	Jin et al. (2009)
Pineapple	MeJA solution (10^{-3}, 10^{-4}, and 10^{-5} M)	10°C and 85% RH for 21 days	Reduced CI by reducing electrolyte leakage	Nilprapruck et al. (2008)
Pome-granate	MeJA (0.01 mM)	2°C and 90% RH for 84 days + 20°C for 4 days	Alleviated CI by increasing phenolics content and antioxidant capacity	Sayyari et al. (2011b)
Mango-steen	*n*-propyl dihy-drojasmonate (0.39 mM)	7°C	Increased chilling tolerance by endog-enous accumulation of jasmonic acid	Kondo et al. (2004)
Tomato	MeJA (0.01 mM)	5°C for 4 weeks + 20°C for 1 day	Increased chilling tolerance by inducing synthesis of stress proteins, HSPs	Ding et al. (2001, 2002)
Cherry tomato	MeJA vapor (0.05 mM) for 12 h	2°C for 21 days	Improved chilling tolerance by altering arginine catabolism	Zhang et al. (2012)
Sweet peppers	MeJA vapor (22.4 μL L^{-1})	0°C for 14 days + 20°C for 9 days	Reduced CI by acti-vating ROS defense system	Fung et al. (2004)
Zucchini squash	MeJA pressure infiltration (0.5, 1.0 mM)	5°C for 14 days + 20°C for 3 days	Alleviated CI by accumulation of poly-amines (spermidine and spermine) and abscisic acid	Wang (1994b), Wang and Buta (1994)

CI, chilling injury; HSPs, heat-shock proteins; MeJA, methyl jasmonate; PAL, phenylalanine ammonialyase; POD, peroxidase; ROS, reactive oxygen species; SOD, superoxide dismutase.

1.5.7 BRASSINOSTEROIDS

Brassinosteroids (BRs) are the group of plant steroidal hormones present ubiquitously in the plant kingdom (Ogweno et al., 2008). BRs have been reported to be essential for plant growth and development, regulating various physiological processes such as stem elongation, root growth, xylem differentiation, synthesis of nucleic acid and proteins, leaf epinasty, floral initiation, ethylene biosynthesis, etc. (Bishop and Koncz, 2002; Fujioka and Yokota, 2003; Kim and Wang, 2010; Sasse, 2003). Further, its effect in ameliorating various biotic and abiotic stresses such as cold stress, thermal stress, salt injury, oxidative damage, metal stress, and pathogen infection have also been reported (Liu et al., 2009; Bajguz and Hayat, 2009).

Postharvest application of BRs has been found highly effective in alleviating CI in several fruits and vegetables during low-temperature storage (Table 1.7). In mango, BRs upregulated the expression of plasma membrane proteins such as REM (remorin family proteins), ASR (abscisic acid stress ripening-like proteins), TIL (temperature-induced lipocalins), TSD (type II Sk2 dehydrin), and genes encoding these proteins thereby maintain membrane integrity and increased chilling tolerance during low-temperature storage (Li et al., 2012b). Moreover, it also maintained higher membrane lipid fluidity by maintaining higher content of unsaturated fatty acids in membrane lipids during chilling stress. Application of BRs has also been found to provide protection from chilling-induced oxidative stress by enhancing both enzymatic and nonenzymatic antioxidant defense system of cells. In bell pepper, application of BRs induced the activities of antioxidant enzymes such as CAT, POD, APX, and GR which reduced oxidative damage (Wang et al., 2012b). Likewise, increased level of phenolic compounds and accumulation of proline in response to exogenous BRs application have also been linked with improved chilling tolerance in tomato (Aghdam et al., 2012b; Aghdam and Mohammadkhani, 2014), peach (Gao et al., 2016), and bamboo shoots (Liu et al., 2016). In bamboo shoots, application of BRs increased activities of H^+-ATPase, Ca^{2+}-ATPase, succinate dehydrogenase, and cytochrome C oxidase, retarded decline in adenosine triphosphate (ATP) content, and maintained high energy charge, thereby increased chilling tolerance (Liu et al., 2016).

TABLE 1.7 Effects of Brassinosteroids in Inducing Chilling Tolerance of Fruits and Vegetables.

Crop	Treatment	Storage temperature	Reported results	References
Mango	Brassinolide solution 10 µM	5°C and 90% RH for 28 days	Increased chilling tolerance by upregulating plasma membrane proteins and lower phase transition and higher fluidity of membrane lipids	Li et al. (2012b)

TABLE 1.7 *(Continued)*

Crop	Treatment	Storage temperature	Reported results	References
Peach	24-Epibrassino-lide solution 15 μM	1°C for 4 weeks + 25°C for 3 days	Reduced CI by increasing phenolic compounds and proline biosynthesis and reducing oxidative stress	Gao et al. (2016)
Bell pepper	Brassinolide solution 15 μM	3°C and 95% RH for 18 days	Enhanced chilling toler-ance by eliciting antioxidant enzyme activity	Wang et al. (2012b)
Tomato	Brassinolide solution 6 μM	1 ± 0.5°C, 85–90% RH for 3 weeks + 25°C for 3 days	Alleviated CI by inhibiting activities of phospholi-pase D, lipoxygenase and enhancing phenolics and proline content	Aghdam et al. (2012b), Aghdam and Moham-madkhani (2014)
Bamboo shoot	Brassinolide solution 0.5 μM	1°C and 95% RH for 42 days + 25°C for 3 days	Increased chilling tolerance by elevating level of ATP, energy charge, and proline accumulation	Liu et al. (2016)

ATP, adenosine triphosphate; CI, chilling injury.

1.5.8 ULTRAVIOLET RADIATION

The application of ultraviolet (UV) radiation in fresh fruits and vegetables has been reported to exert beneficial effects by delaying microbial growth and senescence of produce during their postharvest storage. Among the different ranges of UV radiation, short treatment of UV-C or UV-B has been found to alleviate CI of fruits and vegetables under low-temperature storage (Table 1.8). In peaches, short-term exposure of UV-C was found beneficial in alleviating CI than its long-term exposure. It induced resistance in CI by increasing endogenous levels of Put, Spd and Spm (González-Aguilar et al., 2004). Moreover, application of UV radiation causes increase in phenyl-alanine ammonia-lyase (PAL) activity which results in higher synthesis of phenolic compounds as reported in banana (Pongprasert et al., 2011). However, Vicente et al. (2005) and Liu et al. (2012) reported that UV treated pepper and tomato contain lower phenolic compounds than control fruits in which severe CI occurred during low-temperature storage. In addition, higher activities of antioxidant enzymes such as SOD, CAT, peroxidase, APX, GR, and expression of HSPs in response to prestorage UV exposure were reported as possible mechanisms involved in alleviating CI (Pong-prasert et al., 2011; Cuvi et al., 2011).

TABLE 1.8   Effect of Ultraviolet Radiation in Inducing Chilling Tolerance.

Crop	UV treatment	Storage temperature	Reported results	References
Peach	UV-C irradiation 3, 5, or 10 min	5°C up to 21 days + 7 days at 20°C	Reduced CI by increasing accumulation of polyamines	González-Aguilar et al. (2004)
Banana	UV-C irradiation (0.02, 0.03, and 0.04 kJ m^{-2})	5°C up to 21 days	Alleviated CI by reducing oxidative stress and inducing heat-shock proteins	Pongprasert et al. (2011)
Cucumber	UV-C irradiation 3, 5, 10, or 15 min	5°C and 10°C up to 21 days	Reduced CI	Kasim and Kasim (2008)
Pepper	UV-C light (7 kJ m^{-2})	0°C up to 15 and 22 days + 4 days at 20°C	Reduced incidence and severity of CI	Vicente et al. (2005)
Red pepper	UV-C light (10 kJ m^{-2})	0°C up to 21 days	Alleviated CI by increasing antioxidant enzyme activity	Cuvi et al. (2011)
Tomato	UV-C radiation (4 kJ m^{-2}) and UV-B radiation (20 kJ m^{-2})	2°C up to 20 days + 10 days at 20°C	Increased tolerance to chilling temperature	Liu et al. (2012)

CI, chilling injury; UV, ultraviolet.

1.5.9   γ-AMINOBUTYRIC ACID

GABA, a nonprotein amino acid is a natural signal molecule involved in regulating a variety of biotic and abiotic stresses in plants (Shelp et al., 1999). Rapid accumulation of GABA in plants in response to stresses has been reported as defense system against them (Shelp et al., 1999). For example, strawberry and tomato when stored at elevated-CO_2 or low-O_2 environment, it increased endogenous accumulation of GABA (Deewatthanawong et al., 2010a, 2010b). In plants, accumulation of proline takes place under diverse stress conditions including CI as an adaptive mechanism (Verbruggen and Hermans, 2008). It has been reported that exogenous application of GABA alleviates CI by inducing endogenous proline and GABA accumulation in cold-stored peach fruit (Yang et al., 2011b; Shang et al., 2011). They have also mentioned that GABA treatment increased activities of glutamate decarboxylase, ornithine δ-aminotransferase, Δ¹-pyrroline-5-carboxylate synthetase (P5CS), and decreased activity of proline dehydrogenase (PDH) (Shang et al., 2011). Furthermore, maintenance of higher ATP content and energy charge in GABA-treated fruit has also been postulated another mode of action in reducing CI (Yang et al., 2011b). In banana, treatment of fruit with 5.0 mM GABA was found to reduce CI during storage at low tempera-ture by endogenous proline accumulation and enhancement of antioxidant capacity (Wang et al., 2014). However, further research is needed to eluci-date the clear mode of action of GABA in reducing CI.

1.5.10 OXALIC ACID

Oxalic acid is a naturally occurring organic acid present ubiquitously in plants (Li et al., 2014). Postharvest treatment of peach and mango with oxalic acid maintained higher levels of ATP and energy charge by enhancing activities of H^+-ATPase, Ca^{2+}-ATPase, succinic dehydrogenase, and cyto-chrome C oxidase enzymes (Jin et al., 2014b; Li et al., 2014). It has been reported that incidence of CI in several fruits is associated with limited energy availability and/or energy production while, higher ATP level and energy charge alleviate CI as it plays an important role in fatty acid synthesis and repair of membranes (Rawyler et al., 1999; Chen and Yang, 2013; Chen et al., 2012). Oxalic acid treatment also maintained higher unSFA/SFA ratio which is beneficial in reducing membrane lipid peroxidation, electrolyte leakage, and thereby susceptibility of produce toward CI (Jin et al., 2014b). In mango, oxalic acid treatment reduced CI by increasing accumulation of proline which was attributed to enhanced P5CS and reduced PDH activity (Li et al., 2014; Xue et al., 2012). Likewise, increased tolerance to CI in response to prestorage oxalic acid treatment has also been reported in pome-granate (Sayyari et al., 2012).

1.5.11 1-METHYLCYCLOPROPENE

Ethylene has been reported to stimulate the incidence of CI in chilling-sensi-tive fruits and vegetables. Manipulation of ethylene in stored produce can be achieved by use of ethylene biosynthesis or ethylene action inhibitors. 1-Methylcyclopropene (1-MCP) is a cyclic olefin that reduces the action of ethylene by specifically and irreversibly binding with ethylene receptors. As a result, it blocks the action of the receptors thereby preventing the signal transduction and activation of ethylene-dependent responses in the cell (Serek et al., 1995; Sisler et al., 1996; Sisler and Serek, 1997). But, the efficiency of 1-MCP against ethylene action is limited as plants synthesize new receptors and thereby recovers its sensitivity toward ethylene. Treatment of persimmon, pineapple, and cantaloupe melon with 1-MCP at 0.3, 0.1, and 1 μL L^{-1} reduced CI during cold storage (Salvador et al., 2004; Selvarajah et al., 2001; Ben-Amor et al., 1999). Likewise, in plum, avocado, and pear 1-MCP treatment at 0.4, 0.5, and 1.0 μL L^{-1} alleviated CI (Pesis et al., 2002; Ekman et al., 2004; Candan et al., 2008). However, opposite results has also been reported on the effect of 1-MCP in the incidence of CI. In orange and apricot prestorage treatment with 0.1 μL L^{-1} of 1-MCP for 6 h and 1.0 μL L^{-1} 1-MCP for 20 h, respectively, increased symptoms of CI during low-temperature storage (Porat et al., 1999; Dong et al., 2002).

1.5.12   6-BENZYLAMINOPURINE

6-Benzylaminopurine (6-BAP) is a synthetic cytokinin that stimulates growth and development of plants. Recently, exogenous application of 6-BAP has been reported to induce cold tolerance in chilling sensitive produce. In cucumber, treatment with 50 mM L^{-1} 6-BAP increased the activities of antioxidant enzymes (SOD, CAT, APX, and GR) thereby reduced production and accumulation of O_2^- and H_2O_2. Furthermore, this treatment maintained higher level of total phenolics, ATP content, and energy charge which alleviated CI of cucumber during low-temperature storage (Chen and Yang, 2013).

1.5.13   2,4-DICHLOROPHENOXYACETIC ACID

2,4-dichlorophenoxyacetic acid (2,4-D) is most widely used as herbicide on dicotyledonous plants (Sterling and Hall, 1997). In mango, prestorage application of 2,4-D at 150 mg L^{-1} reduced incidence of CI during storage at 4°C by increasing activities of APX and GR. Furthermore, it also increased endogenous abscisic acid and gibberellic acid and decreased indole-3-acetic acid levels during cold storage (Wang et al., 2008).

1.5.14 LOW-O_2 OR HIGH-CO_2 TREATMENT

Prestorage treatment with very low O_2 or very high CO_2 has been reported to reduce CI symptoms in fruits and vegetables during storage. Exposing avocado fruits in atmospheres of 3% O_2 for 24 h was effective in alleviating CI up to 3 weeks of storage at 2°C (Pesis et al., 1994). Prestorage treatment of citrus with high CO_2 (40%) in air for 9 days reduced CI symptoms (rind pitting) during storage at 0°C up to 90 days (Bertolini et al., 1991). Likewise in avocado, 20% and 25% CO_2 treatment during or before storage reduced incidence of CI at low temperature (Marcellin and Chaves, 1983; Bower et al., 1989). However, the major problem in the method is development of off-flavor during storage due to anaerobic respiration (Lurie and Pesis, 1992; O'Hare and Prasad, 1993).

1.5.15 STORAGE TECHNOLOGIES

Postharvest storage of fruits and vegetables under controlled atmosphere (CA) or modified atmosphere (MA) with low concentration of oxygen and high concentration of carbon dioxide can also minimize the effects of CI during cold storage. Alleviating CI by controlling gaseous composition in the storage atmosphere of fruits and vegetables has been reported in nectarine, guava, avocado, and bitter gourd (Meir et al., 1995; Retamales et al., 1992; Zong et al., 1995; Singh and Pal, 2008; Saltveit and Morris, 1990). In MA storage, high level of CO_2 and low level of O_2 is maintained by packing the produce in polymeric film having selective permeability of O_2 and CO_2. High CO_2 level reduces ethylene production by inhibiting ACC (1-aminocyclopropane-1-carboxylic acid) synthase activity and also sensitivity of produce toward ethylene. Besides this, maintenance of high relative humidity inside package surrounding the produce is also beneficial in inhibiting development of CI symptoms (Forney and Lipton, 1990). Reduction of CI in produce stored at high relative humidity compared to low RH has been reported long back in grapefruit, limes, cucumber, and sweet pepper (Brooks and McColloch, 1936; Morris and Platenius, 1938; Pantastico et al., 1967; Wardowski et al., 1973). Likewise, increasing CO_2 and reducing O_2 level during storage has shown to reduce CI in grapefruit, avocado, peach, nectarine, pineapple, okra, zucchini squash, and potato (Wang, 1993).

1.6 CONCLUSION

CI causes reduction in postharvest life of tropical and subtropical fruits and vegetables during low-temperature storage. Exposure of produce to a chilling temperature causes changes in physical state of membrane lipids from flexible liquid crystalline to rigid solid gel state and oxidative stress. The specific temperature at which this phase transition of membrane lipid occurs depends upon the degree of fatty acids desaturation in the membrane lipids. Different postharvest treatments are employed nowadays to reduce the effect of CI in chilling sensitive fruits and vegetables such as prestorage temperature conditioning, intermittent warming, and SA, NO, MeJA, PA, BR, UV radiation, or other chemical treatments. These treatments increase chilling tolerance of fruits and vegetables by increasing membrane integrity, antioxidant system activity, HSPs accumulation, arginine pathway, or altering activities of PPO and PAL.

KEYWORDS

- chilling injury
- postharvest
- temperature
- storage
- fruits
- vegetables

REFERENCES

Aghdam, M. S.; Bodbodak, S. Physiological and Biochemical Mechanisms Regulating Chilling Tolerance in Fruits and Vegetables under postharvest Salicylates and Jasmonates Treatments. *Sci. Hortic.* **2013,** *156,* 73–85.

Aghdam, M. S.; Bodbodak, S. Postharvest Heat Treatment for Mitigation of Chilling Injury in Fruits and Vegetables. *Food Bioprocess Technol.* **2014,** *7,* 37–53.

Aghdam, M. S.; Mohammadkhani, N. Enhancement of Chilling Stress Tolerance of Tomato Fruit by Postharvest Brassinolide Treatment. *Food Bioprocess Technol.* **2014,** *7* (3), 909–914.

Aghdam, M. S.; Asghari, M. R.; Moradbeygi, H.; Mohammadkhani, N.; Mohayeji, M.; Reza-pour-Fard, J. Effect of Postharvest Salicylic Acid Treatment on Reducing Chilling Injury in Tomato Fruit. *Romanian Biotechnol. Let.* **2012a,** *17* (4), 7466–7473.

Aghdam, M. S.; Asghari, M.; Farmani, B.; Mohayeji, M.; Moradbeygi, H. Impact of Posthar-vest Brassinosteroids Treatment on PAL Activity in Tomato Fruit in Response to Chilling Stress. *Sci. Hortic.* **2012b,** *144*, 116–120.

Aghdam, M. S.; Asghari, M.; Khorsandi, O.; Mohayeji, M. Alleviation of Postharvest Chilling Injury of Tomato Fruit by Salicylic Acid Treatment. *J. Food Sci. Technol.* **2014,** *51* (10), 2815–2820.

Alverez, A. L. Salicylic Acid in Machinery of Hypersensitive Cell Death and Disease Resis-tance. *Plant Mol. Biol.* **2000,** *44*, 429–442.

Asghari, M.; Aghdam, M. S. Impact of Salicylic Acid on Post-harvest Physiology of Horticul-tural Crops. *Trends Food Sci. Technol.* **2010,** *21*, 502–509.

Bajguz, A.; Hayat, S. Effects of Brassinosteroids on the Plant Responses to Environmental Stresses. *Plant Physiol. Biochem.* **2009,** *47*, 1–8.

Barman, K.; Asrey, R. Salicylic Acid Pre-treatment Alleviates Chilling Injury, Preserves Bioactive Compounds and Enhances Shelf Life of Mango Fruit During Cold Storage. *J. Sci. Indust. Res.* **2014,** *73*, 713–718.

Barman, K.; Asrey, R.; Pal, R. K. Putrescine and Carnauba Wax Pretreatments Alleviate Chilling Injury, Enhance Shelf Life and Preserve Pomegranate Fruit Quality During Cold Storage. *Sci. Hortic.* **2011,** *130* (4), 795–800.

Barman, K.; Asrey, R.; Pal, R. K.; Jha, S. K.; Bhatia, K. Post-harvest Nitric Oxide Treatment Reduces Chilling Injury and Enhances the Shelf-life of Mango (*Mangifera indica* L.) Fruit during Low-Temperature Storage. *J. Hortic. Sci. Biotechnol.* **2014,** *89* (3), 253–260.

Bassal, M.; El-Hamahmy, M. Hot Water Dip and Preconditioning Treatments to Reduce Chilling Injury and Maintain Postharvest Quality of Navel and Valencia Oranges During Cold Quarantine. *Postharvest Biol. Technol.* **2011,** *60*, 186–191.

Beligni, M. V.; Lamattina, L. Nitric Oxide in Plants: The History is just Beginning. *Plant Cell Environ.* **2001,** *24*, 267–278.

Ben-Amor, M.; Flores, B.; Latche, A.; Bouzayen, M.; Pech, J. C.; Romojaro, F. Inhibition of Ethylene Biosynthesis by Antisense ACC Oxidase RNA Prevents Chilling Injury in Charentais Cantaloupe Melons. *Plant Cell Environ.* **1999,** *22*, 1579–1586.

Bertolini, P.; Lanza, G.; Tonini, G. Effect of Pre-storage Carbon Dioxide Treatments and Storage Temperatures on Membranosis of 'Femminello comune' Lemons. *Sci. Hortic.* **1991,** *46*, 89–95.

Bishop, G. J.; Koncz, C. Brassinosteroids and Plant Steroid Hormone Signaling. *Plant Cell* **2002,** *14*, S97–S110.

Bower, J. P.; Cutting, J. G. M.; Truter, A. B. Modified Atmosphere Storage and Transport of Avocados—What does it Mean. *South African Avocado Growers' Assn. Yrbk.* **1989,** *12*, 17–20.

Brooks, C.; McColloch, L. P. Some Storage Diseases of Grapefruit. *J. Agric. Res.* **1936,** *52*, 319–351.

Cai, C.; Xu, C.; Shan, L.; Li, X.; Zhou, C. H.; Zhang, W. S.; Ferguson, I.; Chen, K. S. Low Temperature Conditioning Reduces Postharvest Chilling Injury in Loquat Fruit. *Posthar-vest Biol. Technol.* **2006,** *41*, 252–259.

Cai, Y.; Cao, S.; Yang, Z.; Zheng, Y. MeJA Regulates Enzymes Involved in Ascorbic Acid and Glutathione Metabolism and Improves Chilling Tolerance in Loquat Fruit. *Postharvest Biol. Technol.* **2011,** *59,* 324–326.

Candan, A. P.; Graell, J.; Larrigaudiere, C. Roles of Climacteric Ethylene in the Development of Chilling Injury in Plums. *Postharvest Biol. Technol.* **2008,** *47,* 107–112.

Cao, S. F.; Hu, Z. C.; Wang, H. O. Effect of Salicylic Acid on the Activities of Anti-oxidant Enzymes and Phenylalanine Ammonia-Lyase in Cucumber Fruit in Relation to Chilling Injury. *J. Hortic. Sci. Biotechnol.* **2009,** *84* (2), 125–130.

Cao, S.; Hu, Z.; Zheng, Y.; Lu, B. Synergistic Effect of Heat Treatment and Salicylic Acid on Alleviating Internal Browning in Cold Stored Peach Fruit. *Postharvest Biol. Technol.* **2010,** *58,* 93–97.

Cao, S.; Cai, Y.; Yang, Z.; Zheng, Y. MeJA Induces Chilling Tolerance in Loquat Fruit by Regulating Proline and γ-Aminobutyric Acid Contents. *Food Chem.* **2012,** *133,* 1466–1470.

Chen, B. X.; Yang, H. Q. 6-benzylaminopurine Alleviates Chilling Injury of Postharvest Cucumber Fruit Through Modulating Antioxidant System and Energy Status. *J. Sci. Food Agric.* **2013,** *93,* 1915–1921.

Chen, J. Y.; He, L. H.; Jiang, Y. M.; Wang, Y.; Joyce, D. C.; Ji, Z. L.; Lu, W. J. Role of Phenylalanine Ammonia-lyase in Heat Pretreatment Induced Chilling Tolerance in Banana Fruit. *Physiol. Plant* **2008,** *132,* 318–328.

Chen, J. J.; Jin, P.; Li, H. H.; Cai, Y. T.; Zhao, Y. Y.; Zheng, Y. H. Effects of Low Temperature Storage on Chilling Injury and Energy Status in Peach Fruit. *Trans. Chin. Soc. Agric. Eng.* **2012,** *28,* 275–281.

Clark, J.; Durner, D.; Navarre, A.; Klessig, D. F. Nitric Oxide Inhibition of Tobacco Catalase and Ascorbate Peroxidase. *Mol. Plant–Microbe Int.* **2000,** *13,* 1380–1384.

Clarke, S. M.; Mur, L. A. J.; Wood, J. E.; Scott, I. M. Salicylic Acid Dependent Signaling Promotes Basal Thermotolerance but is not Essential for Acquired Thermotolerance in *Arabidopsis thaliana. Planta* **2004,** *38,* 432–447.

Cohen, E. Commercial use of Long-term Storage of Lemon with Intermittent Warming. *HortScience* **1988,** *23,* 400.

Concellón, A.; Añón, M. C.; Chaves, A. R. Effect of Chilling on Ethylene Production in Eggplant Fruit. *Food Chem.* **2005,** *92,* 63–69.

Correa-Aragunde, N.; Graziano, M.; Lamattina, L. Nitric Oxide Plays a Central Role in Determining Lateral Root Development in Tomato. *Planta* **2004,** *218,* 900–905.

Creelman, R. A.; Mullet, J. E. Biosynthesis and Action of Jasmonates in Plants. *Annu. Rev. Plant Physiol. Plant Mol. Biol.* **1997,** *48,* 355–381.

Cuvi, M. J. A.; Vicente, A. R.; Concellón, A.; Chaves, A. R. Changes in Red Pepper Antioxidants as Affected by UV-C Treatments and Storage at Chilling Temperatures. *LWT–Food Sci. Technol.* **2011,** *44,* 1666–1671.

Deewatthanawong, R.; Nock, J. F.; Watkins, C. B. γ-Aminobutyric Acid (GABA) Accumulation in Four Strawberry Cultivars in Response to Elevated CO_2 Storage. *Postharvest Biol. Technol.* **2010a,** *57,* 92–96.

Deewatthanawong, R.; Rowell, P.; Watkins, C. B. γ-Aminobutyric Acid (GABA) Metabolism in CO_2 Treated Tomatoes. *Postharvest Biol. Technol.* **2010b,** *57,* 97–105.

Ding, C. K.; Wang, C. Y. The Dual Effects of Methyl Salicylate on Ripening and Expression of Ethylene Biosynthetic Genes in Tomato Fruit. *Plant Sci.* **2003,** *164,* 589–596.

Ding, C. K.; Wang, C. Y.; Gross, K. C.; Smith, D. L. Reduction of Chilling Injury and Transcript Accumulation of Heat Shock Protein Genes in Tomatoes by Methyl Jasmonate and Methyl Salicylate. *Plant Sci.* **2001**, *161*, 1153–1159.

Ding, C. K.; Wang, C. Y.; Gross, K. C.; Smith, D. L. Jasmonate and Salicylate Induce the Expression of Pathogenesis-Related-Protein Genes and Increase Resistance to Chilling Injury in Tomato Fruit. *Planta* **2002**, *214*, 895–901.

Ding, Z. S.; Tian, S. P.; Zheng, X. L.; Zhou, Z. W.; Xu, Y. Responses of Reactive Oxygen Metabolism and Quality in Mango Fruit to Exogenous Oxalic Acid or Salicylic Acid Under Chilling Temperature Stress. *Physiol. Plant.* **2007**, *130*, 112–121.

Dong, L.; Lurie, S.; Zhou, H. W. Effect of 1-Methylcyclopropene on Ripening of 'Canino' Apricots and 'Royal Zee' Plums. *Postharvest Biol. Technol.* **2002**, *24*, 135–145.

Durzan, D. J.; Pedroso, M. C. Nitric Oxide and Reactive Nitrogen Oxide Species in Plants. *Biotechnol. Genet. Eng. Rev.* **2002**, *19*, 293–337.

Ekman, J. H.; Clayton, M.; Biasi, W. V.; Mitcham, E. J. Interactions Between 1-MCP Concentration, Treatment Interval and Storage Time for 'Barlett' Pears. *Postharvest Biol. Technol.* **2004**, *31*, 127–136.

Evans, P. T.; Malmberg, R. L. Do Polyamines Have Roles in Plant Development? *Annu. Rev. Plant Physiol. Plant Mol. Biol.* **1989**, *40*, 235–269.

Farias-Eisner, R.; Chaudhuri, G.; Aeberhard, E.; Fukuto, J. M. The Chemistry and Tumoricidal Activity of Nitric Oxide/Hydrogen Peroxide and the Implications to Cell Resistance/Susceptibility. *J. Biol. Chem.* **1996**, *271*, 6144–6151.

Flores, H. E.; Protacio, C. M.; Signs, M. W. Primary and Secondary Metabolism of Polyamines in Plants. *Recent Adv. Phytochem.* **1989**, *23*, 329–393.

Flores, B. F.; Sanchez-Bel, P.; Monika, V.; Felix, R. M.; Isabel, E. M. Effects of a Pre-treatment with Nitric Oxide on Peach (*Prunus persica* L.) Storage at Room Temperature. *Eur. Food Res. Technol.* **2008**, *227*, 1599–1611.

Forney, C. F.; Lipton, W. J. Influence of Controlled Atmospheres and Packaging on Chilling Sensitivity. In *Chilling Injury of Horticultural Crops*; Wang, C. Y., Ed.; CRC Press: Boca Raton, FL, 1990; pp 257–268.

Fujioka, S.; Yokota, T. Biosynthesis and Metabolism of Brassinosteroids. *Annu. Rev. Plant Biol.* **2003**, *54*, 1371–1464.

Fung, R. W. M.; Wang, C. Y.; Smith, D. L.; Gross, K. C.; Tian, M. MeSA and MeJA Increase Steady-state Transcript Levels of Alternative Oxidase and Resistance against Chilling Injury in Sweet Peppers (*Capsicum annuum* L). *Plant Sci.* **2004**, *166*, 711–719.

Galston, A. W.; Sawhney, K. Polyamine and Plant Physiology. *Plant Physiol.* **1990**, *94*, 406–410.

Gao, H.; Zhang, Z. K.; Lv, X. G.; Cheng, N.; Peng, B. Z.; Cao, W. Effect of 24-Epibrassinolide on Chilling Injury of Peach Fruit in Relation to Phenolic and Proline Metabolisms. *Postharvest Biol. Technol.* **2016**, *111*, 390–397.

Ghasemnezhad, M.; Marsh, K.; Shilton, R.; Babalar, M.; Woolf, A. Effect of Hot Water Treatments on Chilling Injury and Heat Damage in 'Satsuma' Mandarins: Antioxidant Enzymes and Vacuolar ATPase, and Pyrophosphatase. *Postharvest Biol. Technol.* **2008**, *48*, 364–371.

Gómez-Galindo, F.; Herppich, W.; Gekas, V.; Sjöholm, I. Factors Affecting Quality and Postharvest Properties of Vegetables: Integration of Water Relations and Metabolism. *Crit. Rev. Food Sci. Nutr.* **2004**, *44*, 139–154.

González-Aguilar, G. A.; Gayosso, L.; Cruz, R.; Fortiz, J.; Báez, R.; Wang, C. Y. Polyamines Induced by Hot Water Treatments Reduce Chilling Injury and Decay in Pepper Fruit. *Postharvest Biol. Technol.* **2000**, *18*, 19–26.

González-Aguilar, G. A.; Tiznado-Hernández, M. E.; Savaleta-Gatica, R.; Martineze, M. A. Methyl Jasmonate Treatments Reduce Chilling Injury and Activate the Defense Response of Guava Fruits. *Biochem. Biophys. Res. Commun.* **2004**, *313*, 694–701.

Hardenburg, R. E.; Watada, A. E.; Wang, C. Y. *The Commercial Storage of Fruits, Vegetables, and Florist and Nursery Stocks* (USDA Handbook 66); USDA: Washington, DC, 1986.

Harper, J. R.; Balke, N. E. Characterization of the Inhibition of K^+ Absorption in Oats Roots by Salicylic Acid. *Plant Physiol.* **1981**, *68*, 1349–1353.

Hatton, T. T. Reduction of Chilling Injury with Temperature Manipulation. In *Chilling Injury of Horticultural Crops*; Wang, C. Y., Ed.; CRC Press: Boca Raton, FL, 1990; pp 269–280.

He, Y.; Tang, R. H.; Hao, Y.; Stevens, R. D.; Cook, C. W.; Ahn, S. M. Jing, L.; Yang, Z.; Chen, L.; Guo, F.; Fiorani, F.; Jackson, R. B.; Crawford, N. M.; Pei, Z. M. Nitric Oxide Represses the Arabidopsis Floral Transition. *Science* **2004**, *305*, 1968–1971.

He, L. H.; Chen, J. Y.; Kuang, J. F.; Lu, W. J. Expression of Three sHSP Genes Involved in Heat Pretreatment-Induced Chilling Tolerance in Banana Fruit. *J. Sci. Food Agric.* **2012**, *92*, 1924–1930.

Hernández, M. L.; Padilla, M. N.; Sicardo, M. D.; Mancha, M.; Martínez-Rivas, J. M. Effect of Different Environmental Stresses on the Expression of Oleate Desaturase Genes and Fatty Acid Composition in Olive Fruit. *Phytochemistry* **2011**, *72*, 178–187.

Hodges, D. M.; Lester, G. E.; Munro, K. D.; Toivonen, P. Oxidative Stress: Importance for Postharvest Quality: Oxidative Stress: Postharvest Fruits and Vegetables. *HortScience* **2004**, *39*, 924–929.

Jin, P.; Wang, K.; Shang, H.; Tong, J.; Zheng, Y. Low-Temperature Conditioning Combined with Methyl Jasmonate Treatment Reduces Chilling Injury of Peach Fruit. *J. Sci. Food Agric.* **2009**, *89*, 1690–1696.

Jin, P.; Duan, Y.; Wang, L.; Wang, J.; Zheng, Y. Reducing Chilling Injury of Loquat Fruit by Combined Treatment with Hot Air and Methyl Jasmonate. *Food Bioprocess Technol.* **2014a**, *7*, 2259–2266.

Jin, P.; Zhu, H.; Wang, L.; Shan, T.; Zheng, Y. Oxalic Acid Alleviates Chilling Injury in Peach Fruit by Regulating Energy Metabolism and Fatty Acid Contents. *Food Chem.* **2014b**, *161*, 87–93.

Kasim, R.; Kasim, M. U. The Effect of Ultraviolet Irradiation (UV-C) on Chilling Injury of Cucumbers During Cold Storage. *J. Food Agric. Environ.* **2008**, *6* (1), 50–54.

Khan, W.; Prithviraj, B.; Smith, D. L. Photosynthetic Responses of Corn and Soybean to Foliar Application of Salicylates. *J. Plant Physiol.* **2003**, *160*, 485–492.

Kim, T. W.; Wang, Z. Y. Brassinosteroid Signal Transduction from Receptor Kinases to Transcription Factors. *Annu. Rev. Plant Biol.* **2010**, *61*, 681–704.

Klessig, D. F.; Malamy, J. The Salicylic Acid Signal in Plants. *Plant Mol. Biol.* **1994**, *26*, 1439–1458.

Kondo, S.; Jitatham, A. Relationship Between Jasmonates and Chilling Injury in Mangosteens are Affected by Spermine. *HortScience* **2004**, *39* (6), 1346–1348.

Kramer, G. F.; Wang, C. Y. Correlation of Reduced Chilling Injury with Increased Spermine and Spermidine Levels in Zucchini Squash. *Physiol. Plant* **1989**, *76*, 479–484.

Lara, M. V.; Borsani, J.; Budde, C. O.; Lauxmann, M. A.; Lombardo, V. A.; Murray, R.; Andreo, C. S.; Drincovich, M. F. Biochemical and Proteomic Analysis of 'Dixiland' Peach Fruit (*Prunus persica*) upon Heat Treatment. *J. Exp. Bot.* **2009**, *60*, 4315–4333.

Lederman, I. E.; Zauberman, G.; Weksler, A.; Rot, I.; Fuchs, Y. Ethylene-Forming Capacity During Cold Storage and Chilling Injury Development in "Keitt" Mango Fruit. *Postharvest Biol. Technol.* **1997**, *10*, 107–112.

Leshem, Y. Y. *Nitric Oxide in Plants: Occurrence, Function and Use*; Kluwer: Dordrecht, The Netherlands, 2000.

Leshem, Y. Y.; Wills, R. B. H.; Ku, V. V. V. Evidence for the Function of the Free Radical Gas Nitric Oxide (NO) as an Endogenous Maturation and Senescence Regulating Factor in Higher Plants. *Plant Physiol. Biochem.* **1998**, *36*, 825–833.

Li, W.; Shao, Y.; Chen, W.; Jia, W. The Effects of Harvest Maturity on Storage Quality and Sucrose-Metabolizing Enzymes during Banana Ripening. *Food Bioprocess Technol.* **2011**, *4*, 1273–1280.

Li, D. M.; Guo, Y. K.; Li, Q.; Zhang, J.; Wang, X. J.; Bai, J. G. The Pre-treatment of Cucumber with Methyl Jasmonate Regulates Antioxidant Enzyme Activities and Protects Chloroplast and Mitochondrial Ultrastructure in Chilling-Stressed Leaves. *Sci. Hortic.* **2012a**, *143*, 135–143.

Li, B. Q.; Zhang, C. F.; Cao, B. H.; Qin, G. Z.; Wang, W. H.; Tian, S. P. Brassinolide Enhances Cold Stress Tolerance of Fruit by Regulating Plasma Membrane Proteins and Lipids. *Amino Acids* **2012b**, *43* (6), 2469–2480.

Li, P.; Zheng, X.; Liu, Y.; Zhu, Y. Pre-storage Application of Oxalic Acid Alleviates Chilling Injury in Mango Fruit by Modulating Proline Metabolism and Energy Status Under Chilling Stress. *Food Chem.* **2014**, *142*, 72–78.

Liu, Y. J.; Zhao, Z. G.; Si, J.; Di, C. X.; Han, J.; An, L. Z. Brassinosteroids Alleviate Chilling-Induced Oxidative Damage by Enhancing Antioxidant Defense System in Suspension Cultured Cells of *Chorispora bungeana*. *Plant Growth Regul.* **2009**, *59*, 207–214.

Liu, C.; Jahangir, M. M.; Ying, T. Alleviation of Chilling Injury in Postharvest Tomato Fruit by Preconditioning with Ultraviolet Irradiation. *J. Sci. Food Agric.* **2012**, *92*, 3016–3022.

Liu, Z.; Li, L.; Luo, Z.; Zeng, F.; Jiang, L.; Tang, K. Effect of Brassinolide on Energy Status and Proline Metabolism in Postharvest Bamboo Shoot During Chilling Stress. *Postharvest Biol. Technol.* **2016**, *111*, 240–246.

Los, D. A.; Murata, N. Membrane Fluidity and Its Roles in the Perception of Environmental Signals. *Biochim. Biophys. Acta* **2004**, *1666*, 142–157.

Luo, Z.; Chen, C.; Xie, J. Effect of Salicylic Acid Treatment on Alleviating Postharvest Chilling Injury of 'Qingnai' Plum Fruit. *Postharvest Biol. Technol.* **2011**, *62* (2), 115–120.

Lurie, S. Postharvest heat treatments. *Postharvest Biol. Technol.* **1998**, *14*, 257–269.

Lurie, S.; Handros, A.; Fallik, E.; Shapira, R. Reversible Inhibition of Tomato Fruit Gene Expression at High Temperature (Effects on Tomato Fruit Ripening). *Plant Physiol.* **1996**, *110*, 1207–1214.

Lurie, S.; Pesis, E. Effect of Acetaldehyde and Anaerobiosis as Postharvest Treatments on the Quality of Peaches and Nectarines. *Postharvest Biol. Technol.* **1992**, *1*, 317–326.

Lyons, J. M. Chilling Injury in Plants. *Ann. Rev. Plant Physiol.* **1973**, *24*, 445–466.

Lyons, J. M.; Breidenbach, R. W. Chilling Injury. In *Postharvest Physiology of Vegetables*; Weichmann, J., Ed.; Marcel Dekker: New York, 1987; pp 305–326.

Lyons, J. M.; Raison, J. K. Oxidative Activity of Mitochondria Isolated from Plant Tissues Sensitive and Resistant to Chilling Injury. *Plant Physiol.* **1970**, *45*, 386–389.

Manjunatha, G.; Lokesh, V.; Neelwarne, B. Nitric Oxide in Fruit Ripening: Trends and Opportunities. *Biotechnol. Adv.* **2010,** *28,* 489–499.

Mao, L. C.; Wang, G. Z.; Zhu, C. G.; Pang, H. Q. Involvement of Phospholipase D and Lipoxygenase in Response to Chilling Stress in Postharvest Cucumber Fruits. *Plant Sci.* **2007a,** *172,* 400–405.

Mao, L.; Pang, H.; Wang, G.; Zhu, C. Phospholipase D and Lipoxygenase Activity of Cucumber Fruit in Response to Chilling Stress. *Postharvest Biol. Technol.* **2007b,** *44,* 42–47.

Marcellin, P.; Chaves, A. Effect of Intermittent High CO_2 Treatment on Storage Life of Avocado Fruit in Relation to Respiration and Ethylene Production. *Acta Hortic.* **1983,** *138,* 155–163.

Martînez-Téllez, M. A.; Ramos-Clamont, M. G.; Gardea, A. A.; Vargas-Arispuro, I. Effect of Infiltrated Polyamines on Polygalacturonase Activity and Chilling Injury Responses in Zucchini Squash (*Cucurbita pepo* L.). *Biochem. Biophys. Res. Commun.* **2002,** *295,* 98–101.

Matsuo, T.; Ide, S.; Shitida, M. Correlation Between Chilling Sensitivity of Plant Tissues and Fatty Acid Composition of Phosphatidylglycerols. *Phytochemistry* **1992,** *31,* 2289–2293.

McGlasson, W. B; Scott, K. J.; Mendoza, J. D. B. The Refrigerated Storage of Tropical and Subtropical Products. *Int. J. Refrig.* **1979,** *2,* 199–206.

Meir, S.; Akerman, M.; Fuchs, Y.; Zauberman, G. Further Studies on the Controlled Atmosphere Storage of Avocados. *Postharvest Biol. Technol.* **1995,** *5,* 323–330.

Mirdehghan, S. H.; Rahemi, M.; Serrano, M.; Guillén, F.; Martínez-Romero, D.; Valero, D. Pre-storage Heat Treatment to Maintain Nutritive and Functional Properties During Postharvest Cold Storage of Pomegranate. *J. Agric. Food Chem.* **2006,** *54,* 8495–8500.

Mirdehghan, S. H.; Rahemi, M.; Martínez-Romero, D.; Guillén, F.; Valverde, J. M.; Zapata, P. J.; et al. Reduction of Pomegranate Chilling Injury During Storage After Heat Treatment: Role of Polyamines. *Postharvest Biol. Technol.* **2007a,** *44,* 19–25.

Mirdehghan, S. H.; Rahemi, M.; Castillo, S.; Martínez-Romero, D.; Serrano, M.; Valero, D. Pre-storage Application of Polyamines by Pressure or Immersion Improves Shelf-Life of Pomegranate Stored at Chilling Temperature by Increasing Endogenous Polyamine Levels. *Postharvest Biol. Technol.* **2007b,** *44* (1), 26–33.

Mirdehghan, S. H.; Rahemi, M.; Serrano, M.; Guillen, F.; Martínez-Romero, D.; Valero, D. The Application of Polyamines by Pressure or Immersion as a Tool to Maintain Functional Properties in Stored Pomegranate Arils. *J. Agric. Food Chem.* **2007c,** *55,* 755–760.

Mittler, R. Oxidative Stress, Antioxidants and Stress Tolerance. *Trends Plant Sci.* **2002,** *7,* 405–410.

Möller, I. M. Plant Mitochondria and Oxidative Stress: Electron Transport, NADPH Turnover, and Metabolism of Reactive Oxygen Species. *Annu. Rev. Plant Physiol. Plant Mol. Biol.* **2001,** *52,* 561–591.

Morris, L. L.; Platenius, H. Low Temperature Injury to Certain Vegetables. *Proc. Am. Soc. Hortic. Sci.* **1938,** *36,* 609–613.

Morris, K.; Mackerness, S. A. H.; Page, T.; John, C. F.; Murphy, A. M.; Carr, J. P.; Buchanan-Wollaston, V. Salicylic Acid has a Role in Regulating Gene Expression During Leaf Senescence. *Planta* **2000,** *23,* 677–685.

Murata, N.; Ishizakinishizawa, Q.; Higashi, S.; Hayashi, H.; Tasaka, Y.; Nishida, I. Genetically Engineered Alteration in the Chilling Sensitivity of Plants. *Nature* **1992,** *356,* 710–713.

Nakamura, R.; Inaba, A.; Ito, T. Effect of Cultivating Conditions and Postharvest Stepwise Cooling on the Chilling Sensitivity of Eggplant and Cucumber Fruits. *Sci. Rpt., Okayama Univ., Japan, Sci. Rpt.* **1985,** *66,* 19–29.

Nayyar, H.; Chander, S. Protective Effects of Polyamines against Oxidative Stress Induced by Water and Cold Stress in Chickpea. *J. Agric. Crop Sci.* **2004,** *190,* 355–365.

Neil, S. J.; Desikan, R.; Hancock, J. T. Nitric Oxide Signalling in Plants. *New Phytol.* **2003,** *159,* 11–35.

Nilprapruck, P.; Pradisthakarn, N.; Authanithee, F.; Keebjan, P. Effect of Exogenous Methyl Jasmonate on Chilling Injury and Quality of Pineapple (*Ananas comosus* L.) cv. Pattavia. *Silpakorn. U. Sci. Technol. J.* **2008,** *2* (2), 33–42.

O'Hare, T. J.; Prasad, A. The Effect of Temperature and Carbon Dioxide on Chilling Symptoms in Mango. *Acta Hortic.* **1993,** *343,* 244–250.

Ogweno, J. O.; Song, X. S.; Shi, K.; Hu, W. H.; Mao, W. H.; Zhou, Y. H.; Yu, J. Q.; Nogues, S. Brassinosteroids Alleviate Heat-induced Inhibition of Photosynthesis by Increasing Carboxylation Efficiency and Enhancing Antioxidant Systems in *Lycopersicon esculentum. Plant Growth Regul.* **2008,** *27,* 49–57.

Pantastico, E. B.; Grierson, W.; Soule, J. Chilling Injury in Tropical Fruits: I. Bananas (*Musa paradisiaca* var. Sapientum cv. Lacatan). *Proc. Am. Soc. Hortic. Sci.* **1967,** *11,* 83–91.

Parkin, K. L.; Kuo, S. J. Chilling-induced Lipid Degradation in Cucumber (*Cucumis sativa* L. cv. Hybrid C) Fruit. *Plant Physiol.* **1989,** *90,* 1049–1056.

Patterson, B. D.; Reid, M. S. Genetic and Environmental Influences on the Expression of Chilling Injury. In *Chilling Injury of Horticultural Crops*; Wang, C. Y., Ed.; CRC Press: Boca Raton, FL, 1990; pp 87–112.

Paull, R. E. Chilling Injury of Crops of Tropical and Subtropical Origin. In *Chilling Injury of Horticultural Crops*; Wang, C. Y., Ed.; CRC Press: Boca Raton, FL, 1990; pp 17–36.

Pesis, E.; Marinansky, R.; Zauberman, G.; Fuchs, Y. Pre-storage Low Oxygen Atmosphere Treatment Reduces Chilling Injury Symptoms in 'Fuerte' Avocado Fruit. *HortScience* **1994,** *29,* 1042–1046.

Pesis, E.; Ackerman, M.; Ben-Arie, R.; Feygenberg, O.; Feng, X.; Apelbaum, A.; Goren, R.; Prusky, D. Ethylene Involvement in Chilling Injury Symptoms of Avocado During Cold Storage. *Postharvest Biol. Technol.* **2002,** *24,* 171–181.

Pongprasert, N.; Sekozawa, Y.; Sugaya, S.; Gemma, H. The Role and Mode of Action of UV-C Hormesis in Reducing Cellular Oxidative Stress and the Consequential Chilling Injury of Banana Fruit Peel. *Int. Food Res. J.* **2011,** *18,* 741–749.

Porat, R.; Weiss, B.; Cohen, L.; Daus, A.; Goren, R.; Droby, S. Effects of Ethylene and 1-Methylcyclopropene on the Postharvest Qualities of 'Shamouti' Oranges. *Postharvest Biol. Technol.* **1999,** *15,* 155–163.

Prado, A. M.; Porterfield, D. M.; Feijo, J. A. Nitric Oxide is Involved in Growth Regulation and Reorientation of Pollen Tubes. *Development* **2004,** *131,* 2707–2714.

Purvis, A. C. Free Proline in Peel of Grapefruit and Resistance to Chilling Injury During Cold Storage. *HortScience* **1981,** *16,* 160–161.

Purvis, A. C.; Grierson, W. Accumulation of Reducing Sugar and Resistance of Grapefruit Peel to Chilling Injury as Related to Winter Temperatures. *J. Amer. Soc. Hortic. Sci.* **1982,** *107,* 139–142.

Raison, J. K.; Orr, G. R. Proposals for a Better Understanding of the Molecular Basis of Chilling Injury. In *Chilling Injury of Horticultural Crops*; Wang, C. Y., Ed.; CRC Press: Boca Raton, FL, 1990; pp 145–164.

Rajjou, L.; Belghazi, M.; Huguet, R.; Robin, C.; Moreau, A.; Job, C.; Job, D. Proteomic Investigation of the Effect of Salicylic Acid on Arabidopsis Seed Germination and Establishment of Early Defense Mechanisms. *Plant Physiol.* **2006**, *141*, 910–923.

Rawyler, A.; Pavelic, D.; Gianinazzi, C.; Oberson, J.; Braendle, R. Membrane Lipid Integrity Relies on a Threshold of ATP Production Rate in Potato Cell Cultures Submitted to Anoxia. *Plant Physiol.* **1999**, *120*, 293–300.

Retamales, J.; Cooper, T.; Streif, J.; Kama, J. C. Preventing Cold Storage Disorders in Nectarines. *J. Hortic. Sci.* **1992**, *67*, 619–626.

Rozenzvieg, D.; Elmaci, C.; Samach, A.; Lurie, S.; Porat, R. Isolation of Four Heat Shock Protein cDNAs from Grapefruit Peel Tissue and Characterization of Their Expression in Response to Heat and Chilling Temperature Stresses. *Physiol. Plant.* **2004**, *121*, 421–428.

Rui, H.; Cao, S.; Shang, H.; Jin, P.; Wang, K.; Zheng, Y. Effects of Heat Treatment on Internal Browning and Membrane Fatty Acid in Loquat Fruit in Response to Chilling Stress. *J. Sci. Food Agric.* **2010**, *90*, 1557–1561.

Saba, M. K.; Arzani, K.; Barzegar, M. Postharvest Polyamine Application Alleviates Chilling Injury and Affects Apricot Storage Ability. *J. Agric. Food Chem.* **2012**, *60*, 8947–8953.

Sabehat, A.; Lurie, S.; Weiss, D. Expression of Small Heat-shock Proteins at Low Temperatures. A Possible Role in Protecting Against Chilling Injuries. *Plant Physiol.* **1998a**, *117*, 651–658.

Sabehat, A.; Weiss, D.; Lurie, S. Heat-shock Proteins and Cross-tolerance in Plants. *Physiol. Plant.* **1998b**, *103*, 437–441.

Safizadeh, M.; Rahemi, M.; Aminlari, M. Effect of Postharvest Calcium and Hot-water Dip Treatments on Catalase, Peroxidase and Superoxide Dismutase in Chilled Lisbon lemon fruit. *Int. J. Agric. Res.* **2007**, *2*, 440–449.

Saltveit, M. E. Jr.; Morris, L. L. Overview of Chilling Injury of Horticultural Crops. In *Chilling Injury of Horticultural Crops*; Wang, C. Y., Ed.; CRC Press: Boca Raton, FL, 1990; pp 3–15.

Salvador, A.; Arnal, L.; Monterde, A.; Cuquerella, J. Reduction of Chilling Injury Symptoms in Persimmon Fruit cv. 'Rojo Brillante' by 1-MCP. *Postharvest Biol. Technol.* **2004**, *33*, 285–291.

Sapitnitskaya, M.; Maul, P.; McCollum, G. T.; Guy, C. L.; Weiss, B.; Samach, A.; Porat, R. Postharvest Heat and Conditioning Treatments Activate Different Molecular Responses and Reduce Chilling Injuries in Grapefruit. *J. Exp. Bot.* **2006**, *57*, 2943–2953.

Sasse, J. M. Physiological Actions of Brassinosteroids: An Update. *J. Plant Growth Regul.* **2003**, *22*, 276–288.

Sayyari, M.; Babalare, M.; Kalantarie, S.; Serranoc, M.; Valero, D. Effect of Salicylic Acid Treatment on Reducing Chilling Injury in Stored Pomegranates. *Postharvest Biol. Technol.* **2009**, *53*, 152–154.

Sayyari, M.; Castillo, S.; Valero, D.; Díaz-Mula, H. M.; Serrano, M. Acetyl Salicylic Acid Alleviates Chilling Injury and Maintains Nutritive and Bioactive Compounds and Antioxidant Activity during Postharvest Storage of Pomegranates. *Postharvest Biol. Technol.* **2011a**, *60* (2), 136–142.

Sayyari, M.; Babalar, M.; Kalantari, S.; Martínez-Romero, D.; Guillén, F.; Serrano, M.; Valero, D. Vapour Treatments with Methyl Salicylate or Methyl Jasmonate Alleviated Chilling Injury and Enhanced Antioxidant Potential During Postharvest Storage of Pomegranates. *Food Chem.* **2011b**, *124*, 964–970.

Sayyari, M.; Valero, D.; Babalar, M.; Kalantari, S.; Zapata, P. J.; Serrano, M. Prestorage Oxalic Acid Treatment Maintained Visual Quality, Bioactive Compounds, and Antioxidant Potential of Pomegranate After Long-Term Storage at 2°C. *J. Agric. Food Chem.* **2012**, *58*, 6804–6808.

Scandalios, J. G. Oxygen Stress and Superoxide Dismutases. *Plant Physiol.* **1993**, *101*, 7–12.

Selvarajah, S.; Bauchot, A. D.; John, P. Internal Browning in Cold Stored Pineapples is Suppressed by a Postharvest Application of 1-Methylcyclopropene. *Postharvest Biol. Technol.* **2001**, *23*, 167–170.

Serek, M.; Tamari, G.; Sisler, E. C.; Borochov, A. Inhibition of Ethylene Induced Cellular Senescence Symptoms by 1-Methylcyclopropene, a New Inhibitor of Ethylene Action. *Physiol. Plant* **1995**, *94*, 229–232.

Sevillano, L.; Sanchez-Ballesta, M. T.; Romojaro, F.; Floresc, F. B. Physiological, Hormonal and Molecular Mechanisms Regulating Chilling Injury in Horticultural Species. Postharvest Technologies Applied to Reduce Its Impact. *J. Sci. Food Agric.* **2009**, *89*, 555–573.

Sevillano, L.; Sola, M. M.; Vargas, A. M. Induction of Small Heat-Shock Proteins in Mesocarp of Cherimoya Fruit (*Annona cherimola* Mill.) Produces Chilling Tolerance. *J. Food Biochem.* **2010**, *34*, 625–638.

Shang, H.; Cao, S.; Yang, Z.; Cai, Y.; Zheng, Y. Effect of Exogenous Gamma-Aminobutyric Acid Treatment on Proline Accumulation and Chilling Injury in Peach Fruit after Long-Term Cold Storage. *J. Agric. Food Chem.* **2011**, *59*, 1264–1268.

Shao, X.; Tu, K. Hot Air Treatment Improved the Chilling Resistance of Loquat Fruit Under Cold Storage. *J. Food Process. Preserv.* **2013**. DOI: 10.1111/jfpp.12019.

Shao, X.; Zhu, Y.; Cao, S.; Wang, H.; Song, Y. Soluble Sugar Content and Metabolism as Related to the Heat-Induced Chilling Tolerance of Loquat Fruit During Cold Storage. *Food Bioprocess Technol.* **2013**. DOI: 10.1007/s11947-012-1011-6.

Shelp, B. J.; Bown, A. W.; McLean, M. D. Metabolism and Functions of Gamma-Aminobutyric Acid. *Trends Plant Sci.* **1999**, *4*, 446–452.

Siboza, X. I.; Bertling, I.; Odindo, A. O. Salicylic Acid and Methyl Jasmonate Improve Chilling Tolerance in Cold-Stored Lemon Fruit (*Citrus limon*). *J. Plant Physiol.* **2014**, *171*, 1722–1731.

Singh, S. P.; Pal, R. K. Controlled Atmosphere Storage of Guava (*Psidium guajava* L.) Fruit. *Postharvest Biol. Technol.* **2008**, *47*, 296–306.

Singh, S. P.; Singh, Z.; Swinny, E. E. Postharvest Nitric Oxide Fumigation Delays Fruit Ripening and Alleviates Chilling Injury During Cold Storage of Japanese Plums (*Prunus salicina* Lindell). *Postharvest Biol. Technol.* **2009**, *53*, 101–108.

Sisler, E. C.; Serek, M. Inhibitors of Ethylene Responses in Plants at the Receptor Level: Recent Developments. *Physiol. Plant* **1997**, *100*, 577–582.

Sisler, E. C.; Serek, M.; Dupille, E. Comparison of Cyclopropene, 1-Methylcyclopropene and 3,3-Dimethylcyclopropene as Ethylene Antagonists in Plants. *Plant Growth Regul.* **1996**, *18*, 169–175.

Slocum, R. D. Polyamine Biosynthesis in Plants. In *The Physiology of Polyamines in Plants*; Slocum, R. D., Flores, H. E., Eds.; CRC Press: Boca Raton, FL, 1991; pp 133–140.

Smith, T. A. Polyamines. *Annu. Rev. Plant Physiol.* **1985**, *36*, 117–143.

Stanley, D. W. Biological Membrane Deterioration and Assisted Quality Losses in Food Tissues. *Crit. Rev. Food Sci. Nutr.* **1991**, *30*, 487–553.

Sterling, T. M.; Hall, J. C. Mechanism of Action of Natural Auxins and the Auxinic Herbicides. In *Herbicide Activity: Toxicology, Biochemistry and Molecular Biology*; Roe, R.

M., Burton, J. D., Kuhr, R. J., Eds.; IOS Press: Amsterdam, The Netherlands, 1997; pp 111–141.

Tabacchi, M. H.; Hicks, J. R.; Ludford, P. M.; Robinson, R. W. Chilling Injury Tolerance and Fatty Acid Composition in Tomatoes. *HortScience* **1979**, *14*, 424.

Tian, S.; Qin, G.; Li, B.; Wang, Q.; Meng, X. Effects of Salicylic Acid on Disease Resistance and Postharvest Decay Control of Fruits. *Stewart Postharvest Rev.* **2007**, *6*, 1–7.

Verbruggen, N.; Hermans, C. Proline Accumulation in Plants: A Review. *Amino Acids* **2008**, *35*, 753–759.

Vicente, A. R.; Pineda, C.; Lemoine, L.; Civello, P. M.; Martinez, G. A.; Chaves, A. R. UV-C Treatment Reduce Decay, Retain Quality and Alleviate Chilling Injury in Pepper. *Postharvest Biol. Technol.* **2005**, *35*, 69–78.

Vigh, L.; Maresca, B.; Harwood, J. L. Does the Membrane's Physical State Control the Expression of Heat Shock and Other Genes? *Trends Biochem. Sci.* **1998**, *23*, 369–374.

Wang, C. Y. Approaches to Reduction of Chilling Injury of Fruits and Vegetables. *Hortic. Rev.* **1993**, *15*, 63–95.

Wang, C. Y. Chilling Injury of Tropical Horticultural Commodities. *HortScience* **1994a**, *29* (9), 986–988.

Wang, C. Y. Reduction of Chilling Injury by Methyl Jasmonate. *Acta Hortic.* **1994b**, *368*, 901–907.

Wang, C. Y. Combined Treatment of Heat Shock and Low Temperature Conditioning Reduces Chilling Injury in Zucchini Squash. *Postharvest Biol. Technol.* **1994c**, *4*, 65–73.

Wang, C. Y. Temperature Preconditioning Affects Ascorbate Antioxidant System in Chilled Zucchini Squash. *Postharvest Biol. Technol.* **1996**, *8*, 29–36.

Wang, C. Y.; Baker, J. E. Effects of Two Free Radical Scavengers and Intermittent Warming on Chilling Injury and Polar Lipid Composition of Cucumber and Sweet Pepper Fruits. *Plant Cell Physiol.* **1979**, *20*, 243–251.

Wang, C. Y.; Buta, J. G. Methyl Jasmonate Reduces Chilling Injury in *Cucurbita pepo* Through Its Regulation of Abscisic Acid and Polyamine Levels. *Environ. Exp. Bot.* **1994**, *34*, 427–432.

Wang, L.; Chena, S.; Kong, W.; Li, S.; Archbold, D. D. Salicylic Acid Pretreatment Alleviates Chilling Injury and Affects the Antioxidant System and Heat Shock Proteins of Peaches During Cold Storage. *Postharvest Biol. Technol.* **2006**, *41*, 244–251.

Wang, B.; Wang, J.; Liang, H.; Yi, J.; Zhang, J.; Lin, L.; Wu, Y.; Feng, X.; Cao, J.; Jiang, W. Reduced Chilling Injury in Mango Fruit by 2,4-Dichlorophenoxyacetic Acid and the Antioxidant Response. *Postharvest Biol. Technol.* **2008**, *48*, 172–181.

Wang, H.; Zhang, Z.; Xu, L.; Huang, X.; Pang, X. The Effect of Delay Between Heat Treatment and Cold Storage on Alleviation of Chilling Injury in Banana Fruit. *J. Sci. Food Agric.* **2012a**, *92*, 2624–2629.

Wang, Q.; Ding, T.; Gao, L.; Pang, J.; Yang, N. Effect of Brassinolide on Chilling Injury of Green Bell Pepper in Storage. *Sci. Hortic.* **2012b**, *144*, 195–200.

Wang, Y.; Luo, Z.; Huang, X.; Yang, K.; Gao, S.; Du, R. Effect of Exogenous γ-Aminobutyric Acid (GABA) Treatment on Chilling Injury and Antioxidant Capacity in Banana Peel. *Sci. Hortic.* **2014**, *168*, 132–137.

Wardowski, W. F.; Grierson, W.; Edwards, G. J. Chilling Injury of Stored Limes and Grapefruit as Affected by Differentially Permeable Films. *HortScience* **1973**, *8*, 173–175.

Wendehenne, D.; Pugin, A.; Klessig, D. F.; Durner, J. Nitric Oxide: Comparative Synthesis and Signalling in Animal and Plant Cells. *Trends Plant Sci.* **2001**, *6*, 177–183.

Whitaker, B. D. Changes in Galactolipid and Phospholipid Levels of Tomato Fruits Stored at Chilling and Non-chilling Temperatures. *Phytochemistry* **1992**, *31*, 2627–2630.

Whitaker, B. D. A Reassessment of Heat Treatment as a Means of Reducing Chilling Injury in Tomato Fruit. *Postharvest Biol. Technol.* **1994**, *4*, 75–83.

Wu, B.; Guo, Q.; Li, Q.; Ha, Y.; Li, X.; Chen, W. Impact of Postharvest Nitric Oxide Treatment on Antioxidant Enzymes and Related Genes in Banana Fruit in Response to Chilling Tolerance. *Postharvest Biol. Technol.* **2014**, *92*, 157–163.

Xu, M.; Dong, J.; Zhang, M.; Xu, X.; Sun, L. Cold-induced Endogenous Nitric Oxide Generation Plays a role in Chilling Tolerance of Loquat Fruit during Postharvest Storage. *Postharvest Biol. Technol.* **2012**, *65*, 5–12.

Xue, X. J.; Li, P. Y.; Song, X. Q.; Shen, M.; Zheng, X. L. Mechanisms of Oxalic Acid Alleviating Chilling Injury in Harvested Mango Fruit Under Low Temperature Stress. *Acta Hortic. Sin.* **2012**, *39*, 2251–2257.

Yang, H.; Wu, F.; Cheng, J. Reduced Chilling Injury in Cucumber by Nitric Oxide and the Antioxidant Response. *Food Chem.* **2011a**, *127*, 1237–1242.

Yang, A. P.; Cao, S. F.; Yang, Z. F.; Cai, Y. T.; Zheng, Y. H. Gamma-aminobutyric Acid Treatment Reduces Chilling Injury and Activates the Defence Response of Peach Fruit. *Food Chem.* **2011b**, *129*, 1619–1622.

Zaharah, S. S.; Singh, Z. Postharvest Nitric Oxide Fumigation Alleviates Chilling Injury, Delays Fruit Ripening and Maintains Quality in Cold-stored 'Kensington Pride' Mango. *Postharvest Biol. Technol.* **2011**, *60*, 202–210.

Zhang, J.; Huang,W.; Pan, Q.; Liu, Y. Improvement of Chilling Tolerance and Accumulation of Heat Shock Proteins in Grape Berries (*Vitis vinifera* cv. Jingxiu) by Heat Pretreatment. *Postharvest Biol. Technol.* **2005**, *38* (1), 80–90.

Zhang, X.; Shen, L.; Li, F.; Meng, D.; Sheng, J. Methyl Salicylate-induced Arginine Catabolism is Associated with Up-regulation of Polyamine and Nitric Oxide Levels and Improves Chilling Tolerance in Cherry Tomato Fruit. *J. Agric. Food Chem.* **2011**, *59*, 9351–9357.

Zhang, X.; Sheng, J.; Li, F.; Meng, D.; Shen, L. Methyl Jasmonate Alters Arginine Catabolism and Improves Postharvest Chilling Tolerance in Cherry Tomato Fruit. *Postharvest Biol. Technol.* **2012**, *64*, 160–167.

Zhang, X.; Shen, L.; Li, F.; Meng, D.; Sheng, J. Amelioration of Chilling Stress by Arginine in Tomato Fruit: Changes in Endogenous Arginine Catabolism. *Postharvest Biol. Technol.* **2013a**, *76*, 106–111.

Zhang, X.; Shen, L.; Li, F.; Meng, D.; Sheng, J. Arginase Induction by Heat Treatment Contributes to Amelioration of Chilling Injury and Activation of Antioxidant Enzymes in Tomato Fruit. *Postharvest Biol. Technol.* **2013b**, *79*, 1–8.

Zhang, X.; Shen, L.; Li, F.; Meng, D.; Sheng, J. Hot Air Treatment-induced Arginine Catabolism is Associated with Elevated Polyamines and Proline Levels and Alleviates Chilling Injury in Postharvest Tomato Fruit. *J. Sci. Food Agric.* **2013c**, *93*, 3245–3251.

Zhang, Y.; Zhang, M.; Yang, H. Postharvest Chitosan-g-Salicylic Acid Application Alleviates Chilling Injury and Preserves Cucumber Fruit Quality During Cold Storage. *Food Chem.* **2015**, *174*, 558–563.

Zhao, Z.; Jiang, W.; Cao, J.; Zhao, Y.; Gu, Y. Effect of Cold-shock Treatment on Chilling Injury in Mango (*Mangifera indica* L. cv. 'Wacheng') Fruit. *J. Sci. Food Agric.* **2006**, *86*, 2458–2462.

Zhao, R.; Sheng, J.; Lv, S.; Zheng, Y.; Zhang, J.; Yu, M.; Shen, L. Nitric Oxide Participates in the Regulation of *LeCBF1* Gene Expression and Improves Cold Tolerance in Harvested Tomato Fruit. *Postharvest Biol. Technol.* **2011**, *62*, 121–126.

Zhu, S.; Lina, S.; Mengchen, L.; Jie, Z. Effect of Nitric Oxide on Reactive Oxygen Species and Antioxidant Enzymes in Kiwi Fruit During Storage. *J. Sci. Food Agric.* **2008**, *88*, 324–331.

Zhu, L. Q.; Zhou, J.; Zhu, S. H. Effect of a Combination of Nitric Oxide Treatment and Intermittent Warming on Prevention of Chilling Injury of 'Feicheng' Peach Fruit During Storage. *Food Chem.* **2010**, *121*, 165–170.

Zong, R. J.; Morris, L.; Cantwell, M. Postharvest Physiology and Quality of Bitter-melon (*Momordica charantia* L). *Postharvest Biol. Technol.* **1995**, *6*, 65–72.

CHAPTER 2

HEAT TREATMENTS FOR ENHANCING STORABILITY OF FRUITS AND VEGETABLES

SUSAN LURIE*

Department of Postharvest Science, Agricultural Research Organization, Volcani Center, Bet Dagan, Israel

E-mail: slurie43@agri.gov.il

ABSTRACT

Heat treatment before postharvest storage or intermittent during storage assists insect disinfestation, disease control, mitigates chilling injury symptoms and is favorable for retention of fruit quality. Optimum temperature and exposure duration vary with the commodity in question and the purpose for which it is employed. Vapor heat treatments and hot water immersion techniques are integrated in the commercial fruit industry for varied purposes like disinfestation, maintenance of fruit quality, lowering chilling stress, and extension of storage life. Its wide usage is credited to ease in application, efficacy, and low cost. It prompts resistance to cold stress and impedes development of chilling injury by modulation of ripening process. In this chapter, techniques of hot water treatments, responses of heat treatments on commodity physiology, fungal invasions and insect infestations of different fruits and vegetables are discussed.

2.1 INTRODUCTION

Postharvest heat treatments of fruit are used for insect disinfestation, disease control, to modify fruit responses to cold stress, and to maintain fruit quality during storage. The temperature and exposure duration for a particular crop for a particular purpose are usually determined empirically. Conditioning treatments at temperatures from 30°C to 40°C in hot air (HAT) for times ranging from hours to days have been developed to affect commodity quality and storability. Higher temperatures are normally used for either insect or microorganism control. Some treatments which have commercial approval for insect control involve either vapor heat or hot water immersion of the fruit until the core temperature reaches 47°C, which takes a number of hours unless radiofrequency (RF) is involved (Fig. 2.1). Other treatments, involving hot water dips (HWT), are for a few minutes at temperatures of 50–56°C to control fungal pathogens. In addition, there is a short (10–25 s) treatment of hot water rinsing and brushing (HWRB) which may involve temperatures of up to 63°C.

The application of heat in various forms has also been utilized to prevent the development of physiological alterations triggered by cold storage, thereby extending the shelf life of fruit and vegetables. In fresh commodities, heat treatments affect the respiration, ripening process, ethylene production, and changes in the conformation of macromolecules, including protein aggregation and membrane fluidity, hence affecting the quality of horticultural produce. The temperature and duration of different types of heat treatment may also affect the nutritional and quality attributes of produce. The rate of heating will be different in air or water as shown in Figure 2.1. HWT is widely utilized in many countries for decay control because it is relatively easy to apply and is cost-effective when compared to HAT. Both hot water and hot air have been developed for insect disinfestation protocols and for affecting fresh produce quality, both by inducing resistance to low-temperature damage and by modulating ripening processes.

The number of studies published on the uses of heat treatment on harvested produce has been increasing over the past 25 years. In many cases, the results are interesting but do not lead to a commercial treatment. This chapter will provide tables of the commodities tested since the year 2000, but it should be recognized that it can only be a partial list. In addition to the experiments aimed to control physiological disorders, there is an increasing research on the effects of a heat treatment on gene expression in the fruit or vegetable and how the changes in gene expression lead to the observed effects of the treatment. This chapter will not deal with this aspect of heat treatment, but the reader can refer to Lurie and Pedreschi (2014) for some of the findings.

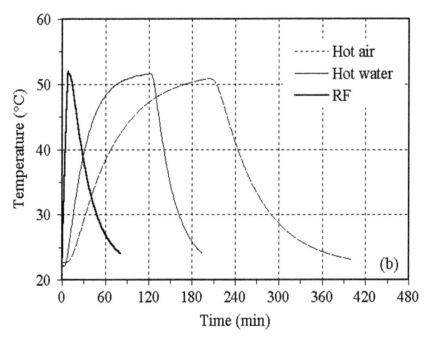

FIGURE 2.1 Comparative kinetics of produce heating and cooling by radiofrequency (RF), hot water, and hot air.

2.2 HOT WATER RINSING AND BRUSHING

In the 1990s, a hot water rinse brushing system (HWRB) was developed at the Agricultural Research Organization in Israel, and this machine has been upgraded and improved in the years (Fallik, 2004; Sivakumar and Fallik, 2013). The fresh produce is first rinsed with cold water and then rolls over brushes into a pressurized recycled hot water rinse at temperatures between 48°C and 63°C for 10–25 s (Fig. 2.2). The machine can easily be incorporated into the commercial packing lines. It was originally developed to clean peppers grown in the desert areas of Israel, and which were harvested covered with dust and sand. Early studies on the effects of HWRB showed that it altered the surface wax to give the fruit a high gloss and sealed small cracks through which pathogens could enter (Fallik et al., 1999). The spray washed off surface pathogens and the high water temperature killed them. It has been used to control pathogens on a number of fruits and vegetables including: *Penicillium expansum* on apple (Fallik et al., 2001), *Penicillium digitatum* on citrus (Porat et al., 2000a), *Alternaria alternata* on mango

(Prusky et al., 1999), *Alternaria alternata* and *Fusarium solani* on melon (Fallik et al., 2000), and *Botrytis cinerea* on tomato (Fallik et al., 2002). It has been adapted in California for the citrus industry (Smilanick et al., 2003).

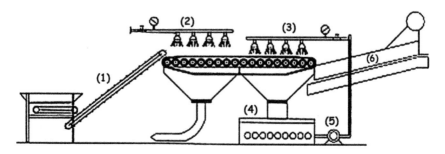

FIGURE 2.2   Schematic of the hot water rinse and brush (HWRB) machine. (1) conveyer; (2) rinse in tap water; (3) hot water spray of the produce on brush rollers; (4) hot water container; (5) water pump for recycling of hot water; and (6) drying the produce with forced air.

2.3   FUNGAL CONTROL

The efficacy of a heat treatment on a fungal pathogen is measured by the reduction in viability of the organism, either in spore germination or mycelial growth. A list of pathogens controlled by heat treatment is presented in the review of Barkai-Golan and Phillips (1991). The ability to control pathogens by high temperature is a function of the time and temperature of exposure. A comparison of dipping grapefruit in hot water (HWT at 53°C for 2 min) or giving HWRB for 30 s at 60°C found that both controlled natural decay development (Porat et al., 2000b). In addition, examining different temperatures of HWRB from 53°C to 62°C for 20 s on inoculated citrus fruit found that the higher temperatures controlled decay development better than the lower temperatures (Porat et al., 2000a).

In addition to the HWRB machine in use to wash and disinfect fresh produce, a steam unit was developed at the Agricultural Research Center in Israel to treat root crops (Gan-Mor et al., 2011). The crops are steamed for 2–3 s. Laboratory tests showed that the treatment reduced carrot and sweet potato rots that develop in storage (Afek and Orenstein, 2002). Recently, this treatment was incorporated into a systems approach for controlling black root rot caused by the fungus *Thielaviopsis basicola* on carrots. The steam is applied and then the carrot receives an application of stabilized hydrogen peroxide or a biological control agent (Eshel et al., 2009). The use of one than more treatment in series was also found to benefit mango. HWRB reduced *Alternaria alternata* but led to lenticel discoloration. The lenticel damage was eliminated when the hot water spray was performed without brushing, and a field wash with NaOH was added for additional fungal control (Feygenberg et al., 2014).

Other crops also benefit from a systems approach to control disease development. Control of *Penicillium expansum* on peaches was best when HWRB at 60°C for 20 s was combined with biological control agent yeast, while *Monilinia fructicola* was controlled by HSRB alone (Karabulut et al., 2002). Combining HWRB at 62°C for 20 s with either sodium bicarbonate or biological control agent yeast controlled decay by *Penicillium digitatum* on inoculated grapefruit in an additive manner (Porat et al., 2002). Table 2.1 includes other crops which have been treated for decay control with a combination of treatments, as well as single heat treatments to control decay.

A novel heat treatment is used to control *Alternaria* spp. on summer basil grown in greenhouses. The fungus infects the leaves during growth and black spots develop on the leaves during postharvest shipping. By closing the greenhouse 3 days before harvest and allowing the temperature to rise above 42°C for several hours, the latent infection was eliminated (Kenigsbuch et al., 2009).

TABLE 2.1 Heat Treatments Tested on Different Commodities to Control Fungal Decay.

Commodity	Fungus	Treatment	Results	References
Apple				
	Various inoculated fungi	HWT 50–54°C, 3 min to 25 s	Controlled fungal development in storage, 65°C caused heat damage	Maxin et al. (2012)
	Penicillium expansum	HWT 45°C, 10 min	Inhibited decay, examined gene expression with microarray	Spadoni et al. (2015)
Banana				
Plantain	*Fusarium verticillioides*	HWT 53°C, 3 min; 50°C, 5 min	Effective decay control on all stages of banana, HAT or steam heat less effective	Aborisade and Akomolafe (2011)
Citrus				
"Clementine" mandarin	*Penicillium digitatum, P. italicum*	HWT 45 or 50°C, 60 or 150 s SC	HWT alone did not control inoculated fruit, SC at 50°C was best	Palou et al. (2002)

TABLE 2.1   *(Continued)*

Commodity	Fungus	Treatment	Results	References
Lemon	*P. digitatum*, *P. italicum*, *Geotrichum citri-aurantii*, phomopsis, and diplodia stem-end rot	HWT 50°C, potassium sorbate, SC, potassium phosphite, hydrogen peroxide	HWT alone did not control inoculated fruit, combinations of the salts could control most of the different fungi	Cerioni et al. (2013)
"Tarocco" orange	Natural decay	HWT 53, 56, or 59°C for 60, 30, and 15 s, TBZ	TBZ at 300 ppm at 53 or 56°C prevented decay after simulated storage and shipment	Palma et al. (2013)
Lemon and orange	*P. digitatum*, *G. citri-aurantii*, *P. italicum*	HWRB 48.9, 54.4, 60.0, or 62.8°C 15 or 30 s	Treatment decreased decay of inoculated fruit with *Penicillium*, *Geotrichum* needed SC	Smilanick et al. (2003)
Loquat	*Colletotrichum acutatum*	HAT 38°C, 36 h + biocontrol	Decreases natural and inoculated decay, antioxidant activity increased after treatment and defense enzymes PAL and β-1,3-glucanase	Liu et al. (2010)
Papaya				
	Colletotrichum gloeosporioides	HWT 54°C, 4 min	Inhibited anthracnose and stem-end rot, delayed softening, enhanced color, induced defense genes	Li et al. (2013)
	C. gloeosporioides	Dry HAT 48.5°C, 4 h	Reduced decay and chilling injury, if moist air used caused heat damage	Perez-Carrillo et al. (2003)
Peach				
	Monilinia fructicola	HAT 50°C 2 h, chitosan, or *Bacillus subtilis* CPA-8	HAT did not stop reinfection during storage. Adding chitosan or biocontrol decreased decay	Casals et al. (2012)
	M. fructicola	HWT 40°C 5 or 10 min	Inhibited spore germination and germ tube elongation, induced defense genes in fruit	Liu et al. (2012)

TABLE 2.1 *(Continued)*

Commodity	Fungus	Treatment	Results	References
	M. laxa	HWT 60°C for 20 s	No decay in fruit inoculated up to 48 h before treatment. Gene expression of β-GAL, PL, PG and PME decreased and PAL, CHI, HSP70 and ROS scavenging genes increased in heated fruit.	Spadoni et al. (2014)
	Monilinia spp.	RF at 27.12 MHz, 17 mm between fruit and upper electrode, 18 min exposure	Decreased both inoculated and natural decay	Casals et al. (2010c)
	Monilinia spp.	RF 4 min of fruit in 40°C water	Decreased both inoculated and natural decay, did not affect fruit firmness	Sisquella et al. (2014)
Peach and nectarine				
	M. laxa, M. fructicola	HAT 40, 45 and 50°C, 30 min to 6 h 60%, 80%, 90%, and 99% RH	Lowest decay on inoculated fruit at 50°C, 99% RH for 2 h. Lower RH and T needed longer times, ripe fruit had less decay control	Casals et al. (2010a, 2010b)
	M. laxa, M. fructicola, and *Monilinia fructigena*	HWT 60°C for 30 and 60 s	Good control of natural decay in commercial trials	Spadoni et al. (2013)
	M. fructicola	MW, 17.5 kW for 50 s and 10 kW for 95 s	Decreased natural decay, some heat damage around fruit pit	Sisquella et al. (2013a, 2013b)
Tomato	Natural infection	HWT 50°C, 2 min; 40°C, 30 min	Eliminated natural infection, maintained fruit firmness	Pinheiro et al. (2012)

β-GAL, beta-galactosidase; CHI, chitinase; HAT, hot air treatment; HSP, heat shock proteins; HWRB, hot water rinse and brush; HWT, hot water treatment; MW, microwave; PAL, phenylalanine ammonia lyase; PG, polygalacturonase; PI, pectin lyase; PME, pectin methyl esterase; RF, radio frequency; RH, relative humidity; ROS, reactive oxygen species; SC, sodium bicarbonate; TBZ, thiabendazole.

2.4   EFFECTS ON COMMODITIES

High-temperature treatments, even the HWRB treatment of a few seconds, have been found to induce stress reactions in commodities. These include reversible inhibition of ripening processes and induction of synthesis of stress and defense compounds such as heat-shock proteins (which can protect against low temperatures) and pathogenesis responsive proteins (which can inhibit pathogens). Peaches held in 40°C water for 10 min had increased defense proteins and the heat treatment inhibited the growth of *Monilinia fructicola* (though the fungus was not killed at this temperature) (Liu et al., 2012). The HWRB treatments of grapefruit also induced accumulation of stress and defense-related proteins in grapefruit, and the fruit became resistant to chilling injury (Porat et al., 2000c; Pavoncello et al., 2001). This same resistance to chilling injury due to a heat temperature has been found in all tropical and subtropical crops examined so far (Table 2.2).

TABLE 2.2   Heat Treatments Tested on Different Commodities to Affect Physiology.

Commodity	Treatment	Results	References
Apple			
"Jonagold" and "Cortland" apples	HAT 46°C 4, 8 or 12 h	Heat injury higher in "Cortland" than in "Jonagold." Injury (flesh browning) found in 8 or 12 h treated apples after storage	Fan et al. (2011)
"Ultima-Gala" apples	HWT 45°C 10 min	Used apple microarray to examine gene expression 1 and 4 h after treatment. Found upregulation of heat-shock proteins, heat-shock cognate protein, and heat-shock transcription factor genes	Spadoni et al. (2015)
Avocado			
"Geada," "Quintal," "Fortuna"	HWT 38°C 60 or 90 min	CI decreased by treatment in "Geada" and "Quintal" but not in "Fortuna" cultivars	Donadon et al. (2012)
Banana			
"Sucrier" banana	HAT 42°C, 78% RH for 6, 12, 18, or 24 h	Inhibited peel spotting. 24 h treatment affected taste, but 18 h fruit were acceptable. PAL, PPO, and LOX activities were lower. Total phenols and dopamine were higher	Kamdee et al. (2009)

TABLE 2.2 *(Continued)*

Commodity	Treatment	Results	References
"Berangan" banana	HWT 50°C 10 or 20 min	Ripening slowed, color changes and softening delayed, respiration and ethylene lower. No effect on pulp	Mirshekari et al. (2013)
Broccoli			
	HWT 50°C 2 h	Inhibited yellowing. Chlorophyllase, Chl-degrading peroxidase (POX), isoperoxidase activities lower than control	Funamoto et al. (2002, 2003)
"Legacy" broccoli	HAT 55°C 3 h	Inhibited yellowing and POX activity, reduced respiration, and maintained chlorophyll *a* and *b* levels, however, caused softening	Guerrero-Arteaga et al. (2014)
Citrus			
"Satsuma" mandarin	HWT 50°C 2 min	Reduced CI (peel pitting), suppressed anaerobic products, ethylene evolution and respiration, decreased POX activity and maintained CAT activity	Ghasemnezhad et al. (2008)
"Fortune" mandarin	HAT 37°C, 95% RH 1, 2, or 3 days	Prevented CI without affecting fruit internal quality. Phenolic compounds identified but no difference between control or HAT	Lafuente et al. (2010)
"Navel" and "Valencia" orange	HWT 41°C 20 min or 50°C 5 min	Reduced CI in cold quarantine 1°C, 90% RH for 20 d for fruit fly. Enhanced POX and catalase (CAT) activities in both fruit peel and juice, and the level of free phenols in juice. Not affect organoleptic qualities	Bassal and El-Hamahmy (2011)
"Tarocco" orange	HWT 52°C 180 s and 56°C 20 s	Increased the level of alcohols, esters, and aliphatic (fatty) aldehydes, did not affect taste, decreased decay	Strano et al. (2014)

TABLE 2.2   *(Continued)*

Commodity	Treatment	Results	References
"Valencia" oranges	HAT 37°C 48 h, 90% RH	Proteomics and metabolomics. Proetins affected were cell rescue, defense, and virulence and metabolism. SOD and POX and sugars higher in heated fruit	Perotti et al. (2011)
"Valencia" oranges	HAT 37°C 48 h, 90% RH	Carbon metabolism different in flavedo-heated fruit along with higher synthesis of sucrose from organic acids.	Perotti et al. (2015)
"Star Ruby" grapefruit	HWT 53°C 2 min, HWRB 60°C 30 s, HAT 36°C 3 days	All treatments decreased CI 80–90% compared to control fruit. HWRB decreased decay. HAT increased weight loss, color change, increased SSC/TA ratio	Porat et al. (2000b, 2000c)
"Marsh" grapefruit	HWT 48°C 120 min	Reduced CI with lower proline levels and higher total soluble and nonreducing sugar levels in flavedo, and less TA in juice.	Ezz et al. (2004)
"Paan" lime	HWT 50°C 5 min	Delayed color change and inhibited chlorophyllase, Chl-degrading POX and pheophytinase as well as Mg-dechelation activity. Decreased respiration and ethylene, maintained TA and delayed increase in SSC	Kaewsuksaeng et al. (2015)
Garlic			
	HWT 55–60°C for 10–2.5 min	Inhibited sprout and root growth, respiration decreased, no effect on firmness or pungency	Cantwell et al. (2003)
	HAT 35°C till water content reached 64%	Weight loss and decay lower than control. HAT of 45°C or higher caused damage	

TABLE 2.2 *(Continued)*

Commodity	Treatment	Results	References
Kiwifruit			
"Hongyang" kiwifruit	HWT 35, 45, 55°C 10 min	35°C and 45°C decreased CI, while 55°C caused damage. 45°C fruit had higher firmness and SSC, and lower malondialdehyde (MDA) content, lipoxygenase (LOX) activity and ethylene production rate, and CBF expression highest.	Ma et al. (2014)
Onion			
	HAT 30°C, 98% RH for up to 9 days	The onion neck became 52% narrower, color of outer skin became dark reddish brown, reduced weight loss by 30%, and rot by 80%.	Eshel et al. (2014)
Papaya			
"Frangi" papaya	HWT 42°C 30 min, then 49°C 20 min	CI reduced and ascorbate POX activity enhanced	Shadmani et al. (2015)
"Maradol" papaya	HAT 48.5°C, 50% RH 4 h	Decreased CI and decay. Moist hot air caused fruit damage. Sugar content maintained and no effect on fruit color	Perez-Carrillo et al. (2003)
Pomegranate			
	HWT 45°C 4 min	Reduced CI (increase in skin browning and electrolyte leakage), increased free putrescine and spermidine and maintained the unsaturated/saturated fatty acid ratio.	Mirdehghan et al. (2007)
	HWT 53°C 3 min	Increased phenol content during storage after treatment.	Onursal et al. (2010)
Strawberry			
"Selva" strawberry	HAT 45°C 3 h	Treated fruit had lower decay, lower respiration, lower levels of H_2O_2, higher antioxidant capacity. APX and SOD increased during storage in these fruit.	Vicente et al. (2003, 2006)

TABLE 2.2 *(Continued)*

Commodity	Treatment	Results	References
Stone fruit			
Peach	HAT 38°C 12 h + 1 mM salisylic acid	Alleviated CI (flesh browning), increased antioxidant enzymes and polyamines	Cao et al. (2010)
Peach	HAT 37°C 3 days	Examined genes for ascorbic acid metabolism. Genes sharply induced immediately after treatment then lower during storage. H_2O_2 levels follow this.	Wang et al. (2014)
"Flavorcrest" peach	HAT 39°C 24 h + 5% O_2 and 15 or 20% CO_2	Control fruit were leathery after 4 weeks at 1°C. Treated fruit ripened normally. HAT caused flesh reddening which CA alleviated	Murray et al. (2007)
"Dixiland" peach	HAT 39°C 24 and 36 h, HWT 39, 41, 44°C 30, 60, 180 min	HAT increased flesh reddening and decreased TA. HWT did not affect fruit	Budde et al. (2006)
"Blackstar" plum	HWT 45°C 10 min	HWT prior to mechanical damage minimized damage development. Wound-induced ethylene and respiration rates were lower in treated plums	Serrano et al. (2004)
Tomato			
"Zinac" tomato	HWT 40–60°C 60 to 2 min	Microbial loads decreased. Fruit firmer after HWT up to 55°C and less firm when treated above 55°C. PME activity lower in all treatments. POX activity lower up to 55°C.	Pinheiro et al. (2012)
"Micro-Tom" tomato	HWT 40°C 7 min	Metabolic profiling showed that treated and cold stored fruit were similar to nonstored fruit and different from cold stored fruit (that developed CI). Arabinose, fructose-6-phosphate, valine, and shikimic acid are associated with treatment protection from CI	Luengwilai et al. (2012)

TABLE 2.2   *(Continued)*

Commodity	Treatment	Results	References
"Messina" tomato	HAT 38°C 12 h	HAT decreased CI index, electrolyte leakage and malondialdehyde content. Enhanced the transcript of LeARG1 and LeARG2, the two genes encoding arginase, and arginase activity. An arginase inhibitor decreased effect of heat indicating arginase involvement in CI tolerance.	Zhang et al. (2013)

APX, ascorbate peroxidase; CAT, catalase; CBF, C-repeat binding factor; CI, chilling injury; HAT, hot air treatment; HWRB, hot water rinse and brush; HWT, hot water treatment; LOX, lipoxygenase; MDA, malondialdehyde; PAL, phenylalanine ammonia lyase; PPO, polyphenol oxidase; POX, peroxidase; RH, relative humidity; SOD, superoxide dismutase; SSC, soluble solids content; TA, titratable acidity.

High-temperature conditioning, or curing, has been found to affect commodities in different ways. In citrus, early season fruits are given a degreening treatment with ethylene to improve the peel color. When this treatment is given at 30°C instead of the usual 20°C, the fruit taste is also improved (Tietel et al., 2010). At the elevated temperature, respiration is enhanced and organic acids are metabolized more rapidly. This leads to an improvement in the sugar to acid ratio, one of the criteria for export quality fruit. Previously, it was found that degreening lemons at 30°C rather than 20°C prevented the red blotch disorder that occurs on the fruit as a result of exposure to ethylene (Cohen et al., 1988).

Onions are normally cured before storage, which causes drying of the outer layers of the onions and reduces decay and water loss in subsequent storage. Curing the onions at 30°C and 98% RH for up to 9 days reduced weight loss and decay in posttreatment stored onions by 30% and 80%, respectively, compared to control bulbs (Eshel et al., 2014).

Persimmons are treated with an atmosphere of high CO_2 to remove the astringency of the fruit. This treatment is generally done at ambient temperature, but some of the new cultivars, such as cv. 123, require extended times at this temperature to achieve full deastringency. Studies by Haya Friedman (personal communication) found that raising the temperature to 35°C allowed deastringency to occur in 1 or 2 days, rather than 4 or more days required at lower temperatures.

2.5 INSECT CONTROL

There are a number of heat treatments approved by APHIS in the United States for insect control, including hot water immersion and vapor heat. However, worldwide most quarantine treatments involve cold treatment, particularly with regard to fly pests, including the Mediterranean fruit fly (*Ceratitis capitata* (Wiedemann)). Since many of the commodities that this insect infects are sensitive to low-temperature storage, heat treatment has been used to increase the resistance of the product to low temperature. Peppers exported to countries which require a quarantine cold treatment of 1.5°C for 3 weeks suffer from chilling injury. HWRB (55°C for 15 s), followed by plastic packaging reduced both chilling injury and decay development in bell peppers given a quarantine treatment (Fallik et al., 2009). For citrus, neither hot water dips nor 6 to 8 h of hot air curing at 38–45°C gave adequate protection against chilling injury. However, hot water dips in combination with shrink packaging reduced decay incidence and chilling injury in citrus fruit (Rodov et al., 2001). This allows the citrus to be given a cold quarantine treatment during their sea shipment.

As an alternative to cold quarantine, studies were conducted to determine the sensitivity to high temperature of different developmental stages of the Mediterranean fruit fly. Time and temperature conditions from 44°C to 52°C were tested (Gazit et al., 2004). As the temperature increased the time needed to kill the insects decreased from over 1 h at 44°C to 1 min at 53°C. The most resistant stage of insect development was the third instar, even more resistant than eggs. Based on the results of the thermal death kinetics of the Mediterranean fruit fly a number of different citrus cultivars were tested using either radiofrequency or hot-air heating. "Navel" and "Valencia" oranges were heated with radiofrequency and held for 15 min in 48°C hot water, then cooled and stored for 10 days at 4°C (Birla et al., 2005). These short times of heating of the fresh produce do not cause any damage to the produce. "Oroblanco" citrus fruit was heated in hot air to either 48°C or 44°C. The fruit developed heat damage at 48°C because of the long time needed (200 min) to raise the temperature in hot air compared to radiofrequency. However, fruit heated to 44°C (which took 140 min) and then held for 30 min had no heat damage. When infested fruits were held at 44°C for 30 min in low oxygen, the insect larvae were killed without any fruit damage (Lurie et al., 2004). Another study on two orange cultivars, "Olinda" and "Campbell" also found that these fruits were not damaged after 100 min at 44°C or 50 min at 46°C (Schirra et al., 2005). Taste tests and quality of the fruits after the treatment and following a week of marketing found no difference between treated or untreated fruit.

A similar study was performed on apples to control the quarantine pest San Jose scale (*Quadraspidiotus perniciosus*). The temperatures tested were lower than that for the citrus (38–46°C) and the time needed for control were longer. San Jose scale was controlled by 10 h heating to 46°C, while the apples had no heat damage (Lurie et al., 1998). Table 2.3 gives other treatments to control insect pests. In addition, a recent review provides a thorough overview of progress in the use of radiofrequency heating for control of postharvest insect pests (Hou et al., 2016).

TABLE 2.3   Heat Treatments Tested on Different Commodities to Control Insect Infestation.

Commodity	Fungus	Treatment	Results	Reference
Apple				
	Codling moth (*Cydia pomonella* L.)	HWT 45°C 30 min then RF 48°C 15 min	Killed all stages of moth with no bad effects on apples	Wang et al. (2006)
Carambola				
	Oriental fruit fly (*Bactocera carambola*)	VHT 46.5°C for 20–30 min	Killed all stages of fly; did not affect fruit quality	Hasbullah et al. (2013)
Citrus				
"Orobanco" grapefruit	Mediterranean fruit fly (*Ceratitis capitata*)	VHT 43°C in low O_2 or 44°C in air for 30 min	Killed first instar larvae (most heat resistant stage) without affecting fruit	Lurie et al. (2004)
Sapote mamey				
	Caribbean fruit fly (*Anastrepha serpentine*)	HAT 40, 43, 45, 46, or 50°C for 120, 150, or 180 min	40°C for 30 min killed larvae but not eggs; 43°C for 30 min killed both and had some effect on fruit ripening. 50°C caused fruit injury	Yahia et al. (2003)
Stone fruit				
Cherry	Queensland fruit fly (*Bactrocera tryoni*)	RF 37–60°C then 0.5°C for 5 d	RF alone did not kill fly, but RF to 50°C or 60°C plus cold storage did. Treatment caused some stem browning	Ekman and Pristijono (2010)

HAT, hot air treatment; HWT, hot water treatment; RF, radiofrequency; VHT, vapor heat treatment.

2.6　CONCLUSIONS

There have been many high-temperature treatments developed in the past quarter of a century to maintain commodity quality after harvest and to control insects and pathogens. The particular treatment depends on the commodity being treated and the purpose of the treatment; that is, decay control, resistance to low-temperature storage, inhibition of ripening, insect control, etc. Details on many of the treatments developed are presented only in research reports and have not been transformed into commercial postharvest practices. However, many do have potential commercial benefit, and in the future may be implemented.

KEYWORDS

- postharvest
- hot water treatment
- vapor heat treatment
- fruits
- vegetables
- shelf life

REFERENCES

Aborisade, A. T.; Akomolafe, O. M. Control of Plantain (*Musa paradisiaca*) Fruit Rot Caused by *Fusarium verticillioides* Using Heat Treatment. *Acta Hortic.* **2011**, *906*, 155–159.

Afek, U.; Orenstein, J. Disinfecting Potato Tubers Using Steam Treatments. *Can. J. Plant Pathol.* **2002**, *24*, 36–39.

Bassal, M.; El-Hamahmy, M. Hot Water Dip and Preconditioning Treatments to Reduce Chilling Injury and Maintain Postharvest Quality of Navel and Valencia Oranges during Cold Quarantine. *Postharvest Biol. Technol.* **2011**, *60*, 186–191.

Barkai-Golan, R.; Phillips, D. J. Postharvest Heat Treatment of Fresh Fruits and Vegetables for Decay Control. *Plant Dis.* **1991**, *75*, 1085–1089.

Birla, S. L.; Wang, S.; Tang, J.; Fellman, J. K.; Mattinson, D. S.; Lurie, S. Quality of Oranges as Influenced by Potential Radio Frequency Heat Treatments against Mediterranean Fruit Flies. *Postharvest Biol. Technol.* **2005**, *38*, 69–79.

Budde, C. O.; Polenta, G.; Lucangeli, C. D.; Murray, R. E. Air and Immersion Heat Treatments Affect Ethylene Production and Organoleptic Quality of 'Dixiland' Peaches. *Postharvest Biol. Technol.* **2006**, *41*, 32–37.

Cantwell, M. I.; Kang, J.; Hong, G. Heat Treatments Control Sprouting and Rooting of Garlic Cloves. *Postharvest Biol. Technol.* **2003**, *30*, 57–65.

Cao, S.; Hu, Z.; Zheng, Y.; Lu, B. Synergistic Effect of Heat Treatment and Salicylic Acid on Alleviating Internal Browning in Cold-Stored Peach Fruit. *Postharvest Biol. Technol.* **2010**, *58*, 93–97.

Casals, C; Teixidó, N.; Viñas, I.; Llauradó, S.; Usall J. Control of Monilinia spp. on Stone Fruit by Curing Treatments: Part I. The Effect of Temperature, Exposure Time and Relative Humidity on Curing Efficacy. *Postharvest Biol. Technol.* **2010a**, *56*, 19–25.

Casals, C.; Teixidó, N.; Viñas, I.; Cambray, J.; Usall, J. Control of *Monilinia* spp. on Stone Fruit by Curing Treatments. Part II: The Effect of Host and *Monilinia* spp. Variables on Curing Efficacy. *Postharvest Biol. Technol.* **2010b**, *56*, 26–30.

Casals, C.; Viñas, I.; Landl, A.; Picouet, P.; Torres, R.; Usall, J. Application of Radio Frequency Heating to Control Brown Rot on Peaches and Nectarines. *Postharvest Biol. Technol.* **2010c**, *58*, 218–224.

Casals, C.; Elmer, P. A. G.; Viclas, I.; Teixidcd, N.; Sisquella, M.; Usall, J. The Combination of Curing with Either Chitosan or *Bacillus subtilis* CPA-8 to Control Brown Rot Infections Caused by *Monilinia fructicola*. *Postharvest Biol. Technol.* **2012**, *64*, 126–132.

Cerioni, L.; Sepulveda, M.; Rubio-Ames, Z.; Volentini, S. I.; Rodríguez-Montelongo, L.; Smilanick, J. L.; Ramallo, J.; Rapisarda, V. A. Control of Lemon Postharvest Diseases by Low-toxicity Salts Combined with Hydrogen Peroxide and Heat. *Postharvest Biol. Technol.* **2013**, *83*, 17–21.

Cohen, E.; Shalom, Y.; Lurie, S. Prevention of Red Blotch in Degreened Lemons. *HortScience* **1988**, *23*, 72–74.

Donadon, J. R.; Durigan, J. F.; Morgado, C. M. A.; Durigan, M. F. B. Chilling Injury in Avocados and Its Prevention with Thermal Treatment. *Acta Hortic.* **2012**, *934*, 747–753.

Ekman, J. H.; Pristijono, P. Combining Radio Frequency Heating and Cool Storage to Disinfest Cherries against Queensland Fruit fly. *Acta Hortic.* **2010**, *877*, 1441–1448.

Eshel, D.; Regev, R.; Orenstein, J.; Droby, S.; Gan-Mor, S. Combining Physical, Chemical and Biological Methods for Synergistic Control of Postharvest Diseases: A Case Study of Black Root Rot of Carrot. *Postharvest Biol. Technol.* **2009**, *54*, 48–52.

Eshel, D.; Teper-Bamnolker, P.; Vinokur, Y.; Saad, I.; Zutahy, Y.; Rodov, V. Fast Curing: A Method to Improve Postharvest Quality of Onions in Hot Climate Harvest. *Postharvest Biol. Technol.* **2014**, *88*, 34–39.

Fallik, E. Pre-storage Hot Water Treatments (Immersion, Rinsing and Brushing). *Postharvest Biol. Technol.* **2004**, *32*, 125–134.

Fallik, E.; Grinberg, S.; Alkalai, S.; Yekutieli, O.; Wiseblum, A.; Regev, R.; Beres, H.; Bar-Lev, E. A Unique Hot Water Treatment to Improve Storage Quality of Sweet Pepper. *Postharvest Biol. Technol.* **1999**, *15*, 25–32.

Fallik, E.; Aharoni, Y.; Copel A.; Rodov, R.; Tuvia-Alkalai, S.; Horev, B.; Yekutieli, O.; Wiseblum, A.; Regev, R. A Short Hot Water Rinse Reduces Postharvest Losses of 'Galia' Melon. *Plant Pathol.* **2000**, *49*, 333–338.

Fallik, E.; Tuvia-Alkalai, S; Feng, X.; Lurie, S. Ripening Characterization and Decay Development of Stored Apples after a Short Prestorage Hot Water Rinsing and Brushing. *Innov. Food Sci. Emerging Technol.* **2001**, *2*, 127–132.

Fallik, E.; Ilic, Z.; Tuvia-Alkalai, S.; Copel, A. Polevaya, Y. A Short Hot Water Rinsing and Brushing Reduces Chilling Injury and Enhances Resistance against *Botrytis cinerea* in Fresh Harvested Tomato. *Adv. Hortic. Sci.* **2002**, *16*, 3–6.

Fallik, E.; Bar-Yosef, A.; Alkalai-Tuvia, S.; Aharon, Z.; Perzelan, Y.; Ilic, Z.; Lurie, S. Prevention of Chilling Injury in Sweet Bell Pepper Stored at 1.5°C by Heat Treatments and Individual Shrink Packaging—3 Years of Research. *Folia Hortic.* **2009**, *21*, 87–97.

Feygenberg, O.; Keinan, A.; Kobiler, I.; Fallik, E.; Pesis, A.; Lers, A.; Prusky, D. Improved Management of Mango Fruit through Orchard and Packinghouse Treatments to Reduce Lenticel Discoloration and Prevent Decay. *Postharvest Biol. Technol.* **2014**, *91*, 128–133.

Funamoto, Y.; Yamauchi, N.; Shigenaga, T.; Shigyo, M. Effects of Heat Treatment on Chlorophyll Degrading Enzymes in Stored Broccoli (*Brassica oleracea* L.). *Postharvest Biol. Technol.* **2002**, *24*, 163–170.

Funamoto, Y.; Yamauchi, N.; Shigyo, M. Involvement of Peroxidase in Chlorophyll Degradation in Stored Broccoli (*Brassica oleracea* L.) and Inhibition of the Activity by Heat Treatment. *Postharvest Biol. Technol.* **2003**, *28*, 39–46.

Gan-Mor, S.; Regev, R.; Levi, A.; Eshel, D. Adapted Thermal Imaging for the Development of Postharvest Precision Steam-Disinfection Technology for Carrots. *Postharvest Biol. Technol.* **2011**, *59*, 265–271.

Gazit, Y.; Wang, S.; Tang, J.; Lurie, S. Thermal Death Kinetics of Egg and Third Instar Mediterranean Fruit fly *Ceratitis capitata* (Weidemann) (Diptera: Tephritidae). *J. Econ. Entomol.* **2004**, *97*, 1540–1546.

Ghasemnezhad, M.; Marsh, K.; Shilton, R.; Babalar, M.; Woolf, A. Effect of Hot Water Treatments on Chilling Injury and Heat Damage in 'Satsuma' Mandarins: Antioxidant Enzymes and Vacuolar ATPase, and Pyrophosphatase. *Postharvest Biol. Technol.* **2008**, *48*, 364–371.

Guerrero-Arteaga, D.; Trejo-Escobar, D.; Mejca-Espacla, D.; Osorio, O. Effect of Hot Air Treatment on the Conservation of Broccoli (*Brassica oleracea* L.), Cultivar 'Legacy'. *Acta Hortic.* **2014**, *1016*, 151–156.

Hasbullah, R.; Rohaeti, E.; Syarief, R. Fruit Fly Disinfestations of Star Fruit (*Averrhoa carambola* L.) Using Vapor Heat Treatment (VHT). *Acta Hortic.* **2013**, *1011*, 147–153.

Hou, L.; Johnson, J. A.; Wang, S. Radio Frequency Heating for Postharvest Control of Pests in Agricultural Products: A Review. *Postharvest Biol. Technol.* **2016**, *113*, 106–118.

Kaewsuksaeng, S.; Tatmala, N.; Srilaong, V.; Pongprasert. Postharvest Heat Treatment Delays Chlorophyll Degradation and Maintains Quality in Thai Lime (*Citrus aurantifolia* Swingle cv. Paan) Fruit. *Postharvest Biol. Technol.* **2015**, *100*, 1–7.

Kamdee, C.; Ketsa, S.; van Doorn, W. G. Effect of Heat Treatment on Ripening and Early Peel Spotting in cv. Sucrier Banana. *Postharvest Biol. Technol.* **2009**, *52*, 288–293.

Karabulut, O. A.; Cohen, L.; Weiss, B.; Daus, A.; Lurie, S.; Droby, S. Control of Brown Rot and Blue Mold of Peach and Nectarine by Short Hot Water Brushing and Yeast Antagonists. *Postharvest Biol. Technol.* **2002**, *24*, 103–111.

Kenigsbuch, D.; Chalupowicz, D.; Aharon, Z.; Maurer, D.; Ovadia, A.; Aharon, N. Preharvest Solar Heat Treatment for Summer Basil (*Ocimum basilicum*) Affects Decay During Shipment and Shelf Life. *Acta Hortic.* **2009**, *880*, 161–166.

Lafuente, M. T.; Calejero, J.; Ballester, A. R.; Zacarias, L.; Gonzalez-Candelas, L. Effect of Heat-conditioning Treatments on Quality and Phenolic Composition of 'Fortune' Mandarin Fruit. *Acta Hortic.* **2010**, *877*, 1333–1340.

Li, X.; Zhu, X.; Zhao, N.; Fu, D.; Li, J.; Chen, W.; Chen. W. Effects of Hot Water Treatment on Anthracnose Disease in Papaya Fruit and Its Possible Mechanism. *Postharvest Biol. Technol.* **2013**, *86*, 437–446.

Liu, F.; Tu, K.; Shao, X.; Zhao, Y.; Tu, S.; Su, J.; Hou, Y.; Zou, X. Effect of Hot Air Treatment in Combination with *Pichia guilliermondii* on Postharvest Anthracnose Rot of Loquat Fruit. *Postharvest Biol. Technol.* **2010**, *58*, 65–71.

Liu, J.; Sui, Y.; Wisniewski, M.; Droby, S.; Tian, S.; Norelli, J.; Hershkovitz, V. Effect of Heat Treatment on Inhibition of *Monilinia fructicola* and Induction of Disease Resistance in Peach Fruit. *Postharvest Biol. Technol.* **2012**, *65*, 61–68.

Luengwilai, K.; Saltveit, M.; Beckles, D. M. Metabolite Content of Harvested Micro-Tom Tomato (*Solanum lycopersicum* L.) Fruit is Altered by Chilling and Protective Heat-shock Treatments as Shown by GCMS Metabolic Profiling. *Postharvest Biol. Technol.* **2012**, *63*, 116–122.

Lurie, S.; Pedreschi, R. Fundamental Aspects of Postharvest Heat Treatments. *Hortic. Res.* **2014**, *1*, 14–30.

Lurie, S.; Fallik, E.; Klein, J. D.; Kozar, F.; Kovacs, K. Postharvest Heat Treatment of Apples to Control San Jose Scale (*Quadraspidiotus perniciosus* Comstock), Blue Mold (*Penicillium expansum* Link), and Maintain Fruit Firmness. *J. Am. Soc. Hortic. Sci.* **1998**, *123*, 110–114.

Lurie, S.; Jemric, T.; Weksler, A.; Akiva, R.; Gazit, Y. Heat Treatment of 'Oroblanco' Citrus Fruit to Control Insect Infestation. *Postharvest Biol. Technol.* **2004**, *34*, 321–329.

Ma, Q.; Suo, J.; Huber, D. J.; Dong, X.; Han, Y.; Zhang, Z.; Rao, J. Effect of Hot Water Treatments on Chilling Injury and Expression of a New C-repeat Binding Factor (CBF) in 'Hongyang' Kiwifruit During Low Temperature Storage. *Postharvest Biol. Technol.* **2014**, *97*, 102–110.

Maxin, P.; Weber, R.; Lindhard Pedersen, H.; Williams, M. Control of a Wide Range of Storage Rots in Naturally Infected Apples by Hot-Water Dipping and Rinsing. *Postharvest Biol. Technol.* **2012**, *70*, 25–31.

Mirdehghan, S. H.; Rahemi, M.; Martinez-Romero, D.; Guillen, F.; Valverde, J. M.; Zapata, P. J.; Serrano, M.; Valero, D. Reduction of Pomegranate Chilling Injury during Storage after Heat Treatment: Role of Polyamines. *Postharvest Biol. Technol.* **2007**, *44*, 19–25.

Mirshekari, A.; Ding, P.; Kadir, J.; Mohd Ghazali, H. Combination of Hot Water Dipping and Fungicide Treatment to Prolong Postharvest Life of 'Berangan' Banana. *Acta Hortic.* **2013**, *1012*, 551–557.

Murray, R.; Lucangeli, C.; Polenta, G.; Budde, C. Combined Pre-storage Heat Treatment and Controlled Atmosphere Storage Reduced Internal Breakdown of 'Flavorcrest' Peach. *Postharvest Biol. Technol.* **2007**, *44*, 116–121.

Onursal, C. E.; Gozlekci, S.; Erkan, M.; Yildirim, I. The Effects of UV-C and Hot Water Treatments on Total Phenolic Compounds of Juice, Peel and Seed Extracts of Pomegranates (*Punica granatum* L.). *Acta Hortic.* **2010**, *8773*, 1505–1509.

Palma, A.; D'Aquino, S.; Vanadia, S.; Angioni, A.; Schirra, M. Cold Quarantine Responses of 'Tarocco' Oranges to Short Hot Water and Thiabendazole Postharvest Dip Treatments. *Postharvest Biol. Technol.* **2013**, *78*, 24–33.

Palou, L.; Usall, J.; Munoz, J. A.; Smilanick, J. L.; Vinas, I. Hot Water, Sodium Carbonate, and Sodium Bicarbonate for the Control of Postharvest Green and Blue Molds of Clementine Mandarins. *Postharvest Biol. Technol.* **2002**, *24*, 93–96.

Pavoncello, D.; Lurie, S.; Droby, S.; Porat, R. A Hot Water Treatment Induces Resistance to *Penicillium digitatum* and Promotes the Accumulation of Heat Shock and Pathogenesis-Related Proteins in Grapefruit Flavedo. *Physiol. Plant.* **2001**, *111*, 17–22.

Perez-Carrillo, E.; Yahia, E.; Ariza, R.; Misael Cornejo, G. Forced Hot Air Treatment at Low Relative Humidity is Effective in Reducing Chilling Injury and Decay Development in Papaya Fruit. *Acta Hortic.* **2003**, *6042*, 697–702.

Perotti, V. E.; Del Vecchio, H. A.; Sansevich, A.; Meier, G.; Bello, F.; Cocco, M.; Garran, S. M.; Anderson, C.; Vazquez, D.; Podesta, F. E. Proteomic, Metabalomic, and Biochemical Analysis of Heat Treated Valencia Oranges during Storage. *Postharvest Biol. Technol.* **2011**, *62*, 97–114.

Perotti, V. E.; Moreno, A. S.; Tripodi, K.; Del Vecchio. H. A.; Meier, G.; Bello, F.; Cocco, M.; Vazquez, D.; Podesta, F. E. Biochemical Characterization of the Flavedo of Heat-Treated Valencia Orange during Postharvest Cold Storage. *Postharvest Biol. Technol.* **2015**, *99*, 80–87.

Pinheiro, J.; Silva, C. L. M.; Alegria, C.; Abreu, M.; Sol, M.; Goncalves, E. M. Impact of Water Heat Treatment on Physical–Chemical, Biochemical and Microbiological Quality of Whole Tomato (*Solanum lycopersicum*) Fruit. *Acta Hortic.* **2012**, *934*, 1269–1276.

Porat, R.; Daus, A.; Weiss, B.; Cohen, L.; Fallik, E.; Droby, S. Reduction of Postharvest Decay in Organic Citrus Fruit by a Short Hot Water Brushing Treatment. *Postharvest Biol. Technol.* **2000a**, *18*, 151–157.

Porat, R.; Pavoncello, D.; Peretz, J.; Ben-Yehoshua, S.; Lurie, S. Effects of Various Heat Treatments on the Induction of Cold Tolerance and on the Postharvest Qualities of 'Star Ruby' Grapefruit. *Postharvest Biol. Technol.* **2000b**, *18*, 159–165.

Porat, R.; Pavonchello, D.; Peretz, J.; Weiss, B.; Daus, A.; Cohen, L.; Ben-Yehoshua, S.; Fallik, E.; Droby, S.; Lurie, S. Induction of Resistance to *Penicillium digitatum* and Chilling Injury in 'Star Ruby' Grapefruit by a Short Hot-water Rinse and Brushing Treatment. *J. Hortic. Sci. Biotechnol.* **2000c**, *75*, 428–432.

Porat, R.; Daus, A.; Weiss, B.; Cohen, L.; Droby, S. Effects of Combining Hot Water, Sodium Bicarbonate and Biocontrol on Postharvest Decay of Citrus. *J. Hortic. Sci. Biotechnol.* **2002**, *77*, 441–445.

Prusky, D.; Fuchs, Y.; Kobiler, I.; Roth, I.; Weksler, A.; Shalom, Y; Falik, E.; Zuaberman, G.; Pesis, E.; Akerman, M.; Yekutiely, O.; Weisblum, A.; Regev, R.; Artes, L. Effect of Hot Water Brushing, Procloraz Treatment and Waxing on the Incidence of Black Spot Decay Caused by *Alternaria alternata* in Mango Fruit. *Postharvest Biol. Technol.* **1999**, *22*, 271–277.

Rodov, V.; Agar, T.; Peretz, Y.; Nafussi, B.; Kim, J. J.; Ben-Yehoshua, S. Effect of Combined Application of Heat Treatments and Plastic Packaging on Keeping Quality of 'Oroblanco' Fruit (*Citrus grandis* L. × *Citrus paradisi* Macf.). *Postharvest Biol. Technol.* **2001**, *20*, 287–294.

Schirra, M.; Mulas, M.; Fadda, A.; Mignani, I.; Lurie, S. Chemical and Quality Traits of 'Olinda' and 'Campbell' Oranges after Heat Treatment at 44 and 46°C for Fruit Fly Disinfestation. *LWT* **2005**, *38*, 519–527.

Serrano, M.; Martinez-Romero, D.; Castillo, S.; Guillen, F.; Valero, D. Role of Calcium and Heat Treatments in Alleviating Physiological Changes Induced by Mechanical Damage in Plum. *Postharvest Biol. Technol.* **2004**, *34*, 155–167.

Shadmani, N.; Ahmad, S. H.; Saari, N.; Tajidin, N. E. Chilling Injury Incidence and Anti-oxidant Enzyme Activities of *Carica papaya* L. 'Frangi' as Influenced by Postharvest Hot Water Treatment and Storage Temperature. *Postharvest Biol. Technol.* **2015**, *99*, 114–119.

Sisquella, M.; Casals, C.; Picouet, P.; Viñas, I.; Torres, R.; Usall, J. Immersion of Fruit in Water to Improve Radio Frequency Treatment to Control Brown Rot in Stone Fruit. *Postharvest Biol. Technol.* **2013a**, *80*, 31–36.

Sisquella, M.; Viñas, I.; Teixidó, N.; Picouet, P.; Usal, I. Continuous Microwave Treatment to Control Postharvest Brown Rot in Stone Fruit. *Postharvest Biol. Technol.* **2013b**, *86*, 1–7.

Sisquella, M.; Viñas, I.; Picouet, P.; Torres, R.; Usall, J. Effect of Host and *Monilinia* spp. Variables on the Efficacy of Radio Frequency Treatment on Peaches. *Postharvest Biol. Technol.* **2014**, *87*, 6–12.

Sivakumar, D.; Fallik, E. Influence of Heat Treatment on Quality Retention of Fresh and Fresh-Cut Produce. *Food Rev. Int.* **2013**, *29*, 294–320.

Smilanick, J. L.; Sorenson, D.; Mansour, M.; Aieyabei, J.; Plaza, P. Impact of a Brief Postharvest Hot Water Drench Treatment on Decay, Fruit Appearance, and Microbe Populations of California Lemons and Oranges. *HortTechnology* **2003**, *13*, 333–338.

Spadoni, A.; Neri, F.; Bertolini, P.; Mari, M. Control of Monilinia Rots on Fruit Naturally Infected by Hot Water Treatment in Commercial Trials. *Postharvest Biol. Technol.* **2013**, *86*, 280–284.

Spadoni, A.; Guidarelli, M.; Marianna Sanzani, S.; Ippolito, A.; Mari, M. Influence of Hot Water Treatment on Brown Rot of Peach and Rapid Fruit Response to Heat Stress. *Postharvest Biol. Technol.* **2014**, *94*, 66–73.

Spadoni, A.; Guidarelli, M.; Phillips, J.; Mari, M.; Wisniewski, M. Transcriptional Profiling of Apple Fruit in Response to Heat Treatment: Involvement of a Defense Response during *Penicillium expansum* Infection. *Postharvest Biol. Technol.* **2015**, *101*, 37–48.

Strano, M. C.; Calandra, M.; Aloisi, V.; Rapisarda, P.; Strano, T.; Ruberto, G. Hot Water Dipping Treatments on Tarocco Orange Fruit and Their Effects on Peel Essential Oil. *Postharvest Biol. Technol.* **2014**, *94*, 26–34.

Tietel, Z.; Weiss, B.; Lewinsohn, E.; Fallik E.; Porat, R. Improving Taste and Peel Color of Early Season Satsuma Mandarins by Combining High Temperature Conditioning and Degreening Treatments. *Postharvest Biol. Technol.* **2010**, *57*, 1–5.

Vicente, A. R.; Chaves, A. R.; Civello, P. M.; Martinez, G. A. Effects of Combination of Heat Treatments and Modified Atmospheres on Strawberry Fruit Quality. *Acta Hortic.* **2003**, *6001*, 197–199.

Vicente, A. R.; Martinez, G. A.; Chaves, A. R.; Civello, P. M. Effect of Heat Treatment on Strawberry Fruit Damage and Oxidative Metabolism During Storage. *Postharvest Biol. Technol.* **2006**, *40*, 116–122.

Wang, S; Birla, S. L.; Tang, J,; Hansen, J. D. Postharvest Treatment to Control Codling Moth in Fresh Apples Using Water Assisted Radio Frequency Heating. *Postharvest Biol. Technol.* **2006**, *40*, 89–96.

Wang, K.; Shao, X.; Gong, Y.; Xu, F.; Wang, H. Effects of Postharvest Hot Air Treatment on Gene Expression Associated with Ascorbic Acid Metabolism in Peach Fruit. *Plant Mol. Biol. Rep.* **2014**, *32*, 881–887.

Yahia, E.; Ariza, R. Postharvest Hot Air Treatments Effect on Insect Mortality and Quality of Sapote Mamey Fruit (*Pouteria sapota*). *Acta Hortic.* **2003**, *6042*, 691–695.

Zhang, X. S.; Li, L.; Meng, F.; Jiping, D. S. Arginase Induction by Heat Treatment Contributes to Amelioration of Chilling Injury and Activation of Antioxidant Enzymes in Tomato Fruit. *Postharvest Biol. Technol.* **2013**, *79*, 1–8.

CHAPTER 3

CALCIUM: AN INDISPENSABLE ELEMENT AFFECTING POSTHARVEST LIFE OF FRUITS AND VEGETABLES

VIJAY YADAV TOKALA[1] and B. V. C. MAHAJAN[2*]

[1]*Curtin Horticulture Research Laboratory, School of Molecular and Life Sciences, Curtin University, Perth, WA, Australia*

[2]*Punjab Horticultural Postharvest Technology Centre, Punjab Agricultural University, Ludhiana, Punjab, India*

[*]*Corresponding author. E-mail: mahajanbvc@gmail.com*

ABSTRACT

Calcium is essential macronutrient involved in numerous biochemical and physiological processes in the plants. It acts as intracellular messenger and is responsible for maintaining the plant cell wall integrity. These properties instigate significant impact on the postharvest fruit quality, ripening, senescence, decay, and physiological disorders. Pre- and postharvest application of calcium has reportedly reduced incidence of different disorders, maintained quality and extended postharvest life in horticulture produce. In the present chapter an attempt is made to provide an overview about different aspects of calcium application, mechanism of action and its impacts on the postharvest life of the fruits and vegetables.

3.1 INTRODUCTION

Fruits and vegetables have an important role in the human diet and their consumption helps to lower the risk of chronic diseases and to maintain a healthy weight (McGuire, 2011). During recent decades, the global demand for year-round availability of fruits and vegetables has increased, which

may be ascribed to the change in dietary habits, increase in health awareness, taste preferences, and the lifestyle of present-day consumers (Yuk et
al., 2006; Pollack, 2001). The continuous increase in population would also
increase the global demand for food, but on the other hand, the growing
competition for land, water, and energy, coupled with overexploitation of
natural resources, will affect our ability to produce food (Godfray et al.,
2010). Prevention of postharvest losses is increasingly cited as a means to
effectively contribute to available food supplies making it a prime goal to
develop technology to reduce their losses (Kitinoja et al., 2011). Many studies
provide indications that postharvest losses are substantial. It is commonly
cited that one-third of the world's agricultural produce is lost before reaching
to consumers (Gustavsson et al., 2011). Fruits and vegetables are comparatively more perishable due to their high moisture content and these losses
vary greatly among developing countries (Lipinski et al., 2005). The magnitude of losses in fruits and vegetables due to pathological or physiological
factors is estimated at about 25–30%, which is because of lack of awareness
about appropriate postharvest handling and nonavailability of adequate postharvest infrastructure at the farm and/or market level (Sharma et al., 2009).
Therefore, efficient postharvest techniques have become an absolute neces-
sity as it not only reduces the monetary loss estimated at millions per year
but also minimizes wastage of labour, energy, and inputs invested in their
production. The shelf-life extension of any fruit and vegetable is prerequisite for minimizing the postharvest losses. Many techniques are available
in the literature to extend the storage life of produce; however, the calcium
application has a significant impact on the shelf-life of fruit and vegetables. Different benefits of calcium application have been discussed by many
workers, namely, delay in ageing or ripening, reduction in postharvest decay,
increase in the nutritional value, and control of different physiological disorders. The aim of this chapter is to give an in-depth detail about the calcium
applications for enhancing storage life and quality improvement of different
horticultural crops.

3.2 IMPORTANCE OF CALCIUM

Calcium is an essential plant nutrient closely related to quality and firmness
of fruits (Sams, 1999) as the divalent cation (Ca^{2+}) is required for various
structural roles in the cell wall and membranes. It acts as a counter-cation for
inorganic and organic anions in the vacuole, and an intracellular messenger in
the cytosol (Marschner, 1995). Earlier studies regarding the effect of calcium

on fruit and vegetable quality were majorly concerned with its association with physiological disorders (DeLong, 1936). Calcium has received considerable attention in recent years because of its desirable effects in delaying senescence and controlling physiological disorders in fruits and vegetables. For fruits and vegetables to function and adapt efficiently to different environmental conditions, their cells must communicate with one another. It is becoming increasingly evident that calcium ions are important as intracellular messengers in plants. Changes in cell-wall structure, membrane permeability, and enzyme activation are known to influence various aspects of cell physiology. Since the discovery of calmodulin, it has become clear that the calcium messages are often relayed by this ubiquitous calcium-binding protein (Cheung, 1980; Poovaiah, 1985). Calcium ions (Ca^{2+}) are used in the synthesis of new cell walls, particularly the middle lamellae that separate newly divided cells. Calcium is also used in the mitotic spindle during cell division. It is required for the normal functioning of plant membranes and has been implicated as a secondary messenger for various plant responses to both environmental and hormonal signals (Sanders et al., 1999). In its function as a secondary messenger, calcium may bind to calmodulin, a protein found in the cytosol of plant cells. The calmodulin–calcium complex regulates many metabolic processes.

Studies on leaf senescence (Poovaiah and Leopold, 1973; Ferguson, 1984) and fruit ripening (Tingwa and Young, 1974; Poovaiah, 1979; Suwwan and Poovaiah, 1978) have indicated that the rate of senescence often depends on the calcium status of the tissue and that by increasing calcium levels, various parameters of senescence such as respiration (Faust and Shear, 1972; Bangerth et al., 1972), protein, and chlorophyll content (Poovaiah and Leopold, 1973), and membrane fluidity (Paliyath et al., 1984) can be altered. It is a well-known fact that calcium plays an important role in maintaining the quality of the fruits and vegetables (Shear, 1975; Bangerth, 1979; Hopfinger and Poovaiah, 1979; Arteca et al., 1980; Collier and Theodore, 1982; Huber, 1983). Calcium deficiency is rare in nature, but there exists possibility in soils with low base saturation or high acidic depositions due to chemical fertilizer abuse (McLauhglin and Wimmer, 1999). Many physiological disorders afflicting fruits and vegetables are known to be related to the calcium content of the tissue (Shear, 1975). White and Broadley (2003) mentioned that the deficiency symptoms can occur in different parts of the plant, namely, (1) in young expanding leaves, such as in "tip burn" of leafy vegetables; (2) in enclosed tissues, such as in "brown heart" of leafy vegetables or "black heart" of celery; or (3) in tissues fed principally by the phloem rather than the xylem, such as in "blossom-end rot" of watermelon, bell pepper, and tomato fruit, "bitter pit" in apples, and "empty pods" in peanuts. They have specified that these disorders occur as calcium cannot be mobilized from older tissues and redistributed via the phloem, because of which the developing tissues have to depend on the immediate supply of calcium in the xylem, which is in turn dependent on transpiration. Disorders such as "cracking" in tomato, cherry, and apple fruit occur in tissues lacking sufficient calcium upon hypoosmotic shock (following increased humidity or rainfall after a long dry spell), presumably as a result of structural weaknesses in cell walls.

TABLE 3.1 Disorders Due to Calcium Deficiency.

Crop	Disorders	References
Apple	Bitter Pit	Simons (1962) and Ferguson and Watkins (1989)
	Cracking	Shear (1971)
	Internal breakdown	Bramlage and Shipway (1967)
Tomato	Blossom-end Rot	Saure (2001)
	Cracking	Bangerth (1973) and Dickson and McCollum (1964)
Lettuce	Tip Burn	Collier and Theodore (1982)
Celery	Blackheart	Takatori et al. (1961)
Cherry	Cracking	Bullock (1952) and Bangerth (1973)
Prune	Cracking	Cline and Tehrani (1973)
Pears	Cork spot	Woodbridge (1971) and Mason and Welsh (1970)
	Black end	Woodbridge (1971)
Carrot	Cavity spot	Guba et al. (1961) and Maynard et al. (1963)
	Leaf tip burn	Tibbitts and Palzkill (1979)
Mango	Spongy tissue (Alphonso)	Wainwright and Burbage (1989)
Strawberry	Leaf tip burn	Tibbitts and Palzkill (1979)

Increasing the calcium normally decreases the incidence of the disorders. Calcium disorders are troublesome and may occur during both storage and shipping, thus making fruits unusable after they arrive at the market. Calcium chloride can also act upon the quality of processed foods by acting as a darkening inhibitor and effective method for the prevention of after-cooking darkening of potato french-fries (Mazza and Qi, 1991). Calcium deficiency is common in horticultural crops, but the deficiency is rarely due to the insufficient calcium in the soil (Naradisorn, 2013). The disorders that result from inadequate calcium in fruit may be due to a poor calcium distribution rather than low calcium uptake, as in the same plant, the calcium content in the leaves is often higher than that of the fruit (Conway et al., 1994a). Similarly, calcium content was significantly higher in the leaf tissues when compared to petioles, which had significantly more calcium than in fruit tissue (Dunn, 2003).

Calcium is essential for structure and function of cell walls and membranes. There are three types of evidence for the role of calcium in membranes. First, under calcium-deficient conditions, there is a profound deterioration of membranes (Marinos, 1962). Second, calcium alters the actual architecture of membranes; its introduction into natural (Paliyath et al., 1984) or artificial membranes of phospholipids (Gary-Bobo, 1970) results in an enormous change in fluidity and water permeability. Third, calcium can powerfully alter an array of physiological activities which are specifically associated with the membrane function, for example, it can turn on the active transport of some ions through membranes (Hanson, 1983).

Calcium plays a special role in maintaining the cell-wall structure in fruits and other storage organs by interacting with the pectic acid in the cell walls to form calcium pectate. Thus, fruits treated with calcium promote maintenance of the cell-wall structure and membrane integrity by increasing the calcium content of the tissues (Scott and Wills, 1975; Poovaiah et al., 1978; Paliyath et al., 1984; Drake and Spayd, 1983). High calcium-containing tubers are less susceptible to decay (McGuire and Kelman, 1983, 1986) and consequently could be stored for longer period of time. Increase in calcium concentration in the tubers lowers the incidence of the internal defects and maintains its quality (Kleinhenz et al., 1999).

3.3 PHYSIOLOGY OF ACTION

Calcium-related physiological disorders that occur in a wide range of vegetables and fruits are usually associated with loss of membrane integrity with resultant loss of cell compartmentalization and ultimately cell death in their tissues (Christiansen and Foy, 1979; Poovaiah, 1979). Sufficient Ca^{2+} is known to be important for maintaining normal cell-wall structure, thereby reducing the incidence of internal breakdown disorders in horticulture crops (Conway et al., 1992). Maintenance of firmness and resistance to softening resulting from the application of calcium have been attributed to the stabilization of membrane systems and formation of calcium pectates, which increase the rigidity of the middle portion and cell wall of the fruit (Grant et al., 1973; Jackman and Stanley, 1995). The role of calcium in maintaining firmness has been associated with plant disease resistance (Naradisorn, 2013). It is suggested that calcium application to fruits can either enhance the resistance of fruit to postharvest pathogens or reduce susceptibility to postharvest diseases and disorders. Additionally, calcium may also cause a reduction of pathogen conidia germination and germ-tube elongation by limiting nutrients available to pathogens on the fruit surface (Moline, 1994).

During ripening, major changes take place in the pectin-rich middle-lamella region of cells where calcium ions have a role in linking adjacent acidic pectin polymers (Seymour et al., 1993). Under normal conditions, as fruit mature, there is increased availability of nutrients for pathogens on the surface of the fruit. However, calcium may enhance the resistance of fruit to pathogens by interacting with cell-wall components. Postharvest pathogens produce pectolytic enzymes, which cause softening of host tissues (Conway et al., 1994b). Lara et al. (2004) suggested that improved resistance to fungal attack in calcium-treated fruit was associated with the preservation of cell-wall and middle-lamella structure. Calcium ions bind tightly to the pectins in the cell walls and produce cationic bridges between pectic acids, or between pectic acids and other acidic polysaccharides. These bridges make the cell walls less accessible to the action of pectolytic enzymes (Moline, 1994; Conway et al., 1994b).

Apart from the structural and functional role of calcium in plant cell walls and membranes, it is also as important as an intracellular "second messenger" (Zocchi and Mignani, 1995). Calcium ions are commonly involved as intracellular messengers in the transduction by plants of a wide range of biotic stimuli including signals from pathogenic and symbiotic fungi (Kang et al., 2006; Vandelle et al., 2006; Yamniuk and Vogel, 2004; Navazio et al., 2007; Gressel et al., 2002; Uhm et al., 2003; Hu et al., 2002). The defense responses such as increased accumulation and excretion of phytoalexins, calloses, etc., which act against pathogens were produced on an application of calcium salts in onion (Dmitriev et al., 1996; Dyachok et al., 1997), sunflower, and grape (Dmitriev et al., 2003; Poinssot et al., 2003).

3.4 CHEMICALS USED

Calcium, as its chloride salt, has great potential in delaying ripening/aging, extending postharvest shelf-life, improving nutritional value by increasing calcium content, maintaining fruit firmness and quality (Lau et al., 1983; Mason et al., 1975; Conway et al., 1992; Ferguson, 2001), and also helping in reducing the incidence of different physiological disorders, namely, internal breakdown, scald, and water core (Yuen, 1994). Calcium chloride ($CaCl_2$) is natural, inexpensive, edible, and has been approved by the FDA (Food and Drug Administration) for postharvest use. Foliar spray effects of calcium chloride, calcium phosphate, and polyphenolic acid chelate of calcium were compared in "McIntosh" apple fruits with the aim to increase calcium uptake into fruits without causing phytotoxicity to find that both calcium chloride and calcium chelates sprays increased fruit calcium concentrations and reduced senescence breakdown, while calcium phosphate sprays did not increase fruit calcium concentrations but increased phosphorus concentration, which was not associated with reduced breakdown (Bramlage et al., 1985).

Ferguson (2001) points out that though calcium nitrate is also an effective source of calcium, its use can cause russet problems later during fruit development. Subbaiah and Perumal (1990) used different sources of calcium, namely, calcium oxide, calcium chloride, and calcium sulfate, and found that calcium chloride sprays had a better impact in improving quality and shelf-life of tomatoes. Subbiah (1994) reported that calcium chloride has an effective role in increasing firmness of the tomato fruits, while firmness index was lowered by the increase in the levels of added nitrogen; hence, calcium nitrate may not be a desirable calcium source for enhancing the firmness of the fruit. Hong and Lee (1999) also reported that calcium chloride treatments are more effective in enhancing the quality of tomato fruits when compared to calcium nitrate. Singh et al. (1993) reported that two consecutive sprays of calcium nitrate (1% or 2% Ca^{2+}) or calcium chloride (0.6% or 1.2% Ca^{2+}) at 20 and 10 days before harvest on "Dashehari" mango trees improved storage life of fruit while retaining its quality. It was further reported that 0.6% Ca^{2+} calcium chloride proved to be the most favorable treatment.

Sprays of 200 ppm Ca^{2+} chelated with carboxylic acids (Calhard®) once a week at the rate of 200 L per 100 m^2 markedly increased strawberry fruit firmness along with calcium and ascorbic acid contents in the fruit (Conway, 1982). Naradisorn (2013) has reported that some fruits are particularly sensitive to damage by chloride and the calcium sulfate ($CaSO_4$) could be a potential alternative, as it also improves the soil structure and has been associated with an improved root health. Postharvest treatment with calcium lactate was suggested as a potential alternative to calcium chloride for shelf-life extension of fresh-cut cantaloupe and treated fruits were approximately 25–33% firmer than cut and untreated cantaloupe samples, without providing undesirable bitterness (Luna-Guzmna and Barrett, 2000). Dipping of strawberry cultivars "Cardinal" and "Sunrise" in calcium lactate improved berry firmness (Morris et al., 1985). Matchima (2013) found that dipping strawberry cultivar "Selva" in calcium lactate at 3000 ppm Ca was effective in delaying botrytis rot development during 7 days of storage.

3.5 METHODOLOGY OF APPLICATION

Methods of postharvest calcium treatment of fruit vary, and it may include dipping (Mootoo, 1991; Suhardi, 1992; Yeun, 1994), spraying (Sharples and Johnson, 1976; Drake et al., 1979), reduced pressure infiltration (Tirmazi and Wills, 1981; Yeun et al., 1993, 1994), and foliar application of calcium chloride to delay ripening and retard mold development in strawberries (Chéour et al., 1990, 1991) and raspberries (Montealegre and Valdes, 1993).

Postharvest dip in calcium chloride solutions is also combined with the heat treatment. Heat allows the formation of COO^- groups from the pectin content of the fruits or vegetables with which Ca^{2+} ions can form salt-bridge cross-links (Stanley et al., 1995). This makes the cell wall-less accessible to the enzymes that cause softening. This practice controls ripening, softening, and decay at the same time (Sams et al., 1993).

Calcium dips have been used as firming agents to extend postharvest shelf-life of fruits and vegetables such as apples (Mir et al., 1993; Sams et al., 1993), peaches (Postlmayr et al., 1956), pineapples (de Carvalho et al., 1998; Goncalves et al., 2000), oranges (Boas et al., 1998), tomatoes (Floros et al., 1992), sliced pears (Rosen and Kader, 1989), and sliced cantaloupe (Luna-Guzman et al., 1999). Strawberries submerged in 1% calcium chloride solution for 15 min at 25°C and stored at refrigerated conditions showed enhanced shelf-life (Garcia et al., 1996; Rosen and Kader, 1989). The calcium chloride pretreatment in combination with gum acacia, gelatin, proved to be an effective method for the prevention of after-cooking darkening of french-fries (Mazza and Qi, 1991).

Conway and Sams (1984) treated "Golden Delicious" apples for 2 min with 0–12% $CaCl_2$ solutions by dipping, vacuum infiltration (4.83 psi), and pressure infiltration (10 psi) and found that out of all the treatments dipping was least effective in increasing the calcium concentration of the tissue, while vacuum infiltration was superior over dipping and pressure infiltration method was most effective. Vacuum infiltration of 4% calcium chloride has shown a significant rise in the quality parameter of apples over control and water application (Poovaiah, 1986). Hong and Lee (1999) have accomplished vacuum infiltration by applying the desired volume of calcium chloride and calcium nitrate solutions in millilitres on the stem-scar, placing fruit under 250 Torr (4.834 lb sq. in^{-1}) for 10 s and slowly releasing the vacuum, allowing the solution to enter the fruit. Combination of 1% calcium chloride and 0.5% oxygen produces immediate firmness effect on the slices of pear and greater maintenance (Rosen and Kader, 1989). Shorter and Joyce (1998) have vacuum (partial pressure) infiltrated calcium (4 g Ca^{2+} L^{-1} as CaCl$_2$) at reduced pressure levels (−33, −66, and −99 kPa) and found that fruits infiltrated at −66 and −99 kPa exhibited injuries, which included exacerbated lenticel blackening and anaerobic off-odor and taste evident at the end of shelf-life.

Applying supplementary calcium into soil has been suggested as the most efficient way of increasing calcium in plant organs (Conway et al., 1994b). However, research on strawberry indicated that calcium is unlikely to be translocated from roots to fruit. Increasing calcium concentration in soil through the application of calcium sulfate increased leaf calcium content but not that of fruit (Dunn and Able, 2006). To increase calcium content in fruit, it is recommended that calcium should be applied directly to the fruit surface (Garcia et al., 1996).

3.6 CONCENTRATION OF CHEMICALS USED

In some apple varieties, namely, Granny Smith and Yellow Newtown, a single postharvest dip of 60 s in 2% (for Yellow Newtown) or 3% (for Granny Smith) calcium chloride solution in place of in-season sprays have been effective. Time of dip and mixing of the solution need to be carefully managed as longer exposures or a higher percent solution can lead to fruit bum. Using steel or iron tank for dipping is not recommended as it can contribute to fruit damage (Caprile, 2012). Treating carrot shreds, slices, and sticks with 1% calcium chloride solution significantly improved the postharvest quality and shelf-life (Izumi and Watada, 1994). Postharvest dip in 1% calcium chloride solution at 25°C has been found to be the most effective method to reduce decay percentage, increase the firmness, and shelf-life of strawberries and pears (Rosen and Kader, 1989; Garcia et al., 1996; Mahajan et al., 2008) (Fig. 3.1). One percent calcium chloride application at about 3 psi significantly extended the shelf-life and about 30% of the significant commercial value of peaches (Wills and Mahendra, 1989; Mahajan and Sharma, 2000). Application of 6% calcium chloride as a preharvest foliar spray in mandarin showed a higher quality and increased storage life in the harvested fruits (El-Hammady et al., 2000). Foliar application of 0.2%

calcium chloride solution in tomatoes, during the fruit growth period considerably improved the quality of produce, while postharvest application of 1 mL of 8% calcium chloride to mature-green tomato fruits through vacuum infiltration method has enhanced the calcium content in the fruits (Subbiah and Perumal, 1990; Hong and Lee, 1999). Subbiah (1994) further observed that 0.5% calcium chloride spray significantly increased the firmness index of tomato fruits, when compared with calcium nitrate. Matchima (2013) found that dipping strawberry in calcium lactate at 3000 ppm was effective in delaying botrytis rot development during 7 days of storage. Singh et al. (1993) experimented by two consecutive sprays of calcium nitrate (1% or 2% Ca^{2+}) or calcium chloride (0.6% or 1.2% Ca^{2+}) at 20 and 10 days before harvest on "Dashehari" mango trees and found that it enhanced storage life while retaining its quality. They further reported that comparatively 0.6% Ca^{2+} calcium chloride proved to be the most favorable, while 1.2% Ca^{2+} calcium chloride treatment caused scorching of the marginal and lamellar portion of the leaves.

The temperature of the calcium chloride dip had a significant influence on the firmness of the melon slices, where treatments at 60°C showed better results than the dip at 20°C and 40°C (Luna-Guzman et al., 1999). Poovaiah (1986) reported that there was a higher increase in the ascorbic acid content of Golden Delicious apple when 4% calcium chloride was vacuum-infiltrated, in comparison to 2%. A single coating of potato pieces with a combination of 0.5% calcium chloride and 5% pectin and sodium alginate reduced the oil content of the French fries by 40% and retained high moisture content (Khalil, 1999).

3.7 STAGES OF APPLICATION

The difference in growing condition, environmental factors, and fruit development can influence the amount of calcium uptake by fruits (Conway et al., 1994b; Wojcik and Swiechowski, 1999). To increase calcium concentration, foliar sprays should be conducted at the late fruit growth stages (Wojcik, 2001).

Preharvest calcium application may be considered as a cultural practice for maintaining adequate calcium concentration in fruit. Calcium-containing compounds have been applied as supplemental fertilizers in soil amendments or foliar sprays. Calcium chloride ($CaCl_2$) and calcium nitrate ($Ca(NO_3)_2$)

are commonly used for foliar sprays (Bramlage et al., 1985). Application of calcium nitrate through irrigation water reduced the severity of leaf grey mould of tomato plants grown in perlite by 70% (Elad and Volpin, 1993).

Foliar application of calcium chloride to peach trees throughout the growing season improves fruit quality and maintains the quality longer than the nontreated fruits (Robson et al., 1989). Foliar application of calcium chloride at 0–20 kg ha^{-1} between 3 and 9 days before harvest led to increased quality and reduced disease incidence in strawberry fruit (Chéour et al., 1990). "Delicious" and "Golden Delicious" apples from trees sprayed with calcium chloride three to four times per year at the rates of 3.60 g L^{-1} and 4.76 g L^{-1}, respectively, had more fruit calcium concentrations in comparison to control trees (Raese and Drake, 1993). Preharvest foliar sprays of calcium chloride in red dragon fruit have increased fruit firmness and reduced severity of postharvest diseases (Ghani et al., 2010). Similarly, calcium chloride sprays at a rate of 1.5 kg ha^{-1} at every 5 days interval from petal fall stage has improved fruit firmness and disease resistance in strawberries (Wójcik and Lewandowski, 2003). A minimum of three preharvest sprays in apples, applied at monthly intervals beginning in May or June showed positive results in California. Dilute sprays (300–400 gal ac^{-1}) are most effective but good coverage of the fruit is essential (calcium doesn't move rapidly within the fruit or from leaf to fruit. For more severe cases, shorter treatment intervals (every 2 weeks) over the same 3-month period are recommended. There is also evidence to indicate that the earlier treatment, beginning in May or even at one-inch fruit size may be more effective in very susceptible varieties (Caprile, 2012). A preharvest foliar spray of CaCl$_2$ at anthesis not only increased the absorption of calcium in leaves and fruits but also enhanced the yield and shelf-life of tomato (Abbasi et al., 2013).

El-Hammady et al. (2000), when applied 4% and 6% calcium chloride solutions to Mandarin trees at mature-green stage of fruits, found positive results pertaining to physical and nutritional parameters of the fruits. Fruits, when vacuum-infiltrated with calcium chloride soon after harvest and stored at refrigerated conditions, showed a significant increase in ascorbic acid levels till 15 weeks of application (Poovaiah, 1986). The pretreatment of potatoes with calcium chloride before processing in combination with gum acacia, gelatin proved to be an effective method for the prevention of after-cooking darkening of french-fries (Mazza and Qi, 1991).

3.8   EFFECTS OF CALCIUM APPLICATION

3.8.1   EFFECT ON QUALITY PARAMETERS

Increasing the concentration of Ca^{2+} by pressure infiltration has shown to delay softening in the fruits (Mootoo, 1991; Suhardi, 1992; Yeun et al., 1993). Similarly, maximizing calcium concentration, without incurring damage, can reduce risk of disorders and help in maintaining firmness and other desirable quality parameters in apple (Stow, 1993; Ferguson, 2001), avocado (Tingwa and Young, 1974), litchi (Roychoudhury et al., 1992), raspberries (Eaves et al., 1972), and tomato (Wills and Tirmazi, 1979). Postharvest application of calcium chloride increased the flesh calcium concentration of cherries, improved texture (firmness and bio-yield), and incidence of pitting resulting from impact damage (Lidster et al., 1978; 1979). Calcium treatment maintained firmness in sliced strawberries even better than that in the case of whole strawberries and this could be attributed to the availability of more surface area for calcium chloride adsorption (Morris et al., 1985) (Table 3.2). Similarly, calcium chloride-treated carrot shreds had more calcium concentration and firm texture, when compared to carrot slices and sticks exposed to similar treatments (Izumi and Watada, 1994). In red dragon fruit, preharvest calcium chloride sprays increased fruit firmness, but fruit quality such as soluble solids and titratable acidity were not affected by the calcium sprays (Ghani et al., 2010). Foliar application of calcium chloride at rates of 0–20 kg ha^{-1} between 3 and 9 days before harvest increased the calcium content of strawberry and also delayed fruit ripening, although there was no effect observed on acidity, soluble solids and titratable acidity (Chéour et al., 1990). Sprays of Ca^{2+} chelated with carboxylic acids (*Calhard*) once a week increased strawberry fruit firmness and the force required to puncture the skin of calcium-treated strawberries was greater than that of control (Conway, 1982).

Mahajan and Sharma (2000) found that preharvest spray of 1% calcium chloride not only improved quality parameters (size and totals sugars) but also maintained the quality during extended storage life of peaches. It also delayed maturity and improved storability of Mandarin oranges (Hsiung and Iwahori, 1984; Schirra and Mulas, 1994) and mangoes (Tirmazi and Wills, 1981; Mootoo, 1991; Yeun et al., 1993; Singh et al., 1993). Application of 6% calcium chloride as a preharvest foliar spray in Mandarin oranges showed comparatively higher values of ascorbic acid and also recorded less physiological loss in weight of fruits (El-Hammady et al., 2000). Calcium chloride dip improved firmness and calcium content of fresh-cut melon slices while reducing the rate of change in metabolism. The respiration rate (CO_2 production) and the ethylene production rate were reduced, which pertain to increase in shelf-life of fresh-cut melon slices (Luna-Guzman et al., 1999), apples (Ferguson, 1984), avocados (Tingwa and Young, 1974), and mangoes (Mootoo, 1991; Van Eeden, 1992).

Calcium treatment also had a significant effect on the organoleptic acceptance of peaches where calcium-treated fruits not only rated superior in appearance, aroma, flavor, and texture after harvest but also continued the overall acceptance greater than the control fruit even after 4 weeks of storage (Robson et al., 1989). The use of calcium chloride may impart bitterness or flavor differences which results from residual calcium chloride on the surface of the fruit (Morris et al., 1985).

FIGURE 3.1 Effect of calcium chloride on storage life and fruit quality of pear and plum. *Source:* Adapted from Mahajan and Dhatt (2004); Mahajan et al. (2008).

TABLE 3.2 Effects of Calcium Application.

Fruit	Effect	References
Apple	Retains fruit firmness	Poovaiah (1986), Conway and Sams (1984), and Ferguson (1984)
	Increase fruit calcium content	
	Increase vitamin C content	
	Reduce postharvest decay	
	Reduce respiration and ethylene production rates	

TABLE 3.2   *(Continued)*

Fruit	Effect	References
Strawberry and blueberry	Extends shelf-life Slows down the rate of decay Maintains firmness of the fruit for an extended period	Garcia et al. (1996), Rosen and Kader (1989), Morris et al. (1985), and Hernandez-Munoz et al. (2006)
Peach	Improve quality Extend shelf-life Reduces brown rotting and disease index	Wills and Mahendra (1989), Souza et al. (1999), Robson et al. (1989), and Mahajan and Sharma (2000)
Pear	Maintains firmness and freshness of sliced pears for an extended period	Rosen and Kader (1989)
Orange	Extend shelf-life Improve quality Fewer disorders	Hsiung and Iwahori (1984), Schirra and Mulas (1994), and El-Hammady et al. (2000)
Pineapple	Reduce internal browning Reduce phenolics and decay	Goncalves et al. (2000) and de Carvalho et al. (1998)
Cantaloupe	Improve firmness Extend the shelf-life	Luna-Guzman et al. (1999)
Carrot	Improve firmness Extend shelf-life	Izumi and Watada (1994)
Tomato	Improve firmness and quality Increases shelf-life Enhance lycopene and ascorbic acid	Subbiah and Perumal (1990), Subbiah (1994), and Hong and Lee (1999)
Potato	Extend storage life Reduces browning and decay Improves the quality of processed products and reduces oil consumption	Walter et al. (1993), Klein-henz et al. (1999), and Mazza and Qi (1991)

3.8.2   EFFECT ON NUTRITIONAL AND BIOCHEMICAL STATUS

Ascorbic acid levels increased in the Golden Delicious apple fruits with postharvest calcium chloride infiltration during storage at refrigerated conditions. On the other hand, ascorbic acid levels showed a decreasing trend under control with the increase in storage period (Poovaiah, 1986). The ascorbic acid and lycopene levels in the tomato fruits increased considerably with foliar application of 0.2% calcium chloride during fruit growth stage (Subbaiah and Perumal, 1990). Strawberries dipped in calcium chloride and stored at 25°C showed high levels of calcium content which decreased with increase in storage temperature (Garcia et al., 1996). Sprays of Ca^{2+} chelated with carboxylic acids (*Calhard*) once a week increased ascorbic acid and calcium content (Conway, 1982).

Postharvest application of calcium reduced enzyme levels and increased the levels of neutral sugar in peaches (Souza et al., 1999). Postharvest calcium treatment reduced the phenolics (indicative of decaying) content of the pineapples and showed extended shelf-life in comparison to control, regardless of the storage temperature and duration of treatment, which may be due to maintenance of the pectin substances in their cell walls due to calcium absorption (de Carvalho et al., 1998; Goncalves et al., 2000).

3.8.3 EFFECT ON MICROBIAL GROWTH AND PHYSIOLOGICAL DISORDERS

Conway and Sams (1984) have indicated that calcium-enriched tissue develops resistance to fungal attack by stabilizing or strengthening cell walls, thereby making them more resistant to harmful enzymes produced by fungi, and it also delays ageing of fruits. Garcia et al. (1996) have demonstrated that postharvest dip in calcium chloride reduced strawberry decay significantly by improving fruit firmness. Brown rotting and disease index was reduced to the significant percentage in peaches treated with calcium chloride, in comparison to untreated peaches (Souza et al., 1999). The total microbial count was substantially reduced when carrot shreds were treated with 1% calcium chloride solution and increased resistance of tissues to the bacterial infections rather than to a bacterial action (Izumi and Watada, 1994). In red dragon fruit, preharvest spray of calcium chloride reduced the severity of anthracnose and brown-rot diseases (Ghani et al., 2010).

Foliar application of calcium chloride at 0–20 kg ha^{-1} between 3 and 9 days before has reduced grey mould (*Botrytis cinerea*) development (Chéour et al., 1990). Similarly, fruit at harvest were firmer and more resistant to *Botrytis* fruit rot than those in the control when fruits were sprayed with calcium chloride at a rate of 1.5 kg ha^{-1} spray at every 5 days interval from petal fall stage (Wójcik and Lewandowski, 2003). Application of calcium sulfate to strawberry plants showed less incidence of grey mould than fruit harvested from plants that received no calcium for cultivars "Aromas" and "Selva." The shelf-life of "Aromas" and "Selva" increased by about 8% when plants received 500 ppm Ca in comparison with plants that did not received calcium treatment. Elad and Volpin (1993) applied calcium sulfate to soil before planting tomato seedlings and found that it reduced the grey mould disease severity by 30–40%. Fruits harvested from strawberry plants applied with calcium sulfate had a less incidence of grey mould than untreated plants along with enhanced shelf-life (Matchima, 2013). Abbasi et al. (2013) reported that calcium chloride at anthesis reduced occurrence of blossom-end rot, improved quality and shelf-life of the tomato fruits (Table 3.1). Goncalves et al. (2000) reported that post-harvest treatment of pineapple slices with calcium chloride retarded their decay rate and internal browning of the fruit regardless of the temperature and duration of the treatment.

3.9   NEGATIVE EFFECTS OF CALCIUM APPLICATION

The extent of positive responses of fruits to calcium application may vary and can be limited (O'hare and Zauberman, 1992; Suhardi, 1992). The fruit damage such as mesocarp discoloration in avocado (Yeun et al., 1994) and lenticel spotting in mango (Tirmazi and Wills, 1981; O'Hare and Zauberman, 1992) can occur at too high Ca^{2+} concentrations. However, a major disadvantage of postharvest calcium treatments has been the inability to predict potential injury (lenticel pitting and surface discoloration) to the fruits (Conway et al., 1994c; Yeun, 1994). Application of calcium nitrate and calcium chloride can cause phytotoxicity and even relatively low concentrations of calcium chloride can produce serious foliar injury, while calcium nitrate is more likely to produce fruit injury (Bramlage et al., 1985). Injury probably is a phytotoxic response to too much calcium in the fruits (Conway and Sams, 1985). At least part of the calcium-induced injury appears to be due to salt stress since the severity of the injury increases with the concentration (Sharples and Johnson, 1976) and decreases when fruit is rinsed of the surface calcium immediately after calcium treatment (Scott and Wills, 1977). Calcium-induced injury to the apples can also be reduced, but not eliminated, by storing fruits in high humidity (Lidster et al., 1977). In case of mango, calcium-induced injury was reduced by packaging the fruit in polymeric films that modified atmosphere and increased the humidity in the air surrounding the fruit (Yeun et al., 1993).

3.10 CONCLUSION

It is clearly established that calcium plays a wide and important role in pre- and postharvest life of fruit and vegetables. The importance of calcium in the nutrition of plants has long been recognized and application of calcium early in plant development may be important for the later, but direct calcium application to fruit may be more significant in the former. The pre- and postharvest factors could clearly influence the effect of calcium on shelf-life extension, quality improvement, and control of decay of harvested produce.

KEYWORDS

- **calcium**
- **postharvest**
- **shelf-life**
- **fruit quality**
- **decay control**

REFERENCES

Abbasi, N. A.; Zafar, L.; Khan, H. A.; Qureshi, A. A. Effects of Naphthalene Acetic Acid and Calcium Chloride Application on Nutrient Uptake, Growth, Yield and Postharvest Performance of Tomato Fruit. *Pak. J. Bot.* **2013,** *45* (5), 1581–1587.

Arteca, R. N.; Poovaiah, B. W.; Hiller, L. K. Electron Microprobe and Neutron Activation Analysis for the Determination of Elemental Distribution in Hollow Heart Potato Tubers. *Am. Potato J.* **1980,** *57,* 241.

Bangerth, F. Investigations upon Ca related Physiological Disorders. *J. Phytopathol.* **1973,** *77* (1), 20–37.

Bangerth, F. Calcium-Related Physiological Disorders of Plants. *Ann. Rev. Phytopathol.* **1979,** *17* (1), 97–122.

Bangerth, F.; Dilley, D. R.; Dewey, D. H. Effect of Postharvest Calcium Treatments on Internal Breakdown and Respiration of Apple Fruits. *J. Am. Soc. Hortic. Sci.* **1972,** *87,* 679.

Boas, E. V. D. B.; Chitarra, A. B.; Menezes, J. B. Modificações dos Componentes de Parede Celular do Melão'orange Flesh'submetido a Tratamento Pós-colheita com Cálcio (Modification of Cell Wall Components in Oranges Flesh Melons Subjected to Postharvest Calcium Treatment). *Braz. Archiv. Biol. Technol.* **1998,** *41* (4), 467–474.

Bramlage, W. J.; Shipway, M. R. Loss of Watercore and Development of Internal Breakdown During Storage of 'Delicious' Apples as Determined by Repeated Light Transmittance Measurements of Intact Apples. *J. Am. Soc. Hortic. Sci.* **1967**, *90*, 475.

Bramlage, W. G.; Drake, M.; Weis, S. A. Comparisons of Calcium Chloride, Calcium Phosphate, and a Calcium Chelate as Foliar Sprays for 'McIntosh' Apple Trees. *J. Am. Soc. Hortic. Sci.* **1985**, *110*, 786–789.

Bullock, R. M. A Study of Some Inorganic Compounds and Growth Promoting Chemicals in Relation to Fruit Cracking of Bing Cherries at Maturity. *Proc. Am. Soc. Hortic. Sci.* **1952**, *59*, 243–253.

Caprile, J. Maximizing Calcium Uptake in Apples, University of California, Small Farm Center Report. 2012. http://sfp.ucdavis.edu/crops/Calcium/ (accessed May 13, 2018).

Chéour, F.; Willemot, C.; Arul, J.; Desjardin, Y.; Makhlouf, Y.; Charest, P. M.; Gosselin, A. A Foliar Application of Calcium Chloride Delays Postharvest Ripening of Strawberry. *J. Am. Soc. Hortic. Sci.* **1990**, *115*, 789–792.

Chéour, F.; Willemot, C.; Arul, J.; Makhlouf, Y.; Desjardin, Y. Postharvest Response of Two Strawberry Cultivars to Foliar Application of CaCl2. *Hortic. Sci.* **1991**, *26*, 1186–1188.

Cheung, W. Y. Calmodulin Plays a Pivotal Role in Cellular Regulation. *Science* **1980**, *207*, 19.

Christiansen, M. N.; Foy, C. D. Fate and Function of Calcium in Tissue. *Commun. Soil Sci. Plant Anal.* **1979**, *10* (1–2), 427–442.

Cline, R. A.; Tehrani, G. Effects of Boron and Calcium Sprays and of Mulch on Cracking of Italian Prune. *Can. J. Plant Sci.* **1973**, *53*, 827.

Collier, G. F.; Theodore, W. T. Tipburn of Lettuce. *Horticultural Reviews*; Palgrave Macmillan: UK, 1982; pp 49–65.

Conway, W. S. Effect of Postharvest Calcium Treatment on Decay of Delicious Apples. *Plant Dis.* **1982**, *66*, 402–403.

Conway, W. S.; Sams, C. E. Possible Mechanisms by Which Postharvest Calcium Treatment Reduces Decay in Apples. *Phytopathology* **1984**, *74* (2), 208–210.

Conway, W. S.; Sams, C. E. Influence of Fruit Maturity on the Effect of Postharvest Calcium Treatment on Decay of Golden Delicious Apples. *Plant Dis.* **1985**, *69*, 42–44.

Conway, W. S.; Tobias, R. B.; Sams, D. C. E. Reduction of Storage Decay in Apples by Postharvest Calcium Infiltration. *Int. Symp. Pre- Postharvest Physiol. Pome-fruit* **1992**, *326*, 115–122.

Conway, W. S.; Sams, C. E.; Brown, G. A.; Beavers, W. S.; Tobias, R. B.; Kennedy, L. S. Pilot Test for the Commercial Use of the Postharvest Pressure Infiltration of Calcium into Apples to Maintain Fruit Quality in Storage. *Hortic. Technol.* **1994a**, *4*, 239–243.

Conway, W. S.; Sams, C. E.; Kelman, A. Enhancing the Natural Resistance of Plant Tissues to Postharvest Disease through Calcium Applications. *Hortic. Sci.* **1994b**, *29*, 751–754.

Conway, W. S.; Sams, C. E.; Wang, C. Y.; Abbott, J. A. Additive Effects of Postharvest Calcium and Heat Treatments on Reducing Decay and Maintaining Quality in Apples. *J. Am. Soc. Hortic. Sci.* **1994c**, *119*, 49–53.

de Carvalho, V. D. Changes in the Cell Wall Compounds of Pineapple Treated with CaCl2 Solution at Different Temperatures. *Cien. Agrotecnol.* **1998**, *22*, 359–365.

DeLong, W. A. Variations in the Chief as Constituents of Apples Affected with Blotchy Cork. *Plant Physiol.* **1936**, *11*, 453–456.

Dickson, D. B.; McCollum, J. P. The Effect of Calcium on Cracking in Tomato Fruits. *Proc. Am. Soc. Hortic. Sci.* **1964**, *84*, 485.

Dmitriev, A.; Djatsok, J.; Grodzinskii, D. The Role of Ca2+ in Elicitation of Phytoalexin Synthesis in Cell Culture of *Allium cepa*. *Plant Cell Rep.* **1996,** *15,* 945–948.

Dmitriev, A.; Tena, M.; Jorrin, J. Systemic Acquired Resistance in Sunflower (*Helianthus annuus*). *Tsitologiya I Genetika.* **2003,** *37* (3), 9–15.

Drake, S. R.; Spayd, S. E. Influence of Calcium Treatment on 'Golden Delicious' Apple Quality. *J. Food Sci.* **1983,** *48,* 403.

Dunn, J. The Effect of Calcium Nutrition on Post-harvest Strawberry Quality and Botrytis Fruit Rot Development. *Honours Thesis,* Discipline of Plant and Food Science, The University of Adelaide, Adelaide, Australia, 2003.

Dunn, J.; Able, A. J. Pre-harvest Calcium Effects on Sensory Quality and Calcium Mobility in Strawberry. *Acta Hortic.* **2006,** *708,* 307–312.

Dyachok, Y. V.; Dmitriev, A. P.; Grodzinskii, D. M. Ca2+ as a Second Messenger in the Elicitation of Phytoalexin and Callose Synthesis in *Allium cepa* Cell Culture. *Rus. J. Plant Physiol.* **1997,** *44,* 333–338.

Eaves, C. A.; Lockhart, C. L.; Stark, R.; Craig, D. L. Influence of Preharvest Sprays of Calcium Salts and Wax on Fruit Quality of Red Raspberry. *J. Am. Soc. Hortic. Sci.* **1972,** *97,* 706–707.

Elad, Y.; Volpin, H. Reduced Development of Grey Mould (*Botrytis cinerea*) in Bean and Tomato Plants by Calcium Nutrition. *J. Phytopathol.* **1993,** *139,* 146–156.

El-Hammady, A. M.; Abdel-Hamid, N.; Saleh, M.; Salah, A. Effects of Gibberellic Acid and Calcium Chloride Treatment on Delaying Maturity, Quality and Storability of 'Balady' Mandarin Fruits. *Arab. Univ. J. Agric. Sci.* **2000,** *8* (3), 755–766.

Faust, M.; Shear, C. B. The Effect of Calcium on Respiration of Apples. *J. Am. Soc. Hortic. Sci.* **1972,** *97,* 437.

Ferguson, I. In *Calcium in Apple Fruit,* Proceedings of 17th Annual Tree Fruit Postharvest Conference, Washington State University, Wenatchee, WA, 13–14 March, 2001; Vol. 13.

Ferguson, I. B. Calcium in Plant Senescence and Fruit Ripening. *Plant Cell Environ.* **1984,** *7,* 477–489.

Ferguson, I. B.; Watkins, C. B. Bitter Pit in Apple Fruit. *Hortic. Rev.* **1989,** *11,* 289–355.

Floros, J. D.; Ekanayake, A.; Abide, G. P.; Nelson, P. E. Optimization of Diced Tomato Calcification Process. *J. Food Sci.* **1992,** *57,* 1144–1148.

Garcia, J. M.; Herrara, S.; Morilla, A. Effects of Postharvest Dips in Calcium Chloride on Strawberry. *J. Agric. Food Chem.* **1996,** *44,* 30–33.

Gary-Bobo, C. M. Effects of Calcium on the Water and Non-electrolyte Permeability of Phospholipid Membranes. *Nature* **1970,** *228,* 1101.

Ghani, M. A. A.; Awang, Y.; Sijam, K. Disease Occurrence and Fruit Quality of Pre-Harvest Calcium Treated Red Flesh Dragon Fruit (*Hylocereus polyrhizus*). *Afr. J. Biotechnol.* **2010,** *10* (9), 1550–1558.

Godfray, H. C. J.; Beddington, J. R.; Crute, I. R.; Haddad, L.; Lawrence, D.; Muir, J. F.; Pretty, J.; Robinson, S.; Thomas, S. M.; Toulmin, C. Food Security: The Challenge of Feeding 9 Billion People. *Science* **2010,** *327* (5967), 812–818.

Goncalves, N. B.; de Carvalho, V. D.; de Goncalves, J. R. Effect of Calcium Chloride and Hot Water Treatment on Enzyme Activity and Content of Phenolic Compounds in Pineapples. *Pesq. Agropec. Bras.* **2000,** *35* (10), 2075–2081.

Grant, G. T.; Morris, E. R.; Rees, D. A.; Smith, P. J. C.; Thom, D. Biological Interaction Between Polysaccharides and Divalent Cations: The Egg-Box Model. *FEBS Lett.* **1973,** *32,* 195–198.

Gressel, J.; Michaeli, D.; Kampel, V.; Amsellem, Z.; Warshawsky, A. Ultra Low Calcium Requirements of Fungi Facilitate Use of Calcium Regulating Agents to Suppress Host Calcium-dependent Defenses, Synergizing Infection by a Mycoherbicide. *J. Agric. Food Chem.* **2002,** *50* (22), 6353–6360.

Guba, E. F.; Young, R. E.; Ui, T. Cavity Spot Disease of Carrot and Parsnip Roots. *Plant Dis. Rep.* **1961,** *45* (2), 102.

Gustavsson, J.; Cederberg, C.; Sonesson, U.; Van Otterdijk, R.; Meybeck, A. *Global Food Losses and Food Waste.* Food and Agriculture Organization of the United Nations: Rome, 2011.

Hanson, J. B. The Roles of Calcium in Plant Growth. In *Current Topics in Plant Biochemistry and Physiology*; Randall, D. D., Blevins, D. G., Larson, R., Eds.; University of Missouri: Columbia, 1983; vol 1, p 1.

Hernandez-Munoz, P.; Almenar, E.; Ocio, M. J.; Gavara, R. Effect of Calcium Dips and Chitosan Coatings on the Postharvest Life of Strawberries (*Fragaria × ananassa* Duch.). *Postharvest Biol. Technol.* **2006,** *39* (3), 247–253.

Hong, J. H.; Lee, S. K. Effect of Calcium Treatment on Tomato Fruit Ripening. *J. Korean Soc. Hortic. Sci.* **1999,** *40* (6), 638–642.

Hopfinger, J. A.; Poovaiah, B. W. Calcium and Magnesium Gradients in Apples with Bitter Pit. *Commun. Soil Sci. Plant Anal.* **1979,** *10*, 57.

Hsiung, T. C.; Iwahori, S. Prevention of Abscission of Ponkan, (*Citrus reticulate* Blanco) Leaves by Various Calcium Salts. *Mem. Fac. Agric., Kangoshima Univ.* **1984,** *20*, 55–62. (Hortic. Abst. 35, 9886).

Hu, X.; Neill, S. J.; Fang, J.; Cai, W.; Tang, Z. The Mediation of Defense Responses of Ginseng Cells to an Elicitor from Cell Walls of *Colletotrichum lagerarium* by Plasma Membrane NAD (P) H Oxidases. *Acta Bot. Sin.* **2002,** *45* (1), 32–39.

Huber, D. J. Role of Cell Wall Hydrolases in Fruit Softening. *Hortic. Rev.* **1983,** *5*, 169.

Izumi, H.; Watada, A. E. Calcium Treatments Effect Storage Quality of Shredded Carrots. *J. Food Sci.* **1994,** *59*, 106–199.

Jackman, R. L.; Stanley, D. W. Perspectives in the Textural Evaluation of Plant Foods. *Trends Food Sci. Technol.* **1995,** *6*, 187–194.

Kang, S.; Kim, H. B.; Lee, H.; Choi, J. Y.; Heu, S.; Oh, C. J.; Kwon, S. I.; An, C. S. Overexpression in *Arabidopsis* of a Plasma Membrane-Targeting Glutamate Receptor from Small Radish Increases Glutamate-Mediated Ca2+ Influx and Delays Fungal Infection. *Mol. Cells* **2006,** *21* (3), 418–427.

Khalil, A. H. Quality of French Fried Potatoes as Influenced by Coating with Hydrocolloids. *Food Chem.* **1999,** *66* (2), 201–208.

Kitinoja, L.; Saran, S.; Roy, S. K.; Kader, A. A. Postharvest Technology for Developing Countries: Challenges and Opportunities in Research, Outreach and Advocacy. *J. Sci. Food Agric.* **2011,** *91*, 597–603.

Kleinhenz, M. D.; Palta, J. P.; Gunter, C. C.; Kelling, K. A. Impact of Source and Timing of Calcium and Nitrogen Applications on 'Atlantic' Potato Tuber at Calcium Concentrations and Internal Quality. *J. Am. Soc. Hortic. Sci.* **1999,** *124* (5), 498–506.

Lara, I.; García, P.; Vendrell, M. Modifications in Cell Wall Composition after Cold Storage of Calcium-Treated Strawberry (*Fragaria× ananassa* Duch.) Fruit. *Postharvest Biol. Technol.* **2004,** *34* (3), 331–339.

Lau, O. L.; Meheriuk, M.; Olsen, K. L. Effects of "Rapid CA", High CO2, and CaCl2 Treatment on Storage Behavior of 'Golden Delicious' Apples. *J. Am. Soc. Hortic. Sci.* **1983,** *108,* 230–233.

Lidster, P. D.; Porritt, S. W.; Eaton, G. W. The effect of Storage Relative Humidity on Calcium Uptake by Spartan Apples. *J. Am. Soc. Hortic. Sci.* **1977,** *102,* 394–396.

Lidster, P. D.; Porritt, S. W.; Tung, M. A. Texture Modification of 'Van' Sweet Cherries by Postharvest Calcium Treatments. *J. Am. Soc. Hortic. Sci.* **1978,** *103* (4), 527–530.

Lidster, P. D.; Tung, M. A.; Yada, R. G. Effects of Preharvest and Postharvest Calcium Treatments on Fruit Calcium Content and the Susceptibility of 'Van' Cherry to Impact Damage. *J. Am. Soc. Hortic. Sci.* **1979,** *104* (6), 790–793.

Lipinski, B.; Hanson, C.; Lomax, J.; Kitinoja, L.; Waite, R.; Searchniger, T. Reducing Food Loss and Waste. *World Resources Institute Working Paper,* June 2013.

Luna-Guzman, I.; Barrett, D. M. Comparison of Calcium Chloride and Calcium Lactate Effectiveness in Maintaining Shelf Stability and Quality of Fresh-cut Cantaloupes. *Postharvest Biol. Technol.* **2000,** *19,* 61–72.

Luna-Guzman, I.; Cantwell, M.; Barrett, D. M. Fresh-cut Cantaloupe: Effects of CaCl2 Dips and Heat Treatments on Firmness and Metabolic Activity. *Postharvest Biol. Technol.* **1999,** *17,* 201–213.

Mahajan, B. V. C.; Sharma, R. C. Effect of Pre-harvest Applications of Growth Regulators and Calcium Chloride on Physico-chemical Characteristics and Storage Life of Peach (*Prunus persica* Batsch) cv. Shan-e-Punjab. *Haryana J. Hortic. Sci.* **2000,** *29* (1/2), 41–43.

Mahajan, B. V. C.; Dhatt A. S. Studies on Postharvest Calcium Chloride Application on Storage Behavior and Quality of Asian Pear During Cold Storage. *J. Food Agric. Environ.* **2004,** *2* (3–4), 157–159.

Mahajan, B. V. C.; Randhawa, J. S.; Kaur, H.; Dhatt, A. S. Effect of Post-harvest Application of Calcium Nitrate and Gibberellic Acid on the Storage Life of Plum. *Indian J. Hortic.* **2008,** *65* (1), 94–96.

Marinos, N. G. Studies on Sub-microscopic Aspects of Mineral Deficiencies. I. Calcium Deficiency in the Shoot Apex of Barley. *Am. J. Bot.* **1962,** *49,* 834.

Marschner, H. *Mineral Nutrition of Higher Plants,* 2nd ed.; London Academic Press: London, **1995,** 285–299.

Mason, J. L.; Welsh, M. F. Cork Spot (Pit) of 'Anjou' Pear Related to Calcium Concentration in Fruit. *Hortic. Sci.* **1970,** *5,* 447.

Mason, J. L.; Jasmin, J. J.; Granger, R. L. Softening of 'McIntosh' Apples by a Post-Harvest Dip in Calcium Chloride Solution Plus Thickener. *Hortic. Sci.* **1975,** *10,* 524–525.

Matchima, N. Effect of Calcium Nutrition on Fruit Quality and Postharvest Diseases. *Int. J. Sci. Innov. Discov.* **2013,** *3* (1), 8–13.

Maynard, D. N.; Gersten, B.; Young, R. E.; Vernell, H. F. The Influence of Plant Maturity and Calcium Level on the Occurrence of the Carrot Cavity Spot. *Proc. Am. Soc. Hortic. Sci.* **1963,** *83,* 506.

Mazza, G.; Qi, H. Control of After-cooking Darkening in Potatoes with Edible Film Forming Products and Calcium Chloride. *J. Agric. Food Chem.* **1991,** *39* (12), 2163–2166.

McLauhglin, S. B.; Wimmer, R. Calcium Physiology and Terrestrial Ecosystem Processes. *New Phytol.* **1999,** *142,* 373–417.

McGuire, S. US Department of Agriculture and US Department of Health and Human Services, *Dietary Guidelines for Americans, 2010.* Washington, DC: US Government Printing Office, January 2011. *Adv. Nutr.: An Int. Rev. J.* **2011,** 2 (3), 293–294.

McGuire, R. G.; Kelman, A. Reduced Severity of Erwinia Soft Rot in Potato Tubers with Increased Calcium Content, Doctoral dissertation, University of Wisconsin, Madison, 1983.

McGuire, R. G.; Kelman, A. Calcium in Potato Tuber Cell Walls in Relation to Tissue Maceration by *Erwinia carotovora pv. atroseptica. Phytopathology* **1986**, *76* (4), 401–406.

Mir, N. A.; Bhat, J. N.; Bhat, A. R. Effect of Calcium Infiltration on Storage Behavior of Red Delicious Apples. *Indian J. Plant Physiol.* **1993**, 36, 65–66.

Moline, H. E. Preharvest Management for Postharvest Biological Control. In *Biological Control of Postharvest Diseases: Theory and Practice.* Wilson, C. L., Wisniewski, M. E., Eds.; CRC Press: Boca Raton, FL, 1994; pp 57–62.

Montealegre, J. R.; Valdés, J. M. Efecto de Aplicaciones de Calcio en Precosecha Sobre la Susceptibilidad de Frutos de Frambuesa a *Botrytis cinerea. Fitopatologia* **1993**, *2*, 93–98.

Mootoo, A. Effect of Postharvest Calcium Chloride Dips on Ripening Changes in 'Julie' Mangoes. *Trop. Sci.* **1991**, *31*, 243–248.

Morris, J. R.; Sistrunk, W. A.; Sims, C. A.; Mian, G. L. Effects of Cultivar, Postharvest Storage Preprocessing Dip Treatments and Style of Pack on the Processing Quality of Strawberries. *J. Am. Soc. Hortic. Sci.* **1985**, *110*, 172–177.

Naradisorn, M. Effect of Calcium Nutrition on Fruit Quality and Postharvest Diseases. *Int. J. Sci. Inno. Discov.* **2013**, *3* (1), 8–13.

Navazio, L.; Baldan, B.; Moscatiello, R.; Zuppini, A.; Woo, S. L.; Mariani, P.; Lorito, M. Calcium-mediated Perception and Defense Responses Activated in Plant Cells by Metabolite Mixtures Secreted by the Biocontrol Fungus *Trichoderma atroviride. BMC Plant Biol.* **2007**, *7* (1), 41.

O'Hare, T. J.; Zauberman, G. Chemical Control of Mango Shelf Life. In *Horticulture Postharvest Group Biennial Review*; Department of Primary Industries: Queensland, 1992; p 33.

Paliyath, G.; Poovaiah, B. W.; Munske, G. R.; Magnuson, J. A. Membrane Fluidity in Senescing Apples: Effects of Temperature and Calcium. *Plant Cell Physiol.* **1984**, *26*, 977.

Poinssot, B.; Vandelle, E.; Bentejac, M.; Adrian, M.; Levis, C.; Brygoo, Y.; Garin, J.; Sicilia,F.; Coutos-Thevenot, P.; Pugin, A. The Endo-polygalacturonase I from *Botrytis cinerea* Activates Grapevine Defense Reactions Unrelated to its Enzymatic Activity. *Mol. Plant-Microbe Interactions* **2003**, *16* (6), 553–564.

Pollack, S. Consumer Demand for Fruit and Vegetables: The US Example. In *Changing Structure of Global Food Consumption and Trade. Economic Research Service*; U.S. Department of Agriculture, Agriculture and Trade Report. WRS-01-1, 2001.

Poovaiah, B. W. Role of Calcium in Ripening and Senescence. *Comm. Soil Sci. Plant Anal.* **1979**, *10* (1–2), 83–88.

Poovaiah, B. W. Role of Calcium and Calmodulin in Plant Growth and Development. *Hortic. Sci.* **1985**, *20*, 347.

Poovaiah, B. W. Role of Calcium in Prolonging Storage Life of Fruits and Vegetables. *Food Tech.* **1986**, *40*, 86–89.

Poovaiah, B. W.; Leopold, A. C. Deferral of Leaf Senescence with Calcium. *Plant Physiol.* **1973**, *52*, 236.

Poovaiah, B. W.; Shekhar, V. C.; Patterson, M. E. Postharvest Calcium and Other Solute Infiltration into Apple Fruits by Pressures and Vacuum Methods. *Hortic. Sci.* **1978**, *13* (3), 357.

Postlmayr, H. L.; Luh, B. S.; Leonard, S. J. Characterization of Pectin Changes in Clingstone Peaches During Ripening and Processing. *Food Technol.* **1956**, *10*, 618–625.

Raese, J. T.; Drake, S. R. Effects of Preharvest Calcium Sprays on Apple and Pear Quality. *J. Plant Nutr.* **1993**, *16*, 1807–1819.

Robson, M. G.; Hopfinger, J. A.; Eck, P. Postharvest Sensory Evaluation of Calcium Treated Peach Fruit. *Acta. Hortic.* **1989**, *254*, 173–177.

Rosen, J. C.; Kader, A. A. Postharvest Physiology and Quality Maintenance of Sliced Pear and Strawberry Fruits. *J. Food Sci.* **1989**, *54*, 656–659.

Roychoudhury, R.; Kabir, J.; Ray, S. K. D.; Dhua, R. S. Effect of Calcium on Fruit Quality of Litchi. *Indian J. Hortic.* **1992**, *49*, 27–30.

Sams, C. E. Preharvest Factors Affecting Postharvest Texture. *Postharvest Biol. Technol.* **1999**, *15*, 249–254.

Sams, C. E.; Conway, S. W.; Abbott, J. A.; Lewis, R. J.; Benshalom, N. Firmness and Decay of Apples Following Postharvest Pressure Infiltration of Calcium and Heat Treatment. *J. Am. Soci. Hortic. Sci.* **1993**, *118*, 623–627.

Sanders, D.; Brownlee, C.; Harper, J. E. Communicating with Calcium. *Plant Cell* **1999**, *11*, 691–706.

Saure, M. C. Blossom-end Rot of Tomato (*Lycopersicon esculentum* Mill.): A Calcium or Stress Related Disorder? *Sci. Hortic.* **2001**, *90*, 193–208.

Schirra, M.; Mulas, M. Storage of 'Montreal' clementines as Affected by CaCl2 and TBZ Postharvest Treatments. *Agric. Mediterr.* **1994**, *124* (4), 238–248.

Scott, K. J.; Wills, R. B. H. Post-harvest Application of Calcium as a Control for Storage Breakdown of Apples. *Hortic. Sci.* **1975**, *10*, 75.

Scott, K. J.; Wills, R. B. H. Vacuum Infiltration of Calcium Chloride: A Method for Reducing Bitter Pit and Senescence of Apples during Storage at Ambient Temperatures. *Hort Sci.* **1977**, *12*, 71–72.

Seymour, G.; Taylor, J.; Tucker, G. Introduction: Texture Changes. *Biochemistry of Fruit Ripening*; Chapman and Hall: London, **1993**, 17–24.

Sharma, R. R.; Singh, D.; Singh, R. Biological Control of Postharvest Diseases of Fruits and Vegetables by Microbial Antagonists: A Review. *Biol. Cont.* **2009**, *50*, 205–221.

Sharples, R. O.; Johnson, D. S. Post-harvest Chemical Treatments for the Control of Storage Disorders of Apples. *Ann. Appl. Biol.* **1976**, *83*, 157–167.

Shear, C. B. Symptoms of Calcium Deficiency on Leaves and Fruit of 'York Imperial' Apple. *J. Am. Soc. Hortic. Sci.* **1971**, *96*, 415.

Shear, C. B. Calcium-Related Disorders of Fruits and Vegetables. *Hortic. Sci.* **1975**, *10*, 361.

Shorter, A. J.; Joyce, D. C. Effect of Partial Pressure Infiltration of Calcium into 'Kensington' Mango Fruit. *Austr. J. Exp. Agric.* **1998**, *38*, 297–294.

Simons, R. K. Anatomical Studies of the Bitter Pit Area of Apples. *Proc. Am. Soc. Hortic. Sci.* **1962**, *81*, 41.

Singh, B. P.; Tandon, D. K.; Kalra, S. K. Changes in Postharvest Quality of Mangoes Affected by Preharvest Application of Calcium Salts. *Sci. Hortic.* **1993**, *54* (3), 211–219.

Souza, A. L. B.; Chittarra, de M. I. F.; Chittarra, A. B.; Machado, J. C. Postharvest Resistance of Peaches (cv. Biuti) to *Monilina fruiticola*: Induction of Biochemical Responses Through the Application of CaCl2 at the Site of the Injury. *Cien. Agrotechnol.* **1999**, *23* (4), 865–875.

Stanley, D. W.; Bourne, M. C.; Stone, A. P.; Wismer, W. V. Low Temperature Blanching Effects on Chemistry, Firmness and Structure of Canned Green Beans and Carrots. *J. Food Sci.* **1995**, *60* (2), 327–333.

Stow, J. Effect of Calcium Ions on Apple Fruit Softening During Storage and Ripening. *Postharvest Biol. Technol.* **1993**, *3*, 1–9.

Subbiah, K. Firmness Index of Tomato as Influenced by Added N, K and CaCl$_2$ Sprays. *Madras Agric. J.* **1994**, *81* (1), 32–33.

Subbiah, K.; Perumal, R. Effect of Calcium Sources, Concentrations, Stages and Number of Sprays on Physio-chemical Properties of Tomato Fruits. *South Indian Hortic.* **1990**, *38* (1), 20–27.

Suhardi, Y. Ripening Retardation of Arumanis Mango. *ASEAN Food J.* **1992**, *7* (4), 207–208.

Suwwan, M.; Poovaiah, B. W. Association between Elemental Content and Fruit Ripening in *rin* and Normal Tomatoes. *Plant Physiol.* **1978**, *61*, 883.

Takatori, F. H.; Lorenz, O. A.; Cannell, G. H. Strontium and Calcium for the Control of Blackheart of Celery. *Proc. Am. Soc. Hortic. Sci.* **1961**, *77*, 406–414.

Tibbitts, T. W.; Palzkill, D. A. Requirement for Root-pressure Flow to Provide Adequate Calcium to Low-transpiring Tissue. *Commun. Soil Sci. Plant Anal.* **1979**, *10* (1–2), 251–257.

Tingwa, P. O.; Young, R. E. The Effect of Calcium on the Ripening of Avocado (*Persea americana* Mill.) fruits. *J. Am. Soc. Hortic. Sci.* **1974**, *99*, 540–542.

Tirmazi, S. I. H.; Wills, R. B. H. Retardation of Ripening of Mangoes by Postharvest Application of Calcium. *Trop. Agric.* **1981**, *58*, 137–141.

Uhm, K. H.; Ahn, I. P.; Kim, S.; Lee, Y. H. Calcium/Calmodulin-Dependent Signaling for Pre-penetration Development in *Colletotrichum gloeosporioides. Phytopathology* **2003**, *93* (1), 82–87.

Van Eeden, S. J. Calcium Treatment as Possible Postharvest Treatment to Increase Storage Potential of Mango Fruit; South African Mango Growers Association Yearbook 1992; 1992, Vol. 12, pp 26–27.

Vandelle, E.; Poinssot, B.; Wendehenne, D.; Bentejac, M.; Pugin, A. Integrated Signaling Network Involving Calcium, Nitric Oxide and Active Oxygen Species but Not Mitogen-activated Protein Kinases in BcPG1-Elicited Grapevine Defenses. *Mol. Plant Microbe Interactions* **2006**, *19* (4), 429–440.

Wainwright, H.; Burbage, M. B. Physiological Disorders in Mango (*Mangifera indica* L.) Fruit. *J. Hortic. Sci.* **1989**, *64*, 125–135.

Walter, Jr., W. M.; Fleming, H. P.; McFeeters, R. F. Base-Mediated Firmness Retention of Sweet-Potato Products. *J. Food Sci.* **1993**, *58* (4), 813–816.

White, P. J.; Broadley, M. R. Calcium in Plants. *Ann. Bot.* **2003**, *92* (4), 487–511.

Wills, R. B. H.; Tirmazi, S. I. H. Effect of Calcium and Other Minerals on Ripening of Tomatoes. *Austr. J. Plant Physiol.* **1979**, *6*, 221–227.

Wills, R. B. H.; Mahendra, M. S. Effect of Postharvest Application of Calcium on Ripening of Peach. *Austr. J. Expt. Agric.* **1989**, *29*, 751–753.

Wojcik, P. Effect of Calcium Chloride Sprays at Different Water Volumes on 'Champion' Apple Calcium Concentration. *J. Plant Nutr.* **2001**, *24* (4–5), 639–650.

Wojcik, P.; Swiechowski, W. Effect of Spraying with Calcium Chloride at Different Water Rates on 'Jonagold' Apple Calcium Concentration. *Acta Agrobot.* **1999**, *52* (1–2), 75–84.

Wójcik, P.; Lewandowski, M. Effect of Calcium and Boron Sprays on Yield and Quality of 'Elsanta' Strawberry. *J. Plant Nutr.* **2003**, *26*, 671–682.

Woodbridge, C. G. Calcium Level of Pear Tissues Affected with Cork and Black End. *Hortic. Sci.* **1971**, *6* (5), 451–453.

Yamniuk, A. P.; Vogel, H. J. Structurally Homologous Binding of Plant Calmodulin Isoforms to the Calmodulin-binding Domain of Vacuolar Calcium-ATPase. *J. Biol. Chem.* **2004**, *279* (9), 7698–7707.

Yeun, C. M. C.; Tan, S. C.; Joyce, D.; Chettri, P. Effect of Postharvest Calcium and Polymeric Films on Ripening and Peel Injury in 'Kensington Pride' Mangoes. *ASEAN Food J.* **1993,** *8,* 110–113.

Yeun, C. M. C.; Caffin, N.; Boonyakiat, D. Effect of Calcium Infiltration on Ripening of Avocadoes at Different Maturities. *Austr. J. Exp. Agric.* **1994,** *34,* 123–126.

Yuen, C. M. C. Calcium and Fruit Storage Potential. *Austr. Ctr. Int. Agric. Res.* **1994,** *50,* 218–227.

Yuk, H. G.; Yoo, M. Y.; Yoon, J. W.; Moon, K. D.; Marshall, D. L.; Oh, D. H. Effect of Combined Ozone and Organic Acid Treatment for Control of *Escherichia coli* O157:H7 and *Listeria monocytogenes* on Lettuce. *J. Food Sci.* **2006,** *71,* M83–M87.

Zocchi, G.; Mignani, I. Calcium Physiology and Metabolism in Fruit Trees. *Acta. Hortic.* **1995,** *383,* 15–23.

CHAPTER 4

EFFECTS OF METHYL JASMONATE TREATMENT ON FRUIT QUALITY PROPERTIES

MARÍA SERRANO[1*], ALEJANDRA MARTÍNEZ-ESPLÁ[2], PEDRO J. ZAPATA[2], SALVADOR CASTILLO[2], DOMINGO MARTÍNEZ-ROMERO[2], FABIÁN GUILLÉN[2], JUAN M. VALVERDE[2], and DANIEL VALERO[2]

[1]*Department of Applied Biology, Miguel Hernández University, Ctra Beniel km 3.2, 03312, Orihuela, Alicante, Spain*

[2]*Department of Food Technology, Miguel Hernández University, Ctra Beniel km 3.2, 03312, Orihuela, Alicante, Spain*

Corresponding author. E-mail: m.serrano@umh.es

ABSTRACT

Jasmonates (JAs), including jasmonic acid (JA) and its derivative methyl jasmonate (MeJA), have been found in a wide range of higher plants and considered hormones acting in the regulation of a wide range of physiological processes in plants, including growth, photosynthesis, reproductive development, and responses to abiotic and biotic stresses. The most studied effect of JAs has been their role as elicitors triggering the plant defense responses against herbivores and pathogens attacks. However, recently it has been also reported that JAs, applied as pre- or postharvest treatments, affect fruit quality parameters either at harvest or during postharvest storage, by increasing the fruit content on bioactive compounds with antioxidant activity, such as phenolics and anthocyanins, as well as by reducing the severity of chilling injury damage. Moreover, MeJA and JA treatments have effects accelerating or delaying the fruit ripening depending on fruit species, applied doses and the fruit developmental stage when the treatment is made. Overall, the application of JA and its derivatives, especially MeJA, has a great potential in the agro-food industry due to their beneficial effect on fruit quality attributes (organoleptic, nutritive, and increased bioactive compounds). These benefits can be obtained by either the application as preharvest or postharvest treatments. Especially, the preharvest MeJA treatment could be considered as simple, economic, and easily applied treatment for practical purposes and is therefore proposed as a recommendable practice to increase fruit quality and the content of specific phenolic compounds with a potential health benefits.

4.1 INTRODUCTION

Jasmonic acid (JA), methyl jasmonate (MeJA), and its other derivatives are composed of a group of plant hormones that have been discovered recently, compared to other plant hormones, such as auxin, abscisic acid, cytokinins, gibberellic acid, and ethylene, and participate in the regulation of diverse processes in plants, including growth, photosynthesis, reproductive development, and responses to abiotic and biotic stresses (Creelman and Mullet, 1997; Wasternack, 2015). One of the first reports of JA as a hormone was the identification of MeJA as the senescence-promoting substance in wormwood (Ueda and Kato, 1980). JA was first of all isolated from the fungal cultures of *Lasiodiplodia theobromae* while MeJA was obtained from the essential oil of *Jasminum grandiflorum* L. and *Rosmarinus officinalis* L. JA and its derivatives, known as jasmonates (JAs), belong to the family of oxygenated fatty acid derivatives collectively called as oxylipins, which are synthesized via the oxidative metabolism of polyunsaturated fatty acids (Dar et al., 2015).

The most studied effect of JAs has been their role as elicitors triggering the plant defense responses against herbivores and pathogens attacks across the plant kingdom, from angiosperms to gymnosperms, suggesting that the signaling machinery underlying JA-mediated secondary metabolite elicitation appeared early in the higher plant evolution. The three major classes of plants' secondary metabolic pathways activated by JAs are those of terpenoids, alkaloids, and phenylpropanoids, the last one leading to the accumulation of phenolic compounds. Several studies have demonstrated that JAs trigger an extensive transcriptional reprogramming of metabolism throughout upregulating the expression of gene-encoding enzymes involved in one particular secondary metabolic pathway and leading to the recognition of so-called transcriptional regulons (De Geyter et al., 2012; Dar et al., 2015; Wasternack 2014, 2015).

In nature, MeJA is a volatile compound and evidence exists about its use as a postharvest treatment with special implications for reducing chilling injury (CI) symptoms or inhibiting fungal infection. The involvement of JAs in regulating developmental processes has not been explored to the same depth as their roles in plant defense or signaling. MeJA is already classified by the United States Food and Drug Administration (FDA) as "generally recognized as safe" (GRAS) substance and having potential for commercial purposes (FDA-EPA, 2013). In this chapter, the main JAs effects, either applied as postharvest or preharvest treatments, on fruit quality attributes at harvest and during storage are reviewed.

4.2 POSTHARVEST MeJA TREATMENT

Currently, most of the knowledge about MeJA effects on fruit quality attributes and ripening is derived from postharvest treatments, which have been focused on reducing a number of stress-induced injuries during the postharvest period such as CI, infection by some pathogens, and mechanical stress among others (Peña-Cortés et al., 2005; Sayyari et al., 2011). In addition, it has been reported that postharvest treatments with MeJA promote climacteric fruit ripening by increasing ethylene production in fruit such as peach, mango, tomato, and apple (Peña-Cortés et al., 2005), as well as in plum (Khan and Singh, 2007), and even in nonclimacteric fruit such as strawberry (Concha et al., 2013). Additionally, MeJA induced increases in flavonoid and anthocyanin compounds during cold storage of bayberry, blackberry, and raspberry (Wang et al., 2008, 2009; De la Peña et al., 2010).

4.2.1 EFFECT OF JAs ON REDUCING CI

Cold storage is widely used as a postharvest treatment to delay ripening and senescence in fruits and vegetables, so upholding their postharvest quality. But the problem of its application to tropical and subtropical fruits and vegetables is the susceptibility of these commodities to suffer CI at temperatures below 12°C. Fruit CI symptoms are mainly skin pitting, internal browning, wooliness, reddening, leatheriness, and abnormal ripening, which lead to an undesirable effects on quality. Cell membrane integrity is the primary cell structure affected by CI due to membrane fatty acid peroxidation, increase of the saturation degree of these fatty acids, degradation of phospholipids and galactolipids, and the rise of the sterol to phospholipid ratio, leading to reduction of membrane fluidity and performance. The membrane damage initiates a cascade of secondary reactions finally leading to the disruption of cellular and subcellular structures (Aghdam et al., 2013). Oxidative stress also occurs in CI, leading to increase in the levels of reactive oxygen species (ROS) which stimulates lipid peroxidation in cell membranes and in turn enhances cell membrane damages. Plant defense against oxidative stress is accomplished in two ways. The first is by activation of the expression of gene-encoding proteins involved in reducing ROS production such as alternative oxidase (AOX), and the second one by inducing the gene expression for ROS scavenger enzymes such as superoxide dismutase (SOD), catalase (CAT), ascorbate peroxidase (APX), glutathione peroxidase (GPX), glutathione-S-transferase (GST), monodehydroascorbate reductase (MDHAR), dehydroascorbate reductase (DHAR), and glutathione reductase (GR). Higher activity of these enzymatic antioxidant system leads to the reduction of ROS, therefore improving membrane integrity and ultimately inducing resistance toward CI (Aghdam and Bodbodak, 2013).

To increase fruit tolerance to CI and to extend their storage life, postharvest protocols such as cold storage coupled with heat treatments, temperature preconditioning, intermittent warming, modified and controlled atmosphere storage, ultraviolet light, salicylates, and JAs treatments have been developed (Valero et al., 2002; Valero and Serrano, 2010; Aghdam et al., 2013).

Postharvest treatment with MeJA can reduce the development of CI symptoms in a number of horticultural crops, including papaya (González-Aguilar et al., 2003), peach (Meng et al., 2009), guava (González-Aguilar et al., 2004), lemon (Siboza et al., 2014), and pomegranate (Sayyari et al., 2011), manifested by inhibiting electrolyte leakage and lipid peroxidation. However, the mode of action of MeJA in reducing CI and quality deterioration has not been clearly elucidated. In this sense, the effect of MeJA treatments (0.1 and 0.01 mM for 16 h at 20°C) on alleviating CI in pomegranate fruit was attributed to maintenance of membrane integrity (Sayyari et al., 2011). On the other hand, MeJA treatment of banana fruit effectively reduced the increase in relative electrolyte leakage and visible CI symptoms, as well as enhanced proline and the endogenous-JA content and expression levels of JA-biosynthetic genes, maintaining quality of banana fruit during cold storage (Zhao et al., 2013). Treatment with MeJA significantly reduced the severity of CI in loquat and maintained significant higher concentrations of proline and γ-aminobutyric acid (GABA) with respect to control fruits, which might contribute to the reduced CI in MeJA-treated fruit (Cao et al., 2012). GABA is a four-carbon, nonprotein amino acid which, in addition to its antichilling function, plays a crucial role in human health due to its antihypertensive effects (Peñas et al., 2015). So, in addition to alleviating CI, the increase of endogenous levels of GABA in fruits and vegetables by MeJA postharvest treatments could have beneficial effects on human health.

Accordingly, in cherry tomato fruit the expression of genes encoding for arginase activity increased significantly in fruit pretreated with MeJA compared with the control fruit during cold storage (Zhang et al., 2012). Since arginase hydrolyses arginine to urea and ornithine and the last one is the precursor of putrescine (Put) and proline, the increase in arginase activity may be involved in chilling tolerance induced by throughout increases in proline content. Moreover, the increased Put concentration in MeJA-treated cherry tomato fruit (Zhang et al., 2012) may also contribute to the increase of chilling tolerance in MeJA-treated fruit, given the known effect of polyamines in reducing CI of sensitive fruits (Valero and Serrano, 2010). On the other hand, MeJA could also decrease incidence of CI by enhancing the activities of SOD, CAT, and APX and lowering the activity of lipoxygensase (LOX), as it has been proposed in loquat fruit (Cao et al., 2009). In a similar way, decreases in LOX activity have been proposed as the reason for the reduction of CI in MeJA-treated avocado together with increases in fatty acid desaturase activity and heat shock proteins when MeJA treatment was combined with low-temperature conditioning (Sivankalyani et al., 2015). Finally, in peach, the reduction of CI symptoms by MeJA treatment was attributed to higher activity of peroxidase (Meng et al., 2009) as well as to maintain cell energy status, ATP, and adenylate energy charge through increase in H^+-ATPase, Ca^{2+}-ATPase, succinic dehydrogenase (SDH), and cytochrome C oxidase (CCO) activities which are key players in cell energy and ATP production (Jin et al., 2013). Moreover, recently it has been reported that soluble sugar metabolism is also involved in chilling resistance of MeJA-treated peach fruit, since sucrose and sorbitol levels were significantly higher in treated fruit than in control during storage, paralleled by higher gene expression and activity of sucrose phosphate synthase and lower expression and activity of both acid invertase and sorbitol dehydrogenase (Yu et al., 2016). Thus, higher levels of sorbitol and especially of sucrose could regulate osmotic pressure, contribute to membrane stability, activate the antioxidization system to eliminate free radicals, and enhance chilling tolerance in fruit.

4.2.2 ROLE OF MeJA ON INDUCING FRUIT RESISTANCE TO FUNGAL DECAY

Several reports have shown that MeJA could be a useful technique in reducing postharvest diseases in horticultural products, possibly by directly suppressing pathogen growth, and/or indirectly inducing disease resistance, probably due to the systemic acquired resistance (SAR) as a signal molecule (Beckers and Spoel, 2005). Thus, the application of postharvest MeJA treatments effectively suppressed anthracnose rot caused by *Colletotrichum acutatum* in loquat (Cao et al., 2008) and reduced postharvest brown rot caused by *Monilinia fructicola* (Yao and Tian, 2005) and blue mould decay caused by *Penicillium expansum* (Wang et al., 2015a) in sweet cherry. Pathogenesis-related (PR) proteins are a class of defense molecules that are induced by pathogen attack and involved in plant defense mechanisms against fungal infection. Among PR proteins, chitinase (CHI) and β-1,3-glucanase (GLU), which hydrolyze polymers of fungal cell walls, are increased by MeJA treatments in sweet cherries (Wang et al., 2015a). Postharvest treatments with JA and MeJA also reduced decay of grapefruit caused by the green mold *Penicillium digitatum* after either artificial inoculation of wounded fruit at 24°C or natural decay in cold-stored fruit, the most effective concentration being 10 μM (Droby et al., 1999).

In addition, MeJA treatments increased disease resistance to *Botrytis cinerea* in tomato due to an accumulation of H_2O_2, concomitant with enhanced expression of the genes codifying for Cu–Zn superoxide dismutase (Cu–Zn SOD), CAT, and APX. In turn, ascorbate (ASC) and glutathione (GSH) content were increased in MeJA-treated fruits, being beneficial for scavenging excess ROS and alleviating oxidative damage of proteins (Zhu and Tian, 2012). Resistance to *B. cinerea* was also increased by MeJA treatments in grape berries (Wang et al., 2015b), in which low concentration of MeJA (10 μM) triggered priming defense mechanism, whereas higher concentrations of MeJA (50 or 100 μM) directly activated defense responses. Thus, the expression of the defensive VvNPR1.1 gene, which plays a key role to upregulate the expression of PR1 and PR2 (β-1,3-glucanase), and the accumulation of the phytoalexins trans-resveratrol and its oligomer (trans-)ε-viniferin in 10 μM/L MeJA-treated grape berries were significantly enhanced only upon inoculating the berries with *B. cinerea*, whereas the 50 or 100 μM MeJA treatment directly induced these defense responses (Wang et al., 2015b). Accordingly, MeJA significantly reduced the anthracnose (caused by *Colletotrichum musae*) disease index, and lesion diameter of banana fruit, along with enhanced accumulation of endogenous SA or JA contents, and higher expression levels of some pathogenesis-related (PRs) genes (Tang et al., 2013).

On the other hand, the application of the JA derivative *n*-propyl dihydro-jasmonate (PDJ) to berries of grapevines decreased the lesion diameters of the inoculated fungus *Glomerella cingulata* by increasing endogenous ABA and JA concentrations and thus suggesting that defense mechanism in grape berries may depend on the synergistic effect of JA and ABA (Wang et al., 2015c).

Moreover, *in vitro* experiment showed that MeJA had a direct and transient effect on inhibiting spore germination and germ tube elongation of *P. expansum* (Wang et al., 2015a). Accordingly, it has been reported that MeJA significantly inhibited spore germination, germ tube elongation, and mycelial growth of *C. acutatum* (Cao et al., 2008). Zhu and Tian (2012) also found that MeJA negatively affected plasma membrane integrity of *B. cinerea* spores and showed direct antimicrobial activity in vitro, suggesting that MeJA has a direct fungitoxic property against this pathogen. However, previously, it had been reported that MeJA had no direct inhibitory effect on the spore germination and germ tube elongation of the pathogen *B. cinerea* and the mycelial growth was only significantly inhibited when agar media was supplemented with 600 µM MeJA but not with 200 or 400 µM (Darras et al., 2005). Accordingly, neither JA nor MeJA had any direct antifungal effect on *P. digitatum* spore germination or germ tube elongation (Droby et al., 1999) and MeJA had a little inhibitory effect on mycelial growth and spore germination of *M. fructicola* (Yao and Tian, 2005). Thus, the direct effect of MeJA against fungi is not clear and the contradictory results could be due to different sensitivities of fungal species to MeJA or to different doses and experimental conditions.

4.3 PREHARVEST MeJA TREATMENT

Information about the effect of the application of MeJA as a preharvest treatment on fruit quality attributes at harvest and during postharvest storage is much limited than those of previously commented postharvest treatments, and most of the reports deal on inducing disease resistance, as has been observed on reducing infection of *M. fructicola* in sweet cherry (Yao and Tian, 2005).

4.3.1 MeJA EFFECT ON FRUIT GROWTH

During peach development and ripening, endogenous JA levels are elevated at early stages of fruit development (S1), then gradually decrease (at S2–S3 stages) and start to increase during ripening (S4), these changes being correlated with ethylene biosynthesis (Torrigiani et al., 2012). In a similar way, JA concentration was high in immature apples (24 days after full blossom, DAFB) and decreased as fruit developed, with the lowest endogenous levels being found at the time of harvest, the JA behavior being coincident with those of respiration rate and ethylene production (Fan et al., 1997). In this report, apple fruit was harvested at preclimacteric stage and increase in ethylene production or JA concentration could not be found. Given this JA behavior, some researchers have been focused on the application of JA or MeJA during on-tree fruit development with the aim to increase their endogenous levels at the time of harvest and their impact on fruit physiology.

MeJA applied as preharvest treatment could affect fruit growth, although these effects are dependent on fruit species, cultivar, concentration, and time of application. Thus, Martínez-Esplá et al. (2014) treated "Black Splendor" (BS) and "Royal Rosa" (RR) plum trees with MeJA at concentrations of 0.5, 1.0, and 2.0 mM at three key points of fruit development on tree (pit hardening, initial color changes, and onset of ripening) and found that in BS plums both 0.5 and 1.0 mM MeJA led to a significant increase in fruit volume and weight (Fig. 4.1) while no significant effect was observed for 2.0 mM. However, in RR plums, the effective concentrations in increasing these fruit size parameters were 0.5 and 2.0 mM, whereas 1.0 mM had no significant effect. At harvest time, the major increase in fruit volume was 8% and 18% in 0.5 and 2.0 mM treated plums for BS and RR, respectively (Martínez-Esplá et al., 2014), and in fruit weight, these increases were of 9.2% and 21.6% for BS and RR, respectively (Fig. 4.1). We have found similar effects of MeJA preharvest treatments on increasing fruit size in sweet cherry, as can be observed in Figure 4.1. From the agronomic and commercial point of view, the results obtained would have a great importance since fruits with larger size are more appreciated by consumers and would reach higher prices in the market. Accordingly, fruit volume and size was also increased in "Fuji" apples by MeJA treatments (at 5, 10, and 20 mM) at weekly intervals during the last month of fruit development on tree (Altuntas et al., 2012). On the contrary, it was previously reported in this apple cultivar that MeJA treatments at similar concentrations reduced fruit size and weight due to inhibition of cell expansion or elongation when applied 48 days after full bloom (DAFB), while no effect was observed when these treatments were performed at 119 DAFB (Rudell et al., 2005). In agreement with these findings, a single treatment in peach with 0.8 mM MeJA at 56 DAFB (S2 stage) or 0.2 mM applied at S3 stage did not affect fruit diameter or weight at harvest (Ziosi et al., 2008; Ruiz et al., 2013). However, Martínez-Esplá et al. (2014) reported that MeJA showed positive effects in terms of increasing plum fruit size and weight, probably due to the repeated applications of MeJA at the three key points of the fruit growth curve.

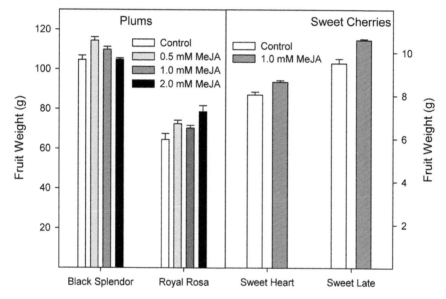

FIGURE 4.1 Effects of preharvest treatments of plum and sweet cherry cultivars with MeJA at three key points of fruit development (pit hardening, initial color changes, and onset of ripening) on fruit weight at harvest.

Source: Adapted from Martínez-Esplá et al. (2014) for plum and unpublished data for sweet cherry.

4.3.2 MeJA AND FRUIT RIPENING

Different experiments have demonstrated that tree treatments with JA or MeJA have different effects on the parameters related to fruit ripening depending on fruit species, applied concentration, and even the fruit developmental stage when the treatment is performed. Thus, in peach, field treatment with MeJA (0.4 mM) at the initial S3 stage of fruit development resulted in a delay in the ripening process, manifested by lower softening, skin color, and total soluble solids (TSS) accumulation, while with later MeJA application, lower effects were observed (Ziosi et al., 2008). However, treatment of peach trees with MeJA at 8.0 mM led to acceleration of fruit ripening and leaf senescence (Janoudi and Flore, 2003). Moreover, the application of MeJA to peach fruits 1 day before harvest stimulated the ethylene biosynthesis and accelerated softening without affecting the content of TSS (Torrigiani et al., 2012). Accordingly, MeJA treatment at an early developmental stage of apples induced a delay of the ripening process, while an acceleration of this process occurred when MeJA was applied at the latest developmental stages (Rudell et al., 2005). However, in "Fortune" plums, MeJA treatment at 5.0 or 10 mM, 2 weeks before harvest, retarded color change and accumulation of TSS during on-tree ripening, while firmness was not affected (Karaman et al., 2013).

On the other hand, in BS and RR plums, treatments with MeJA at three key points of fruit development led to a delay in skin color change and acidity losses with 0.5 mM MeJA treatment, while 1.0 and 2.0 mM doses had no significant effect, showing a delay on the ripening process by the lower doses of MeJA (Martínez-Esplá et al., 2014). However, the rate of softening during on-tree fruit ripening was significantly retarded by MeJA treatments, and at the time of harvest, fruit firmness was higher in plums from all MeJA-treated trees, the main effects being observed with 0.5 mM for both cultivars (Fig. 4.2). Skin and flesh firmness at harvest were also higher in "Fuji" apples treated with MeJA during the last month of fruit development on tree (Altuntas et al., 2012). On the contrary, TSS increased for both cultivars during on-tree ripening, although any effect was observed attributable to MeJA treatments with final values ca. 11 and 10 g/100 g for BS and RR cultivars, respectively, in both control and treated fruits (Fig. 4.2). Accordingly, the application of MeJA at 0.01 or 0.1 mM to blackberry and raspberry plants of several cultivars increased the content of TSS at harvest, the effect being proportional to the applied concentration (Wang and Zheng, 2005; Wang et al., 2008). Thus, MeJA treatment can differentially impact on each of the parameters involved in fruit ripening, depending on doses, time of application, fruit species, and even on cultivar.

The different effects of MeJA treatments on parameters related with on-tree fruit ripening could be due to the effect of MeJA on ethylene biosynthesis, at least in climacteric fruits. In fact, it has been reported that 2.0 mM MeJA treatment accelerated and increased ethylene production in RR plum fruit, while MeJA at 0.5 mM inhibited ethylene production in RR and BS cultivars, and 1.0 mM MeJA did not show significant effects in RR but inhibited ethylene production in BS (Zapata et al., 2014). Accordingly,

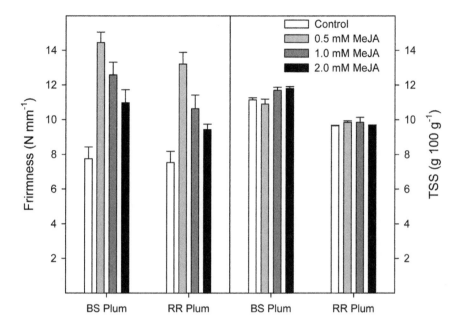

FIGURE 4.2   Effects of preharvest treatments of plum "Black Splendor" (BS) and "Royal Rosa" (RR) cultivars with MeJA at three key points of fruit development (pit hardening, initial color changes, and onset of ripening) on fruit firmness and total soluble solids (TSS) content at harvest.

Source: Adapted from Martínez-Esplá et al. (2014).

preharvest exogenous application of MeJA at low concentration inhibited ethylene production in apple (Fan et al., 1997). Similarly, treatment of peach trees with 0.88 mM MeJA, applied at S3, led to less ripe fruit at harvest and to a delay in ethylene production during fruit ripening on tree (Soto et al., 2012). Accordingly, Ziosi et al. (2008) have reported that 0.22 mM MeJA treatment did not affect ethylene production or ripening, while MeJA at 0.44 mM dramatically inhibited ethylene production at harvest. The mechanism by which MeJA affects ethylene production seems to be regulated at gene level, with stimulation or inhibition of the genes responsible for its biosynthesis being species specific. Thus, the inhibition of ethylene production by MeJA is supported by the reduced expression of 1-aminocyclopropane-1-carboxylic acid (ACC), ACC synthase (ACS), ACC oxidase (ACO), and ethylene receptors (ETR) transcripts in peach (Soto et al., 2012; Ruiz et al., 2013). Contrarily, the postharvest application of PDJ in pear picked at the preclimacteric ripening stage enhanced the expression of ethylene biosynthetic genes for ACC synthase and ACC oxidase, while the accumulations of

ACS1 mRNA decreased in the fruit treated by PDJ at the climacteric stage (Kondo et al., 2007). On the other hand, postharvest 1.0 mM MeJA treatment increased and advanced the ethylene production in three plum cultivars, by increasing ACS and ACO activities (Khan and Singh, 2007).

The evolution of fruit quality parameters during postharvest storage is also affected by preharvest MeJA treatments. Thus, in BS and RR plums, overall quality parameters such as firmness, color, TSS, and total acidity (TA) evolved faster in fruits from 2.0 mM MeJA-treated trees than in controls, while these changes were slower in plums from 0.5-mM treated trees (Zapata et al., 2014), leading to the conclusion that preharvest MeJA treatment at 2.0 mM accelerated and 0.5 mM delayed the ripening process in both plum cultivars, which could be attributed to its effect on accelerating and retarding ethylene production, respectively.

4.3.3 MeJA EFFECT ON FRUIT BIOACTIVE COMPOUNDS

Fruit and vegetables contain a wide range of phytochemical compounds that exhibit antioxidant activity, which vary widely in chemical structure and function, the most common being phenolics (including anthocyanins), carotenoids, and vitamins, which have shown protective effects against several chronic diseases associated with ageing including atherosclerosis, cardiovascular diseases, cancer, cataracts, ulcer, neurodegenerative diseases, brain and immune dysfunction, and even against bacterial and viral diseases (Del Rio et al., 2013; Nile and Park, 2014; Woodside et al., 2015).

In plums and sweet cherry, increase in phenolics, carotenoids, and total antioxidant activity during on-tree fruit ripening has been reported in a wide range of cultivars (Díaz-Mula et al., 2008, 2009). However, the concentration of total phenolic in plum flesh at harvest increased as a consequence of 0.5 mM MeJA tree treatments in BS and RR cultivars. The same effect was observed in total antioxidant activity in the hydrophilic fraction (H-TAA) (Zapata et al., 2014). These effects could be attributed to a stimulation of the phenolic biosynthesis pathway as previously described for Chinese bayberry, for which MeJA increased the phenylalanine ammonialyase (PAL) activity (Wang et al., 2009). Moreover, MeJA treatment increased the effect of blackberry extracts on inhibiting cancer cell proliferation and inducing apoptosis (Wang et al., 2008), showing the health-beneficial effect of fruit with increased phenolic concentration by MeJA treatment.

In addition, the concentration of total phenolics and total anthocyanins, as well as the H-TAA, was found at higher levels in sweet cherries from

1.0 mM MeJA-treated trees than in controls, at harvest and after 20 days of cold storage (Fig. 4.3). However, in "Fortune" plums, preharvest application of MeJA at 10 mM increased total phenolics at harvest, while no effect was found for a 5.0-mM dose. These results were inconsistent with those reported for individual phenolics, since control fruit had 5.30-fold higher concentration of the main phenolic compound (chlorogenic acid) compared with those MeJA-treated fruit (Karaman et al., 2013). Accordingly, preharvest MeJA treatments also induced accumulation of total phenolics, total anthocyanins, and antioxidant activity in raspberry, although no significant qualitative changes were observed in phenolic profile due to MeJA treatment (Wang and Zheng, 2005). Similar effects were observed in MeJA-sprayed raspberries and blackcurrants (at 0.01 and 0.1 mM) in which significant increase in some individual phenolics, such as myricetin, ellagic acid, and quercetin were found as compared with controls, even when low-MeJA concentration was applied (Flores and Ruíz del Castillo, 2014, 2015). Similarly, postharvest MeJA treatments led to increased concentrations of these bioactive compounds during cold storage in pomegranate, strawberry, and blackberry (Chanjirakul et al., 2006, 2007; Sayyari et al., 2011). In grape and wine, phenolic content was also increased in the samples from MeJA treatment in comparison with control samples (Portu et al., 2015).

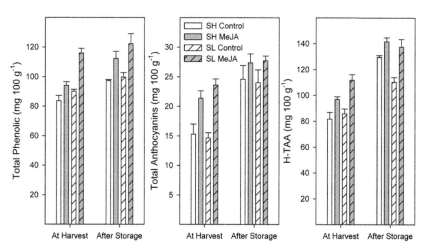

FIGURE 4.3 Effects of preharvest treatments of "Sweet Heart" (SH) and "Sweet Late" (SL) sweet cherry cultivars with 1.0 mM MeJA at three key points of fruit development (pit hardening, initial color changes, and onset of ripening) on total phenolic, total anthocyanins, and hydrophilic total antioxidant activity (H-TAA) at harvest and after 20 days of storage at 1°C.

MeJA sprayed to mango trees (at 80 µL/mL) 20 days before harvesting led to increased anthocyanin concentration in mango peel, due to an enhancement of PAL activity. In addition, this MeJA treatment increased phenolic concentration in mango fruit flesh at harvest and this concentration was higher along storage as compared with control fruit (Muengkaew et al., 2016). Accordingly, 0.25 mM MeJA applied at different crop developmental stages, from flowering to fruit ripening, increased the concentration of total anthocyanins in Chilean strawberries (Saavedra et al., 2016) at harvest and during postharvest storage suggesting a residual effect and activation of the phenylpropanoid pathway. In this respect, previous research has shown improvements in anthocyanin and flavonol content when MeJA was applied before harvest to blackberries (Wang et al., 2008), raspberries (Wang and Zheng, 2005), or apples (Shafiq et al., 2013).

Total anthocyanin, total phenolics, and total antioxidant capacity of MeJA-treated fruits were significantly higher than control fruits. In this sense, high correlations were found in "Fuji" apples between applied MeJA doses and the increase in total anthocyanin, total phenolics, and total antioxidant capacity (Ozturk et al., 2015a). Similar results from these authors were found when MeJA (at 1120 or 2240 mg/L) were sprayed to "Fortune" and "Friar" plums, in which the content of the individual phenolics (chlorogenic acid, caffeic acid, p-coumaric acid, rutin, ferulic acid, quercetin, naringenin, and kaemferol) was always higher in MeJA-treated plums compared with controls when picked at three different ripening stages, especially for 2240 mg/L (Ozturk et al., 2015b).

Moreover, MeJA application to grapevine clusters of Syrah, Monastrell, and Barbera cultivars led to an increase in stilbene and anthocyanin content in grape and wine (Fernández-Marín et al., 2014; Ruíz-García et al., 2012, 2013). Accordingly, foliar application of MeJA to grapevine led to a significant enhancement of anthocyanin synthesis in fruits. Consequently, grapes from MeJA treatment showed higher content of 3-O-glucosides of petunidin and peonidin and trans-p-coumaroyl derivatives of cyanidin and peonidin and total anthocyanin concentration, calculated as the sum of individual anthocyanin contents, was increased 23% due to MeJA foliar treatment (Portu et al., 2015). These higher anthocyanin levels found in grapes from MeJA treatment were reflected in the corresponding wines.

Enhancement of anthocyanin synthesis seems to be explained by the accumulation of different enzymes involved in the phenylpropanoid pathway. In this sense, it has been reported that grapevines may respond to MeJA application by activating gene expression of enzymes responsible for phenolic biosynthesis such as PAL, chalcone synthase, stilbene synthase, and UDP-glucose: flavonoid-O-transferase, proteinase inhibitors, and CHI, with subsequent accumulation of anthocyanins and stilbenes in grapevine cell cultures (Belhadj et al., 2008). Other bioactive compounds such as resveratrol in grapes (Vuong et al., 2014) and glucosinolates in cruciferous (Fritz et al., 2010) were also enhanced by JA treatment. Moreover, in other reports, it has been shown that some lipophilic compounds such as lycopene and vitamin E are increased by MeJA treatments in fruits and vegetables (Wasternack, 2014; Wang and Zheng, 2005). However, in BS and RR plums, no significant effects on carotenoid concentration or lipophilic total antioxidant activity (L-TAA) were obtained due to MeJA treatment (Martínez-Esplá et al., 2014).

Apart from the effect of MeJA on the above commented antioxidant compounds, its role on the behavior of antioxidant enzymes have been recently studied. Thus, preharvest treatment with MeJA at 0.5 mM to plum trees led to the higher activities of POD, CAT, and APX at harvest compared with those obtained in control plums for both BS and RR cultivars (Fig. 4.4). Along storage, a general increase of the enzymatic activities was observed in both control and MeJA-treated plums, although the activities of these antioxidant enzymes were always significantly increased as a consequence of

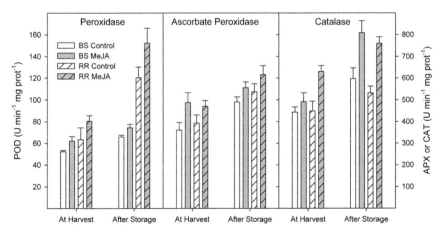

FIGURE 4.4 Effects of preharvest treatments MeJA at 0.5 mM of plum "Black Splendor" (BS) and "Royal Rosa" (RR) cultivars at three key points of fruit development (pit hardening, initial color changes, and onset of ripening) on antioxidant enzymes peroxidase (POD), APX, and CAT at harvest and after 35 days of storage at 1°C.

Source: Adapted from Zapata et al. (2014).

preharvest MeJA treatment. The increase of these enzymes could enhance the tissue capacity to eliminate ROS leading to a delay in ripening and senescence processes. In a similar way, postharvest MeJA treatment led to higher activities of SOD, CAT, and APX during storage in treated than control in control peach and loquat fruit, although this effect was related to a lower CI damage (Cao et al., 2009; Jin et al., 2009). In a nonchilling sensitive commodity, such as raspberry, MeJA treatment induced higher activity on a wide range of antioxidant enzymes and in turn increased the fruit resistance to decay (Chanjirakul et al., 2006).

4.4   CONCLUSIONS

The application of JA and its derivatives, especially MeJA, has a great potential in the agro-food industry due to their beneficial effect on fruit quality attributes (organoleptic, nutritive, and increased bioactive compounds). These benefits can be obtained by either the application as preharvest or postharvest treatments. Especially, the preharvest MeJA treatment could be considered as simple, economic, and easily applied treatment for practical purposes and is therefore proposed as a recommendable practice to increase fruit quality and the content of specific phenolic compounds with a potential health benefits.

KEYWORDS

- ripening
- jasmonates
- phenylpropanoids
- chilling injury
- antioxidant enzymes

REFERENCES

Aghdam, M. S.; Bodbodak, S. Physiological and Biochemical Mechanisms Regulating Chilling Tolerance in Fruits and Vegetables under Postharvest Salicylates and Jasmonates Treatments. *Sci. Hortic.* **2013,** *156,* 73–85.

Aghdam, M. S.; Sevillano, L.; Flores, F. B.; Bodbodak, S. Heat Shock Proteins as Biochemical Markers for Postharvest Chilling Stress in Fruits and Vegetables. *Sci. Hortic.* **2013,** *160*, 54–64.

Altuntas, E.; Ozturk, B.; Özkan, Y.; Yildiz, K. Physico-mechanical Properties and Colour Characteristics of Apple as Affected by Methyl Jasmonate Treatments. *Int. J. Food Eng.* **2012,** *8*, 16.

Beckers, G. J. M.; Spoel, S. H. Fine-tuning Plant Defence Signalling: Salicylate Versus Jasmonate. *Plant Biol.* **2005,** *7*, 1–10.

Belhadj, A.; Telef, N.; Saigne, C.; Cluzet, S.; Barrieu, F.; Hamdi, S.; Mérillon, J. M. Effect of Methyl Jasmonate in Combination with Carbohydrates on Gene Expression of PR Proteins, Stilbene and Anthocyanin Accumulation in Grapevine Cell Cultures. *Plant Physiol. Biochem.* **2008,** *46*, 493–499.

Cao, S.; Zheng, Y.; Yang, Z.; Tang, S.; Jin, P.; Wang, K.; Wang, X. Effect of Methyl Jasmonate on the Inhibition of *Colletotrichum acutatum* Infection in Loquat Fruit and the Possible Mechanisms. *Postharvest Biol. Technol.* **2008,** *49*, 301–307.

Cao, S.; Zheng, Y.; Wang, K.; Jin, P.; Rui, H. Methyl Jasmonate Reduces Chilling Injury and Enhances Antioxidant Enzyme Activity in Postharvest Loquat Fruit. *Food Chem.* **2009,** *115*, 1458–1463.

Cao, S.; Cai, Y.; Yang, Z.; Zheng, Y. MeJA Induces Chilling Tolerance in Loquat Fruit by Regulating Proline and γ-Aminobutyric Acid Contents. *Food Chem.* **2012,** *133*, 1466–1470.

Chanjirakul, K.; Wang, S. Y.; Wang, C. H.; Siriphanich, J. Effect of Natural Volatile Compounds on Antioxidant Capacity and Antioxidant Enzymes in Raspberries. *Postharvest Biol. Technol.* **2006,** *40*, 106–115.

Chanjirakul, K.; Wang, S. Y.; Wang, C. H.; Siriphanich, J. Natural Volatile Treatments Increase Free-radical Scavenging Capacity of Strawberries and Black-berries. *J. Sci. Food Agric.* **2007,** *87*, 1463–1472.

Concha, C. M.; Figueroa, N. E.; Poblete, L. A.; Oñate, F. A.; Schwab, W.; Figueroa, C. R. Methyl Jasmonate Treatment Induces Changes in Fruit Ripening by Modifying the Expression of Several Ripening Genes in *Fragaria chiloensis* Fruit. *Plant Physiol. Biochem.* **2013,** *70*, 433–444.

Creelman, R. A.; Mullet, J. E. Biosynthesis and Action of Jasmonates in Plants. *Annu. Rev. Plant Physiol. Plant Mol. Biol.* **1997,** *48*, 355–381.

Dar, T. A.; Uddin, M.; Khan, M. M. A.; Hakeem, K. R.; Jaleel, H. Jasmonates Counter Plant Stress: A Review. *Environ. Exp. Bot.* **2015,** *115*, 49–57.

Darras, A. I.; Terry, L. A.; Joyce, D. C. Methyl Jasmonate Vapour Treatment Suppresses Specking Caused by *Botrytis cinerea* on Cut *Freesia hybrida* L. Flowers. *Postharvest Biol. Technol.* **2005,** *38*, 175–182.

De Geyter, N.; Gholami, A.; Goormachtig, S.; Goossens, A. Transcriptional Machineries in Jasmonate-Elicited Plant Secondary Metabolism. *Trends Plant Sci.* **2012,** *17*, 349–359.

De la Peña, F.; Blanch, G. P.; Ruiz del Castillo, M. L. (+)-Methyl jasmonate-Induced Bioformation of Myricetin, Quercetin and Kaempferol in Red Raspberries. *J. Agric. Food Chem.* **2010,** *58*, 11639–11644.

Del Rio, D.; Rodriguez-Mateos, A.; Spencer, J. P. E.; Tognolini, M.; Borges, G.; Crozier, A. Dietary (Poly)Phenolics in Human Health: Structures, Bioavailability, and Evidence of Protective Effects Against Chronic Diseases. *Antioxid. Redox Sig.* **2013,** *18*, 1818–1892.

Díaz-Mula, H. M.; Zapata, P. J.; Guillén, F.; Castillo, S.; Martínez-Romero, D.; Valero, D.; Serrano, M. Changes in Physicochemical and Nutritive Parameters and Bioactive

Compounds During Development and On-tree Ripening of Eight Plum Cultivars: A Comparative Study. *J. Sci. Food Agric.* **2008**, *88*, 2499–2507.

Díaz-Mula, H. M.; Castillo, S.; Martínez-Romero, D.; Valero, D.; Zapata, P. J.; Guillén, F.; Serrano, M. Organoleptic, Nutritive and Functional Properties of Sweet Cherry as Affected by Cultivar and Ripening Stage. *Food Sci. Technol. Int.* **2009**, *15*, 535–543.

Droby, S.; Porat, R.; Cohen, L.; Weiss, B.; Shapiro, B.; Philosoph-Hadas, S.; Meir, S. Suppressing Green Mold Decay in Grapefruit with Postharvest Jasmonate Application. *J. Am. Soc. Hortic. Sci.* **1999**, *124*, 184–188.

Fan, X.; Mattheis, J. P.; Fellman, J. K.; Patterson, M. E. Changes in Jasmonic Acid Concentration During Early Development of Apple Fruit. *Physiol. Plant* **1997**, *101*, 328–332.

Fernández-Marín, M. I.; Puertas, B.; Guerrero, R. F.; García-Parrilla, M. C.; Cantos-Villar, E. Preharvest Methyl Jasmonate and Postharvest UVC Treatments: Increasing Stilbenes in Wine. *J. Food Sci.* **2014**, *79*, C310–C317.

Flores, G., Ruiz del Castillo, M. L. Influence of Preharvest and Postharvest Methyl Jasmonate Treatment on Flavonoid Content and Metabolomics Enzyme in Red Raspberry. *Postharvest Biol. Technol.* **2014**, *97*, 77–82.

Flores, G.; Ruiz del Castillo, M. L. Variations in Ellagic Acid, Quercetin and Myricetin in Berry Cultivars after Preharvest Methyl Jasmonate Treatments. *J. Food Comp. Anal.* **2015**, *39*, 55–61.

FDA-EPA. Methyl Jasmonate: Exemption from the Requirement of a Tolerance. Document 78 FR 22789, 2013; pp 22789–22794.

Fritz, V. A.; Justen, V. L.; Bode, A. M.; Schuster, T.; Wang, M. Glucosinolate Enhancement in Cabbage Induced by Jasmonic Acid Application. *HortSci.* **2010**, *45*, 1188–1191.

González-Aguilar, G. A.; Buta, J. G.; Wang, C. Y. Methyl Jasmonate and Modified Atmosphere Packaging (MAP) Reduce Decay and Maintain Postharvest Quality of Papaya 'Sunrise'. *Postharvest Biol. Technol.* **2003**, *28*, 361–370.

González-Aguilar, G. A.; Tiznado-Hernández, M. E.; Zavaleta-Gatica, R.; Martínez-Téllez, M. A. Methyl Jasmonate Treatments Reduce Chilling Injury and Activate the Defense Response of Guava Fruits. *Biochem. Biophys. Res. Commun.* **2004**, *313*, 694–701.

Janoudi, A.; Flore, J. A. Effects of Multiple Applications of Methyl Jasmonate on Fruit Ripening, Leaf Gas Exchange and Vegetative Growth in Fruit Trees. *J. Hortic. Sci. Biotechnol.* **2003**, *78*, 793–797.

Jin, P.; Zheng, Y.; Tang, S.; Rui, H.; Wang, C. Y. A Combination of Hot Air and Methyl Jasmonate Vapor Treatment Alleviates Chilling Injury of Peach Fruit. *Postharvest Biol. Technol.* **2009**, *52*, 24–29.

Jin, P.; Zhu, H.; Wang, J.; Chen, J.; Wang, X.; Zheng, Y. Effect of Methyl Jasmonate on Energy Metabolism in Peach Fruit During Chilling Stress. *J. Sci. Food Agric.* **2013**, *93*, 1827–1832.

Karaman, S.; Ozturk, B.; Genc, N.; Celik, S. M. Effect of Preharvest Application of Methyl Jasmonate on Fruit Quality of Plum (*Prunus salicina* Lindell cv. "Fortune") at Harvest and During Cold Storage. *J. Food Process. Pres.* **2013**, *37*, 1049–1059.

Khan, A. S.; Singh, Z. Methyl Jasmonate Promotes Fruit Ripening and Improves Fruit Quality in Japanese Plum. *J. Hortic. Sci. Biotechnol.* **2007**, *82*, 695–706.

Kondo, S.; Yamada, H.; Setha, S. Effect of Jasmonates Differed at Fruit Ripening Stages on 1-Aminocyclopropane-1-Carboxylate (ACC) Synthase and ACC Oxidase Gene Expression in Pears. *J. Am. Soc. Hortic. Sci.* **2007**, *132*, 120–125.

Martínez-Esplá, A.; Zapata, P. J.; Castillo, S.; Guillén, F.; Martínez-Romero, D.; Valero, D.; Serrano, M. Preharvest Application of Methyl Jasmonate (MeJA) in Two Plum Cultivars. 1. Improvement of Fruit Growth and Quality Attributes at Harvest. *Postharvest Biol. Technol.* **2014**, *98*, 98–105.

Meng, X.; Han, J.; Wang, Q.; Tian, S. Changes in Physiology and Quality of Peach Fruits Treated by Methyl Jasmonate under Low Temperature Stress. *Food Chem.* **2009**, *114*, 1028–1035.

Muengkaew, R.; Chaiprasart, P.; Warrington, I. Changing of Physiochemical Properties and Color Development of Mango Fruit Sprayed Methyl Jasmonate. *Sci. Hortic.* **2016**, *198*, 70–77.

Nile, S. H.; Park, S. W. Edible Berries: Bioactive Components and Their Effect on Human Health. *Nutrition* **2014**, *30*, 134–144.

Ozturk, B.; Yildiz, K.; Ozkan, Y. Effects of Pre-harvest Methyl Jasmonate Treatments on Bioactive Compounds and Peel Color Development of "Fuji" Apples. *Int. J. Food Prop.* **2015a**, *18*, 954–962.

Ozturk, B.; Yildiz, K.; Ozkan, Y. Effect of Pre-harvest Methyl Jasmonate Treatments on Ethylene Production, Water-Soluble Phenolic Compounds and Fruit Quality of Japanese Plums. *J. Sci. Food Agric.* **2015b**, *95*, 583–591.

Peña-Cortés, H.; Barrios, P.; Dorta, F.; Polanco, V.; Sánchez, C.; Sánchez, E.; Ramírez, I. Involvement of Jasmonic Acid and Derivatives in Plant Response to Pathogen and Insects and in Fruit Ripening. *J. Plant Growth Regul.* **2005**, *23*, 246–260.

Peñas, E.; Limón, R. I.; Martínez-Villaluenga, C.; Restani, P.; Pihlanto, A.; Frias, J. Impact of Elicitation on Antioxidant and Potential Antihypertensive Properties of Lentil Sprouts. *Plant Foods Hum. Nutr.* **2015**, *70*, 401–407.

Portu, J.; Santamaría, P.; López-Alfaro, I.; López, R.; Garde-Cerdán, T. Methyl Jasmonate Foliar Application to Tempranillo Vineyard Improved Grape and Wine Phenolic Content. *J. Agric. Food Chem.* **2015**, *63*, 2328–2337.

Rudell, D. R.; Fellman, J. K.; Mattheis, J. P. Preharvest Application of Methyl Jasmonate to 'Fuji' Apples Enhances Red Coloration and Affects Fruit Size, Splitting, and Bitter Pit Incidence. *Hortic. Sci.* **2005**, *40*, 1760–1762.

Ruiz, K. B.; Trainotti, L.; Bonghi, C.; Ziosi, V.; Costa, G.; Torrigiani, P. Early Methyl Jasmonate Application to Peach Delays Fruit/Seed Development by Altering the Expression of Multiple Hormone-related Genes. *J. Plant Growth Regul.* **2013**, *32*, 852–864.

Ruiz-García, Y.; Romero-Cascales, I.; Gil-Muñoz, R., Fernández-Fernández, J. I.; López-Roca, J.M.; Gómez-Plaza, E. Improving grape phenolic content and wine chromatic characteristics through the use of two different elicitors: Methyl jasmonate versus benzothiadiazole. *J. Agric. Food Chem.* **2012**, *60*, 1283–1290.

Ruiz-García, Y.; Gil-Muñoz, R.; López-Roca, J. M.; Martínez-Cutillas, A.; Romero-Cascales, I.; Gómez-Plaza, E. Increasing the Phenolic Compound Content of Grapes by Preharvest Application of Abcisic Acid and a Combination of Methyl Jasmonate and Benzothiadiazole. *J. Agric. Food Chem.* **2013**, *61*, 3978–3983.

Saavedra, G. M.; Figueroa, N. E.; Poblete, L. A.; Cherian, S.; Figueroa, C. R. Effects of Preharvest Applications of Methyl Jasmonate and Chitosan on Postharvest Decay, Quality and Chemical Attributes of *Fragaria chiloensis* Fruit. *Food Chem.* **2016**, *190*, 448–453.

Sayyari, M.; Babalar, M.; Kalantari, S.; Martínez-Romero, D.; Guillén, F.; Serrano, M.; Valero, D. Vapour Treatments with Methyl Salicylate or Methyl Jasmonate Alleviated

Chilling Injury and Enhanced Antioxidant Potential during Postharvest Storage of Pomegranates. *Food Chem.* **2011**, *124*, 964–970.

Shafiq, M.; Singh, Z.; Khan, A. S. Time of Methyl Jasmonate Application Influences the Development of 'Cripps Pink' Apple Fruit Colour. *J. Sci. Food Agric.* **2013**, *93*, 611–618.

Siboza, X. I.; Bertling, I.; Odindo, A. O. Salicylic Acid and Methyl Jasmonate Improve Chilling Tolerance in Cold-Stored Lemon Fruit (*Citrus limon*). *J. Plant Physiol.* **2014**, *171*, 1722–1731.

Sivankalyani, V.; Feygenberg, O.; Maorer, D.; Zaaroor, M.; Fallik, E.; Alkan, N. Combined Treatments Reduce Chilling Injury and Maintain Fruit Quality in Avocado Fruit during Cold Quarantine. *PLoS ONE* **2015**, *10*, e0140522.

Soto, A.; Ruiz, K. B.; Ziosi, V.; Costa, G.; Torrigiani, P. Ethylene and Auxin Biosynthesis and Signaling are Impaired by Methyl Jasmonate Leading to a Transient Slowing Down of Ripening in Peach Fruit. *J. Plant Physiol.* **2012**, *169*, 1858–1865.

Tang, Y.; Kuang, J. F.; Wang, F. Y.; Chen, L.; Hong, K. Q.; Lu, W. J.; Chen, J. Y. Molecular Characterization of PR and WRKY Genes During SA- and MeJA-Induced Resistance against *Colletotrichum musae* in Banana Fruit. *Postharvest Biol. Technol.* **2013**, *79*, 62–68.

Torrigiani, P.; Fregola, F.; Ziosi, V.; Ruiz, K. B.; Kondo, S.; Costa, G. Differential Expression of Allene Oxide Synthase (AOS), and Jasmonate Relationship with Ethylene Biosynthesis in Seed and Mesocarp of Developing Peach Fruit. *Postharvest Biol. Technol.* **2012**, *63*, 67–73.

Ueda, J.; Kato, J. Isolation and Identification of a Senescence-Promoting Substance from Wormwood (*Artemisia absinthium* L.). *Plant Physiol.* **1980**, *66*, 246–249.

Valero, D.; Martínez-Romero, D.; Serrano, M. The Role of Polyamines in the Improvement of the Shelf Life of Fruit. *Trends Food Sci. Technol.* **2002**, *13*, 228–234.

Valero, D.; Serrano, M. Postharvest Biology and Technology for Preserving Fruit Quality; CRC Press: Boca Raton, USA, 2010. pp. 83–87, 103–106,142–146, 178–188.

Vuong, T. V.; Franco, C.; Zhang, W. Treatment Strategies for High Resveratrol Induction in *Vitis vinifera* L. Cell Suspension Culture. *Biotech. Reports.* **2014**, *1–2*, 15–21.

Wang, S. Y.; Zheng, W. Preharvest Application of Methyl Jasmonate Increases Fruit Quality and Antioxidant Capacity in Raspberries. *Int. J. Food Sci. Technol.* **2005**, *40*, 187–195.

Wang, S. Y.; Bowman, L.; Ding, M. Methyl Jasmonate Enhances Antioxidant Activity and Flavonoid Content in Blackberries (*Rubus* sp.) and Promotes Antiproliferation of Human Cancer Cells. *Food Chem.* **2008**, *107*, 1261–1269.

Wang, K.; Jin, P.; Cao, S.; Shang, H.; Yang, Z.; Zheng, Y. Methyl Jasmonate Reduces Decay and Enhances Antioxidant Capacity in Chinese Bayberries. *J. Agric. Food Chem.* **2009**, *57*, 5809–5815.

Wang, L.; Jin, P.; Wang, J.; Jiang, L.; Shan, T.; Zheng, Y. Methyl Jasmonate Primed Defense Responses Against *Penicillium expansum* in Sweet Cherry Fruit. *Plant Mol. Biol. Rep.* **2015a**, *33*, 1464–1471.

Wang, K.; Liao, Y.; Kan, J.; Han, L.; Zheng, Y. Response of Direct or Priming Defense Against *Botrytis cinerea* to Methyl Jasmonate Treatment at Different Concentrations in Grape Berries. *Int. J. Food Microbiol.* **2015b**, *194*, 32–39.

Wang, S.; Takahashi, H.; Saito, T.; Okawa, K.; Ohara, H.; Shishido, M.; Ikeura, H.; Kondo, S. Jasmonate Application Influences Endogenous Abscisic Acid, Jasmonic Acid and Aroma Volatiles in Grapes Infected by a Pathogen (*Glomerella cingulata*). *Sci. Hortic.* **2015c**, *192*, 166–172.

Wasternack, C. Jasmonates: an Update on Biosynthesis, Signal Transduction and Action in Plant Stress Response, Growth and Development. *Ann. Bot.* **2014**, *100*, 681–697.

Wasternack, C. How Jasmonates Earned Their Laurels: Past and Present. *J. Plant Growth Regul.* **2015**, *34*, 761–794.

Woodside, J. V.; McGrath, A. J.; Lyner, N.; McKinley, M. C. Carotenoids and Health in Older People. *Maturitas* **2015**, *80*, 63–68.

Yao, H.; Tian, S. Effects of Pre- and Post-harvest Application of Salicylic Acid or Methyl Jasmonate on Inducing Disease Resistance of Sweet Cherry Fruit in Storage. *Postharvest Biol. Technol.* **2005**, *35*, 253–262.

Yu, L.; Liu, H.; Shao, X.; Yu, F.; Wei, Y.; Ni, Z.; Xu, F.; Wang, H. Effects of Hot Air and Methyl Jasmonate Treatment on the Metabolism of Soluble Sugars in Peach Fruit During Cold Storage. *Postharvest Biol. Technol.* **2016**, *113*, 8–16.

Zapata, P. J.; Martínez-Esplá, A.; Guillén, F.; Díaz-Mula, H. M.; Martínez-Romero, D.; Serrano, M.; Valero, D. Preharvest Application of Methyl Jasmonate (MeJA) in Two Plum Cultivars. 2. Improvement of Fruit Quality and Antioxidant Systems during Postharvest Storage. *Postharvest Biol. Technol.* **2014**, *98*, 115–122.

Zhao, M. L.; Wang, J. N.; Shan, W.; Fan, J. G.; Kuang, J. F.; Wu, K. Q.; Li, X. P.; Chen, W. X.; He, F. Y.; Chen, J. Y.; Lu, W. J. Induction of Jasmonate Signalling Regulators MaMYC2s and Their Physical Interactions with MaICE1 in Methyl Jasmonate-Induced Chilling Tolerance in Banana Fruit. *Plant Cell Environ.* **2013**, *36*, 30–51.

Zhang, X.; Sheng, J.; Li, F.; Meng, D.; Shen, L. Methyl Jasmonate Alters Arginine Catabolism and Improves Postharvest Chilling Tolerance in Cherry Tomato Fruit. *Postharvest Biol. Technol.* **2012**, *64*, 160–167.

Zhu, Z.; Tian S.P. Resistant Responses of Tomato Fruit Treated with Exogenous Methyl Jasmonate to *Botrytis cinerea* Infection. *Sci. Hortic.* **2012**, *142*, 38–43.

Ziosi, V.; Bonghi, C.; Bregoli, A. M.; Trainotti, L.; Biondi, S.; Sutthiwal, S.; Kondo, S.; Costa, G.; Torrigiani, P. Jasmonate-induced Transcriptional Changes Suggest a Negative Interference with the Ripening Syndrome in Peach Fruit. *J. Exp. Bot.* **2008**, *59*, 563–573.

CHAPTER 5

NITRIC OXIDE FOR ENHANCING STORAGE LIFE OF FRUITS AND VEGETABLES

SHUHUA ZHU*, LILI ZHANG, CHANGBAO CHEN, and JIE ZHOU

Laboratory of Chemical Biology, College of Chemistry and Material Science, Shandong Agricultural University, Taian 271018, Shandong, China

Corresponding author. E-mail: shuhua@sdau.edu.cn

ABSTRACT

Fruits and vegetables are important cash crops. However, ripening and senescence, biotic and abiotic stresses usually cause the decrease of postharvest quality and storage life of fruits and vegetables leading to economic damage. As bioactive gas molecule, nitric oxide (NO) can be endogenously produced in higher plants, it is confirmed that NO is involved in plenty cellular processes of plants. Due to its bioactivity, NO was applied to the storage of various fruits and vegetables and showed a favorable fresh-keeping function. In this chapter, the effects of NO on storage quality, respiration and ethylene production, antioxidative defense as well as resistance to chilling injury and diseases of postharvest fruits and vegetables is discussed. Moreover, we have deliberated about few problems related to postharvest nitric oxide applications for fruits and vegetables which need to be studied and clarified in further research.

5.1 INTRODUCTION

The ripening and senescence of harvested fruits and vegetables are important physiological processes causing decrease in postharvest quality and shelf-life leading to economic damage. Many methods are used to delay this process and prolong the storage of fruits and vegetables. Low temperature (Günther et al., 2015; Johnson et al., 2015), modified atmosphere (Viškelis et al., 2011; Forney et al., 2015), pretreatment with radiation (Severo et al., 2015; Topcu et al., 2015), chemical reagents (Valverde et al., 2015; Lim et al., 2016; Palma et al., 2015), and edible coating (Guerreiro et al., 2015) can efficiently devote to the storage of fruits and vegetables. Pretreatments with chemical reagents are popularly used in coordination with cold storage (Palma et al., 2015; Lim et al., 2016; Zhang et al., 2015). It is suggested that parts of phytogenic chemicals are safer than synthetic chemicals for food and human health.

As a bioactive gas molecule, nitric oxide (NO) has been confirmed to be endogenously produced in higher plants and plays important roles in regulating the ripening and senescence of fruits and vegetables (Leshem et al., 1998; Zhang et al., 2011). Increasing evidences confirmed that both exogenous and endogenous NO involve in prolonging the lives, maintaining the qualities, and decreasing the rotten ratio of fruits and vegetables during storage (Ruan et al., 2015; Wang et al., 2015a, 2015b; Wills et al., 2015; Shi et al., 2015; Jing et al., 2016). The biology of subcellular NO in prokaryote cells, plant organelles, and animal cells is well reviewed by Roszer (2012).

5.2 NITRIC OXIDE AND STORAGE QUALITY OF FRUITS AND VEGETABLES

Application of exogenous NO could reduce the decay and prolong the storage life of fruits and vegetables (Wills et al., 2000; Abdollahi et al., 2013). Generally, there are three methods for exogenous NO application: NO gas, NO solution, and NO donors. Fumigation with NO gas in the concentration range of 5–10 μL L^{-1} could produce >50% extension in shelf-life of strawberry (Wills et al., 2000). NO gas (10 μL L^{-1}) could delay the ripening of plums by 3–4 days at 21 $\pm$ 1°C, and alleviate chilling injury (CI) symptoms during cold storage at 0°C for 6 weeks (Singh et al., 2009). Fumigation with 10 μmol L^{-1} NO significantly inhibited the decrease of firmness, accumulation of sugar, and acid:sugar ratio in peach fruit during storage (Sun et al., 2011). NO saturated solution can be produced under anaerobic conditions and diluted to proper concentrations (Lim et al., 2005). Dipping fruits with NO solution at 1 μmol L^{-1} significantly delayed the decrease in vitamins C and E and maintained the content of soluble solids in kiwifruit during storage (Zhu et al., 2008). The NO donors, such as sodium nitroprusside (SNP), N-*tert*-butyl-*a*-phenylnitrone (PBN), S-nitroso-N-acetylpenicillamine (SNAP), 3-morpholinosydnonimine (Sin-I), and diethylenetriamine/nitric oxide (DETANO), are studied in horticultural crops. Spraying with 5 μmol L^{-1} SNP effectively controls decay organisms and maintains fruit quality during storage at 2.5°C for 15 days (Abdollahi et al., 2013). SNP (2.0 mmol L^{-1}) reduces pericarp browning, maintains high anthocyanin content, and preserves bioactive compounds (phenolics and ascorbic acid) and antioxidant capacity in litchi fruit during storage (Barman et al., 2014). SNP at 0.05 mmol L^{-1} delays chlorophyll degradation and enhances antioxidant activity in banana fruits after cold storage (Wang et al., 2015a). SNP keeps high levels of titratable acidity, soluble protein, ascorbic acid, reducing sugar, and reduced loss of weight loss and soluble solid concentration (SSC), retards ripening of glorious oranges (Liu et al., 2012). Similar biological effects are also found by PBN, SNAP, Sin-I, and DETANO (Leshem and Haramaty, 1996; Leshem, 2000; Wills et al., 2007). Integrated application of NO with modified atmosphere packaging also improves quality retention of button mushroom (Jiang et al., 2011). Endogenous NO induced by UV-B irradiation can also involve in enhancing the firmness and delays the ripening of mangoes (Ruan et al., 2015).

NO devotes to maintain proper color of fruits and vegetables during storage. Short-term exposure to NO gas can inhibit the browning on the surface of apple slices (Pristijono et al., 2006; Huque et al., 2013). Fumigation with NO gas or dipping fruits in an aqueous solution of DETANO also inhibits the browning on the surface of cut lettuce slices (Wills et al., 2008; Iakimova and Woltering, 2015). Both exogenous NO solution and simultaneous use of ascorbic acid and NO can inhibit surface browning of fresh-cut peach slices (Zhu et al., 2009a). SNP delays pericarp browning, inhibits the activities of enzymes in relation to phenolic metabolism, and maintains a high total phenol content in longan fruit during storage (Duan et al., 2007). It is suggested that the gaseous form (nitrous oxide) was more effective than a solution (NO) to prevent browning in the pericarp of longkong fruit (Lichanporn and Techavuthiporn, 2013). Exogenous NO can significantly delay the increase in red index and phenolic metabolism of postharvest Chinese winter jujube (Zhu et al., 2009b). NO also delays chlorophyll degradation in banana fruits after cold storage (Wang et al., 2015a).

5.3 NITRIC OXIDE, RESPIRATION AND ETHYLENE

Fruits and vegetables are classified as climacteric and nonclimacteric on the basis of patterns of respiration and of ethylene production during ripening (Manjunatha et al., 2010). Excessive respiration and ethylene production lead to postharvest fruits and vegetables rot and reduced shelf-life. Generally, inhibiting respiration and ethylene production is the first considered thing to prolong storage life of postharvest fruits and vegetables. NO is confirmed to inhibit the respiration and ethylene production in postharvest fruits and vegetables during storage (Zhu and Zhou, 2007; Leshem and Wills, 1998).

Plant mitochondria are the source and target for NO (Igamberdiev et al., 2014). NO could inhibit cytochrome oxidase, the terminal enzyme of the mitochondrial respiratory chain, to regulate mitochondrial respiration and cell functions (Cleeter et al., 1994; Brown, 1995; Sarti et al., 2012). NO inhibits aconitase activity to induction of the alternative oxidase in plant (Gupta et al., 2012). NO significantly inhibits the tricarboxylic acid cycle of postharvest peaches via inhibiting the activities of succinate thiokinase and succinate dehydrogenase, increases the activities of pyruvate dehydrogenase complex and fumarase during the whole storage, inhibits aconitase activity in the first period, and promotes activities of α-ketoglutarate dehydrogenase complex and malate dehydrogenase in the last period of storage (Ma et al., 2011). However, most of evidences about the regulation by NO on respiration focus on animal and plant tissues. The regulation metabolism of NO on respiration of fruit and vegetable still needs further and deep study.

According to Yang cycle of ethylene biosynthesis (Yang and Hoffman, 1984), 1-aminocyclopropane-1-carboxylic acid (ACC) is the substrate of ethylene biosynthesis. ACC synthase catalyzes the formation of ACC and ACC oxidase catalyzes the conversion from ACC to ethylene. ACC synthase and ACC oxidase are regarded as limiting factors in the ethylene production cycle. NO fumigation inhibited ethylene biosynthesis through inhibition of ACC synthase and ACC oxidase activities leading to reduced ACC content in the mango fruit pulp (Zaharah and Singh, 2011a). It is suggested that NO possibly reacts with the $-NH_2$ group of Lys278 of 1-aminocyclopropane-1-carboxylic acid synthase (ACS), which prevents to release ACC by the transaldimination reaction and deactivates ACS (Zhu and Zhou, 2007). In normal processes, the metal center of ACC oxidase combines with ACC and O_2 in the presence of bicarbonate and ascorbate to form the ACC–ACC oxidase–O complex that contains a double bond of Fe=O. The ACC–ACC oxidase–O complex decomposes to produce ethylene (Tierney et al., 2005). However, NO and ACC bind to the iron in the metal center of ACC oxidase, generating a stable ternary ACC–ACC oxidase–NO complex, that could not decompose to produce ethylene (Zhu et al., 2006). In mango fruit, pretreatment with salicylic acid (SA) and NO significantly downregulated *MiACO* and *MiERS1*, whereas *MiETR1* and *MiEIN2* are upregulated, and *MiACS* and *MiERF* exhibit fluctuating expression patterns, suggesting that *MiACO*, *MiACS*, *MiETR1*, *MiERS1*, *MiEIN2*, and *MiERF* might participate in the SA and NO signal induction pathways during ripening (Hong et al., 2014). Besides inhibiting the biosynthesis of ethylene in plants, NO also counters ethylene effects on ripening fruits, which is well reviewed by Manjunatha et al. (2012).

5.4 NITRIC OXIDE AND CHILLING INJURY

Most of fruits and vegetables of tropical and subtropical origin are highly susceptible to CI when stored at low temperatures, and this limits the exploitation of cold storage in extending storage life. Low temperature can induce the production of endogenous NO (Xu et al., 2012) and both endogenous and exogenous NO can alleviate CI of fruits and vegetables during cold storage (Zaharah and Singh, 2011b; Xu et al., 2012; Zhao et al., 2009). NO is regarded as a mediator of cold stress response (Baudouin and Jeandroz, 2015). Cold storage is popularly used to prolong the storage life of postharvest fruits and vegetables. During cold storage of fruits and vegetables, NO is a key element of signaling networks underlying plant response to low temperature (Baudouin and Jeandroz, 2015). Exogenous NO effectively reduces CI of cucumber fruit, improves the antioxidative defense system, and 1,1-diphenyl-2-picryl-hydrazyl (DPPH)-radical scavenging activity of cucumber in cold stress (Yang et al., 2011). Combination of exogenous NO and intermittent warming can prevent CI of peach fruit during storage (Zhu et al., 2010). Exogenous NO from SNP solutions significantly alleviates CI symptoms of plums during entire period of cold storage (Sharma and Sharma, 2015). Application of exogenous NO is most effective in reducing CI of banana fruit via improving the antioxidative defense system (Wu et al., 2014; Wang et al., 2016) and improving proline accumulation (Wang et al., 2013). NO promotes accumulation of polyamines (PAs), γ-aminobutyric acid (GABA), and proline, and it enhances activities of arginine decarboxylase, ornithine decarboxylase, diamine oxidase, and polyamine oxidase. Accumulation of PAs, GABA, and proline contributes to the chilling tolerance in cold-stored banana fruit (Wang et al., 2016). Cold storage can induce the generation of endogenous NO, which plays a critical role in alleviating

CI symptoms by affecting the antioxidative defense systems in loquat fruit during postharvest storage (Xu et al., 2012). Cold tolerance of cucumber fruit induced by yeast saccharide is linked with the induction of endogenous NO accumulation (Dong et al., 2012a). Exogenous NO protects tomatoes from cold injury by inducing endogenous NO accumulation and expression of *LeCBF1* (Zhao et al., 2011). NO can also enhance enzyme activities involved in energy metabolism, maintain the level of adenosine triphosphate (ATP) and energy charge, and alleviate CI in postharvest banana fruit (Wang et al., 2015b). The relationship between cold and NO is complicated. The advance on the responses of NO to low temperature is well reviewed by Puyaubert and Baudouin (2014).

5.5   NITRIC OXIDE AND ANTIOXIDATIVE DEFENSE

Antioxidant defense is a key factor for plant response and tolerance to abiotic oxidative stress (Hasanuzzaman et al., 2012). Production and accumulation of reactive oxygen species (ROS) are considered as important events in fruits during storage and are involved in regulating fruit senescence (Tian et al., 2013). Intercellular levels of ROS depend on the balance between their generation and the capacity to remove them. Maintenance of a high antioxidant capacity to scavenge the toxic ROS has been linked to increased tolerance of plants to these environmental stresses. There are enzymatic and nonenzymatic paths in antioxidative defense system in plant. The enzymatic antioxidants include superoxide dismutase (SOD), catalase (CAT), guaiacol peroxidase (GPX), and enzymes of ascorbate–glutathione (AsA–GSH) cycle such as ascorbate peroxidase (APX), monodehydro-ascorbate reductase, dehydroascorbate reductase, and glutathione reductase (GR) (Sharma et al., 2012). The nonenzymatic antioxidant system includes ascorbic acid (vitamin C), vitamin E (α-tocopherol), glutathione, and polyphenolic compounds (Ahmad et al., 2009).

Exogenous NO significantly reduces the accumulation of superoxide (O_2^-) and hydrogen peroxide (H_2O_2) and increases the activity of SOD and CAT in kiwifruit during storage (Zhu et al., 2008). Fumigation with exogenous NO can also delay the decrease in DPPH radical-scavenging activity in Chinese bayberry fruits, increases in both the rate of O_2^- production and H_2O_2 contents, and increase activities of SOD, CAT, and APX in Chinese bayberry fruits (Wu et al., 2012). Exogenous NO also delays senescence through inhibition of H_2O_2 accumulation in fresh-cut lettuce (Iakimova and Woltering, 2015). Cold triggers a marked increase in NO levels in loquat fruit, and abolition of endogenous NO accumulation exhibits significantly higher superoxide anion production rates and H_2O_2 contents and reduces activities of SOD, CAT, APX, and peroxidase (POD) in the fruit during cold storage (Xu et al., 2012). Increasing evidences confirmed that both exogenous and endogenous NO induced by other pretreatments can result in low content of ROS and high activities of antioxidant enzymes and expression of antioxidant related genes (Wu et al., 2014; Wang et al., 2013). These results indicate that NO could maintain the balance between the formation and detoxification of ROS and improve the antioxidative defense system.

The AsA–GSH cycle is a key antioxidant system involved in the finely tuned regulation of H_2O_2 in cells. The cycle involves the antioxidant metabolites: ascorbate, glutathione, and NADPH and the enzymes linking these metabolites (Noctor and Foyer, 1998). Exogenous NO delays the decrease of vitamin C in Chinese winter jujube fruit and exhibits inhibitory effects on polyphenol oxidase (PPO) and phenylalnine ammonialyase (PAL) activities in a dose-dependent manner (Zhu et al., 2009b) in papaya fruit (Li et al., 2014). Exogenous NO can also increase the levels of glutathione and phenolic compounds in fruit (Toivonen, 2004). NO can modulate the antioxidant capacity of AsA–GSH cycle by posttranslational modifications such as S-nitrosylation and/or tyrosine nitration (Begara-Morales et al., 2015). S-nitrosylation caused by NO can positively regulate APX activity (Yang et al., 2015) and maintain redox balance in plant cells (Correa-Aragunde et al., 2015).

Fumigation with exogenous NO can also delay the decrease in total phenolics contents to prevent enzymatic browning in bayberry fruits (Wu et al., 2012), winter jujube (Zhu et al., 2009b), litchi (Barman et al., 2014), longkong fruit (Lichanporn and Techavuthiporn, 2013), and mushroom (Dong et al., 2012b). High contents of phenolic compounds contribute to enhance antioxidative response in fruits during storage.

5.6 NITRIC OXIDE AND POSTHARVEST DISEASES

Postharvest disease is an important reason causing rot of fruits and vegetables during storage. NO is confirmed to play important and positive roles in activating defense responses and inhibiting postharvest diseases. Exogenous NO solution can also control brown rot disease caused by *Monilinia fructicola* in postharvest peach fruit (Shi et al., 2015; Gu et al., 2014). Exogenous NO induces postharvest disease resistance in citrus fruit to *Colletotrichum gloeosporioides* (Zhou et al., 2016), inhibits anthracnose, and induces defense-related enzymes in mango fruit (Hu et al., 2014). NO released by SNP aqueous solution could effectively retard pericarp reddening of tomato fruit and increase the resistance of tomato fruit to gray mold rot caused by *Botrytis cinerea* (Lai et al., 2011). It is dominated that yeast saccharide-induced decay alleviation and defense responses of peach fruit were dependent on endogenous NO generation (Yu et al., 2012). Endogenous NO generation induced by the elicitor from *Botrytis cinerea* and disease resistance could be blocked by a NOS inhibitor, and NO synthase is suggested to be a postharvest response in pathogen resistance of fruit (Zheng et al., 2011).

5.7 FUTURE PROSPECTIVE

Overall, it is clear that both exogenous and endogenous NO contribute to enhance the storage life of fruits and vegetables via inhibiting respiration and ethylene production. NO also promotes the antioxidative defense by improving enzymatic and nonenzymatic system, inhibits postharvest disease, and reduces CI to maintain proper storage quality of harvested fruits and vegetables. Most works have been done to shed light on the regulation by NO on ethylene production and CI with the methods of molecular biology and transcriptome. NO-based posttranslational modifications such as *S*-nitrosylation, tyrosine nitration, and *S*-glutathionylation are important pathways to modulate physiological and biochemical processes in postharvest fruits and vegetables and should be studied widely in further works. However, it is still not clear how NO signal regulates components of fruits and vegetables such as sugars, acids, volatiles, and softness and how NO signal crosstalk with cold signal to alleviate chilling injury. Modern scientific instruments and omics methods including metabolomics, transcriptomics, proteomics, and regulatomics can be used to solve above issues. There is a long road to clarify the roles of NO on postharvest fruits and vegetables.

ACKNOWLEDGMENT

The work was supported by National Natural Science Foundation of China (31470686, 31770724). We would like to extend our sincere appreciation to Dr. Xiaokang Zhang for his kind help in preparing this manuscript.

KEYWORDS

- nitric oxide
- storage life
- fruits
- vegetables
- antioxidative defense

REFERENCES

Abdollahi, R.; Asghari, M.; Esmaiili, M.; Abdollahi, A. Postharvest Nitric Oxide Treatment Effectively Reduced Decays of Selva Strawberry Fruit. *Nitric Oxide* **2013**, *3*, 54–59.

Ahmad, P.; Jeleel, C.; Azooz, M.; Nabi, G. Generation of ROS and Non-enzymatic Antioxidants during Abiotic Stress in Plants. *Bot. Res. Intern.* **2009**, *2*, 11–20.

Barman, K.; Siddiqui, M. W.; Patel, V. B.; Prasad, M. Nitric Oxide Reduces Pericarp Browning and Preserves Bioactive Antioxidants in Litchi. *Sci. Hortic.* **2014**, *171*, 71–77.

Baudouin, E.; Jeandroz, S. Nitric Oxide as a Mediator of Cold Stress Response: A Transcriptional Point of View. In *Nitric Oxide Action in Abiotic Stress Responses in Plant*; Springer: Cham Heidelberg, Germany, 2015; pp 129–39.

Begara-Morales, J. C.; Sánchez-Calvo, B.; Chaki, M.; et al. Modulation of the Ascorbate– Glutathione Cycle Antioxidant Capacity by Posttranslational Modifications Mediated by Nitric Oxide in Abiotic Stress Situations. In *Reactive Oxygen Species and Oxidative Damage in Plants Under Stress*; Gupta, D. K., Palma, J. M., Corpas, F. J., Eds.; Springer: Cham Heidelberg, Germany, 2015; pp 305–320.

Brown, G. C. Nitric Oxide Regulates Mitochondrial Respiration and Cell Functions by Inhibiting Cytochrome Oxidase. *FEBS Lett.* **1995**, *369*, 136–139.

Cleeter, M. W.; Cooper, J. M.; Darley-Usmar, V. M.; Moncada, S.; Schapira, A. H. Reversible Inhibition of Cytochrome C Oxidase, the Terminal Enzyme of the Mitochondrial Respiratory Chain, by Nitric Oxide. Implications for Neurodegenerative Diseases. *FEBS Lett.* **1994**, *345*, 50–54.

Correa-Aragunde, N.; Foresi, N.; Lamattina, L. Nitric Oxide Is a Ubiquitous Signal for Maintaining Redox Balance in Plant Cells: Regulation of Ascorbate Peroxidase as a Case Study. *J. Exp. Bot.* **2015**, *66*, 2913–2921.

Dong, J.; Yu, Q.; Lu, L.; Xu, M. Effect of Yeast Saccharide Treatment on Nitric Oxide Accumulation and Chilling Injury in Cucumber Fruit during Cold Storage. *Postharvest Biol. Technol.* **2012a**, *68*, 1–7.

Dong, J.; Zhang, M.; Lu, L.; Sun, L.; Xu, M. Nitric Oxide Fumigation Stimulates Flavonoid and Phenolic Accumulation and Enhances Antioxidant Activity of Mushroom. *Food Chem.* **2012b**, *135*, 1220–1225.

Duan, X.; Su, X; You, Y.; Qu, H.; Li, Y.; Jiang, Y. Effect of Nitric Oxide on Pericarp Browning of Harvested Longan Fruit in Relation to Phenolic Metabolism. *Food Chem.* **2007**, *104*, 571–576.

Forney, C. F.; Jamieson, A. R.; Pennell, K. D. M.; Jordan, M. A.; Fillmore, S. A. Relationships Between Fruit Composition and Storage Life in Air or Controlled Atmosphere of Red Raspberry. *Postharvest Biol. Technol.* **2015**, *110*, 121–130.

Günther, C. S.; Marsh, K. B.; Winz, R. A.; Harker, R. F.; Wohlers, M. W.; White, A. Goddard, M. R. The Impact of Cold Storage and Ethylene on Volatile Ester Production and Aroma Perception in 'Hort16A' Kiwifruit. *Food Chem.* **2015**, *169*, 5–12.

Gu, R.; Zhu, S.; Zhou, J.; Liu, N.; Shi, J. Inhibition on Brown Rot Disease and Induction of Defence Response in Harvested Peach Fruit by Nitric Oxide Solution. *Eur. J. Plant Pathol.* **2014**, *139*, 369–378.

Guerreiro, A. C.; Gago, C. M.; Faleiro, M. L.; Miguel, M. G.; Antunes, M. D. The Effect of Alginate-Based Edible Coatings Enriched with Essential Oils Constituents on *Arbutus unedo* L. Fresh Fruit Storage. *Postharvest Biol. Technol.* **2015**, *100*, 226–233.

Gupta, K. J.; Shah, J. K.; Brotman, Y.; Jahnke, K.; Willmitzer, L.; Kaiser, W. M.; Bauwe, H.; Igamberdiev, A. U. Inhibition of Aconitase by Nitric Oxide Leads to Induction of the Alternative Oxidase and to a Shift of Metabolism Towards Biosynthesis of Amino Acids. *J. Exp. Bot.* **2012**, *63*, 1773–1784.

Hasanuzzaman, M.; Hossain, M.; Da Silva, J. T.; Fujita, M. Plant Response and Tolerance to Abiotic Oxidative Stress: Antioxidant Defense is a Key Factor. In *Crop Stress and Its Management: Perspectives and Strategies*; Venkateswarlu, B., Shanker, A. K., Shanker, C., Maheswari, M., Eds.; Springer: Dordrecht, Netherlands, 2012; pp 261–315.

Hong, K.; Gong, D.; Xu, H.; Wang, S.; Jia, Z.; Chen, J.; Zhang, L. Effects of Salicylic Acid and Nitric Oxide Pretreatment on the Expression of Genes Involved in the Ethylene Signaling Pathway and the Quality of Postharvest Mango Fruit. *New Zeal. J. Crop Hortic. Sci.* **2014**, *42*, 205–216.

Hu, M.; Yang, D.; Huber, D. J.; Jiang, Y.; Li, M.; Gao, Z.; Zhang, Z. Reduction of Postharvest Anthracnose and Enhancement of Disease Resistance in Ripening Mango Fruit by Nitric Oxide Treatment. *Postharvest Biol. Technol.* **2014**, *97*, 115–122.

Huque, R.; Wills, R.; Pristijono, P.; Golding, J. Effect of Nitric Oxide (NO) and Associated Control Treatments on the Metabolism of Fresh-Cut Apple Slices in Relation to Development of Surface Browning. *Postharvest Biol. Technol.* **2013**, *78*, 16–23.

Iakimova, E. T.; Woltering, E. J. Nitric Oxide Prevents Wound-induced Browning and Delays Senescence through Inhibition of Hydrogen Peroxide Accumulation in Fresh-Cut Lettuce. *Innov. Food Sci. Emer. Technol.* **2015**, *30*, 157–169.

Igamberdiev, A. U.; Ratcliffe, R. G.; Gupta, K. J. Plant Mitochondria: Source and Target for Nitric Oxide. *Mitochondrion* **2014**, *19*, 329–333.

Jiang, T.; Zheng, X.; Li, J.; Jing, G.; Cai, L.; Ying, T. Integrated Application of Nitric Oxide and Modified Atmosphere Packaging to Improve Quality Retention of Button Mushroom (*Agaricus bisporus*). *Food Chem.* **2011**, *126*, 1693–1699.

Jing, G.; Zhou, J.; Zhu, S. Effects of Nitric Oxide on Mitochondrial Oxidative Defence in Postharvest Peach Fruits. *J. Sci. Food Agric.* **2016**, *96*, 1997–2003.

Johnson, M. C.; Thomas, A. L.; Greenlief, C. M. Impact of Frozen Storage on the Anthocyanin and Polyphenol Contents of American Elderberry Fruit Juice. *J. Agric. Food Chem.* **2015**, *63*, 5653–5659.

Lai, T.; Wang, Y.; Li, B.; Qin, G.; Tian, S. Defense Responses of Tomato Fruit to Exogenous Nitric Oxide during Postharvest Storage. *Postharvest Biol. Technol.* **2011**, *62*, 127–132.

Leshem, Y. Y. The Biological Conquest of Nitric Oxide. In *Nitric Oxide in Plants*; Dordrecht, Netherlands: Springer, 2000; pp 3–23.

Leshem, Y. Y.; Haramaty, E. The Characterization and Contrasting Effects of the Nitric Oxide Free Radical in Vegetative Stress and Senescence of *Pisum sativum* Linn. Foliage. *J. Plant Physiol.* **1996**, *148*, 258–263.

Leshem, Y. Y.; Wills, R. B. H. Harnessing Senescence Delaying Gases Nitric Oxide and Nitrous Oxide: A Novel Approach to Postharvest Control of Fresh Horticultural Produce. *Biol. Plant.* **1998**, *41*, 1–10.

Leshem, Y. Y.; Wills, R. B. H.; Ku, V. V. V. Evidence for the Function of the Free Radical Gas—Nitric Oxide (NO⁻)—As an Endogenous Maturation and Senescence Regulating Factor in Higher Plants. *Plant Physiol. Biochem.* **1998**, *36*, 825–833.

Li, X. P.; Wu, B.; Guo, Q.; Wang, J. D.; Zhang, P.; Chen, W. X. Effects of Nitric Oxide on Postharvest Quality and Soluble Sugar Content in Papaya Fruit during Ripening. *J. Food Process. Preserv.* **2014**, *38*, 591–599.

Lichanporn, I.; Techavuthiporn, C. The Effects of Nitric Oxide and Nitrous Oxide on Enzymatic Browning in Longkong (*Aglaia dookkoo* Griff.). *Postharvest Biol. Technol.* **2013**, *86*, 62–65.

Lim, M. D.; Lorković, I. M.; Ford, P. C. The Preparation of Anaerobic Nitric Oxide Solutions for the Study of Heme Model Systems in Aqueous and Nonaqueous Media: Some Consequences of NOx Impurities. *Methods Enzymol.* **2005**, *396*, 3–17.

Lim, S.; Han, S. H.; Kim, J.; Lee, H. J.; Lee, J. G.; Lee, E. J. Inhibition of Hardy Kiwifruit (*Actinidia aruguta*) Ripening by 1-Methylcyclopropene during Cold Storage and Anticancer Properties of the Fruit Extract. *Food Chem.* **2016**, *190*, 150–157.

Liu, L.; Wang, J.; Qu, L.; Li, S.; Wu, R.; Zeng, K. Effect of Nitric Oxide Treatment on Storage Quality of Glorious Oranges. *Proc. Eng.* **2012**, *37*, 150–154.

Ma, C.; Sun, Z.; Zhou, J.; Zhu, S. Effects of Exogenous Nitric Oxide on Krebs Cycle-Associated Enzyme Activities in Feicheng Peach Fruit. In *The 3rd Conference on Key Technology of Horticulture*; London Science Publishing: Shenyang, China, 2011; pp 275–283.

Manjunatha, G.; Lokesh, V.; Neelwarne, B. Nitric Oxide in Fruit Ripening: Trends and Opportunities. *Biotechnol. Adv.* **2010**, *28*, 489–499.

Manjunatha, G.; Gupta, K. J.; Lokesh, V.; Mur, L.; Neelwarne, B. Nitric Oxide Counters Ethylene Effects on Ripening Fruits. *Plant Signal. Behav.* **2012**, *7*, 476–483.

Noctor, G.; Foyer, C. H. Ascorbate and Glutathione: Keeping Active Oxygen Under Control. *Annu. Rev. Plant Physiol. Plant Mol. Biol.* **1998**, *49*, 249–279.

Palma, F.; Carvajal, F.; Ramos, J. M.; Jamilena, M.; Garrido, D. Effect of Putrescine Application on Maintenance of Zucchini Fruit Quality during Cold Storage: Contribution of GABA Shunt and Other Related Nitrogen Metabolites. *Postharvest Biol. Technol.* **2015**, *99*, 131–140.

Pristijono, P.; Wills, R. B. H.; Golding, J. B. Inhibition of Browning on the Surface of Apple Slices by Short Term Exposure to Nitric Oxide (NO) Gas. *Postharvest Biol. Technol.* **2006**, *42*, 256–259.

Puyaubert, J.; Baudouin, E. New Clues for a Cold Case: Nitric Oxide Response to Low Temperature. *Plant Cell Environ.* **2014**, *37*, 2623–2630.

Roszer, T. *The Biology of Subcellular Nitric Oxide*; Springer: Dordrecht, Netherlands, 2012.

Ruan, J.; Li, M.; Jin, H.; Sun, L.; Zhu, Y.; Xu, M.; Dong, J. UV-B Irradiation Alleviates the Deterioration of Cold-Stored Mangoes by Enhancing Endogenous Nitric Oxide Levels. *Food Chem.* **2015**, *169*, 417–423.

Sarti, P.; Forte, E.; Giuffre, A.; Mastronicola, D.; Magnifico, M. C.; Arese, M. The Chemical Interplay Between Nitric Oxide and Mitochondrial Cytochrome C Oxidase: Reactions, Effectors and Pathophysiology. *Int. J. Cell Biol.* **2012**. DOI: 10.1155/2012/571067.

Severo, J.; Tiecher, A.; Pirrello, J.; Regad, F.; Latche, A.; Pech, J.; Bouzayen, M.; Rombaldi, C. V. UV-C Radiation Modifies the Ripening and Accumulation of Ethylene Response Factor (ERF) Transcripts in Tomato Fruit. *Postharvest Biol. Technol.* **2015**, *102*, 9–16.

Sharma, P.; Jha, A. B.; Dubey, R. S.; Pessarakli, M. Reactive Oxygen Species, Oxidative Damage, and Antioxidative Defense Mechanism in Plants Under Stressful Conditions. *J. Bot.* **2012**. DOI: 10.1155/2012/217037.

Sharma, S.; Sharma, R. R. Nitric Oxide Inhibits Activities of PAL and PME Enzymes and Reduces Chilling Injury in 'Santa Rosa' Japanese Plum (*Prunus salicina* Lindell). *J. Plant Biochem. Biotechnol.* **2015**, *24*, 292–297.

Shi, J. Y.; Liu, N.; Gu, R. X.; Zhu, L. Q.; et al. Signals Induced by Exogenous Nitric Oxide and Their Role in Controlling Brown Rot Disease Caused by *Monilinia fructicola* in Postharvest Peach Fruit. *J. Gen. Plant Pathol.* **2015**, *81*, 68–76.

Singh, S. P.; Singh, Z.; Swinny, E. E. Postharvest Nitric Oxide Fumigation Delays Fruit Ripening and Alleviates Chilling Injury during Cold Storage of Japanese Plums (*Prunus salicina* Lindell). *Postharvest Biol. Technol.* **2009**, *53*, 101–108.

Sun, Z.; Li, Y.; Zhou, J.; Zhu, S. H. Effects of Exogenous Nitric Oxide on Contents of Soluble Sugars and Related Enzyme Activities in 'Feicheng' Peach Fruit. *J. Sci. Food Agric.* **2011**, *91*, 1795–1800.

Tian, S.; Qin, G.; Li, B. Reactive Oxygen Species Involved in Regulating Fruit Senescence and Fungal Pathogenicity. *Plant Mol. Biol.* **2013**, *82*, 593–602.

Tierney, D. L.; Rocklin, A. M.; Lipscomb, J. D.; Que, L.; Hoffman, B. M. ENDOR Studies of the Ligation and Structure of the Non-heme Iron Site in ACC Oxidase. *J. Am. Chem. Soc.* **2005**, *127*, 7005–7013.

Toivonen, P. Postharvest Storage Procedures and Oxidative Stress. *HortScience* **2004**, *39*, 938–942.

Topcu, Y.; Dogan, A.; Kasimoglu, Z.; Sahin-Nadeem, H.; Polat, E.; Erkan, M. The Effects of UV Radiation during the Vegetative Period on Antioxidant Compounds and Postharvest Quality of Broccoli (*Brassica oleracea* L.). *Plant Physiol. Biochem.* **2015**, *93*, 56–65.

Valverde, J. M.; Gimenez, M. J.; Guillen, F.; Valero, D.; Martinez-Romero, D.; Serrano, M. Methyl Salicylate Treatments of Sweet Cherry Trees Increase Antioxidant Systems in Fruit at Harvest and during Storage. *Postharvest Biol. Technol.* **2015**, *109*, 106–113.

Viškelis, P.; Rubinskiene, M.; Sasnauskas, A.; Bobinas, C.; Kvikliene, N. Changes in Apple Fruit Quality during a Modified Atmosphere Storage. *J. Fruit Ornam. Plant Res.* **2011**, *19*, 155–165.

Wang, Y.; Luo, Z.; Du, R.; Liu, Y.; Ying, T.; Mao, L. Effect of Nitric Oxide on Antioxidative Response and Proline Metabolism in Banana during Cold Storage. *J. Agric. Food Chem.* **2013**, *61*, 8880–8887.

Wang, Y.; Luo, Z.; Du, R. Nitric Oxide Delays Chlorophyll Degradation and Enhances Antioxidant Activity in Banana Fruits after Cold Storage. *Acta Physiol. Plant.* **2015a**, *37*, 1–10.

Wang, Y.; Luo, Z.; Khan, Z. U.; Mao, L.; Ying, T. Effect of Nitric Oxide on Energy Metabolism in Postharvest Banana Fruit in Response to Chilling Stress. *Postharvest Biol. Technol.* **2015b**, *108*, 21–27.

Wang, Y.; Luo, Z.; Mao, L.; Ying, T. Contribution of Polyamines Metabolism and GABA Shunt to Chilling Tolerance Induced by Nitric Oxide in Cold-Stored Banana Fruit. *Food Chem.* **2016**, *197*, 333–339.

Wills, R.; Ku, V.; Leshem, Y. Fumigation with Nitric Oxide to Extend the Postharvest Life of Strawberries. *Postharvest Biol. Technol.* **2000**, 18, 75–79.

Wills, R. B. H.; Soegiarto, L.; Bowyer, M. C. Use of a Solid Mixture Containing Diethylenetriamine/Nitric Oxide (DETANO) to Liberate Nitric Oxide Gas in the Presence of Horticultural Produce to Extend Postharvest Life. *Nitric Oxide* **2007**, *17*, 44–49.

Wills, R. B. H.; Pristijono, P.; Golding, J. B. Browning on the Surface of Cut Lettuce Slices Inhibited by Short Term Exposure to Nitric Oxide (NO). *Food Chem.* **2008**, *107*, 1387–1392.

Wills, R.; Pristijono, P.; Golding, J. Nitric Oxide and Postharvest Stress of Fruits, Vegetables and Ornamentals. In *Nitric Oxide Action in Abiotic Stress Responses in Plants*; Springer: Cham Heidelberg, Germany, 2015; pp 221–238.

Wu, F.; Yang, H.; Chang, Y.; Cheng, J.; Bai, S.; Yin, J. Effects of Nitric Oxide on Reactive Oxygen Species and Antioxidant Capacity in Chinese Bayberry during Storage. *Sci. Hortic.* **2012**, *135*, 106–111.

Wu, B.; Guo, Q.; Li, Q.; Ha, Y.; Li, X.; Chen, W. Impact of Postharvest Nitric Oxide Treatment on Antioxidant Enzymes and Related Genes in Banana Fruit in Response to Chilling Tolerance. *Postharvest Biol. Technol.* **2014**, *92*, 157–163.

Xu, M.; Dong, J.; Zhang, M.; Xu, X.; Sun, L. Cold-induced Endogenous Nitric Oxide Generation Plays a Role in Chilling Tolerance of Loquat Fruit during Postharvest Storage. *Postharvest Biol. Technol.* **2012**, *65*, 5–12.

Yang, S. F.; Hoffman, N. E. Ethylene Biosynthesis and Its Regulation in Higher Plants. *Annu. Rev. Plant Physiol.* **1984**, *35*, 155–189.

Yang, H.; Wu, F.; Cheng, J. Reduced Chilling Injury in Cucumber by Nitric Oxide and the Antioxidant Response. *Food Chem.* **2011**, *127*, 1237–1242.

Yang, H.; Mu, J.; Chen, L.; Feng, J.; Hu, J.; Li, L.; Zhou, J. M.; Zuo, J. S-Nitrosylation Positively Regulates Ascorbate Peroxidase Activity during Plant Stress Responses. *Plant Physiol.* **2015**, *167*, 1604–1615.

Yu, Q.; Chen, Q.; Chen, Z.; Xu, H.; Fu, M.; Li, S.; Wang, H.; Xu, M. Activating Defense Responses and Reducing Postharvest Blue Mold Decay Caused by *Penicillium expansum* in Peach Fruit by Yeast Saccharide. *Postharvest Biol. Technol.* **2012**, *74*, 100–107.

Zaharah, S. S.; Singh, Z. Mode of Action of Nitric Oxide in Inhibiting Ethylene Biosynthesis and Fruit Softening during Ripening and Cool Storage of 'Kensington Pride' Mango. *Postharvest Biol. Technol.* **2011a**, *62*, 258–266.

Zaharah, S. S.; Singh, Z. Postharvest Nitric Oxide Fumigation Alleviates Chilling Injury, Delays Fruit Ripening and Maintains Quality in Cold-stored 'Kensington Pride' Mango. *Postharvest Biol. Technol.* **2011b**, *60*, 202–210.

Zhang, L. L.; Zhu, S. H.; Chen, C. B.; Zhou, J. Metabolism of Endogenous Nitric Oxide during Growth and Development of Apple Fruit. *Sci. Hortic.* **2011**, *127*, 500–506.

Zhang, Y.; Zhang, M.; Yang, H. Postharvest Chitosan–Salicylic Acid Application Alleviates Chilling Injury and Preserves Cucumber Fruit Quality during Cold Storage. *Food Chem.* **2015**, *174*, 558–563.

Zhao, M. G.; Chen, L.; Zhang, L. L.; Zhang, W. H. Nitric Reductase-dependent Nitric Oxide Production is Involved in Cold Acclimation and Freezing Tolerance in *Arabidopsis*. *Plant Physiol*. **2009**, *151*, 755–767.

Zhao, R.; Sheng, J.; Lv, S.; Zheng, Y.; Zhang, J.; Yu, M.; Shen, L. Nitric Oxide Participates in the Regulation of LeCBF1 Gene Expression and Improves Cold Tolerance in Harvested Tomato Fruit. *Postharvest Biol. Technol*. **2011**, *62*, 121–126.

Zheng, Y.; Shen, L.; Yu, M.; Fan, B.; Zhao, D.; Liu, L.; Sheng, J. Nitric Oxide Synthase as a Postharvest Response in Pathogen Resistance of Tomato Fruit. *Postharvest Biol. Technol*. **2011**, *60*, 38–46.

Zhou, Y.; Li, S.; Zeng, K. Exogenous Nitric Oxide-induced Postharvest Disease Resistance in Citrus Fruit to *Colletotrichum gloeosporioides*. *J. Sci. Food Agric*. **2016**, *96* (2), 505–512.

Zhu, S. H.; Zhou, J. Effect of Nitric Oxide on Ethylene Production in Strawberry Fruit during Storage. *Food Chem*. **2007**, *100*, 1517–1522.

Zhu, S.; Liu, M.; Zhou, J. Inhibition by Nitric Oxide of Ethylene Biosynthesis and Lipoxygenase Activity in Peach Fruit during Storage. *Postharvest Biol. Technol*. **2006**, *42*, 41–48.

Zhu, S.; Sun, L.; Liu, M.; Zhou, J. Effect of Nitric Oxide on Reactive Oxygen Species and Antioxidant Enzymes in Kiwifruit during Storage. *J. Sci. Food Agric*. **2008**, *88*, 2324–2331.

Zhu, L. Q.; Zhou, J.; Zhu, S. H.; Guo, L. H. Inhibition of Browning on the Surface of Peach Slices by Short-Term Exposure to Nitric Oxide and Ascorbic Acid. *Food Chem*. **2009a**, *114*, 174–179.

Zhu, S.; Sun, L.; Zhou, J. Effects of Nitric Oxide Fumigation on Phenolic Metabolism of Postharvest Chinese Winter Jujube (*Zizyphus jujuba* Mill. cv. Dongzao) in Relation to Fruit Quality. *LWT-Food Sci. Technol*. **2009b**, *42*, 1009–1014.

Zhu, L. Q.; Zhou, J.; Zhu, S. H. Effect of a Combination of Nitric Oxide Treatment and Intermittent Warming on Prevention of Chilling Injury of 'Feicheng' Peach Fruit during Storage. *Food Chem*. **2010**, *121*, 165–70.

CHAPTER 6

NANOTECHNOLOGY IN PACKAGING OF FRESH FRUITS AND VEGETABLES

OVAIS SHAFIQ QADRI[1,2*], KAISER YOUNIS[1],
GAURAV SRIVASTAVA[1], and ABHAYA KUMAR SRIVASTAVA[2]

[1]*Department of Bioengineering, Integral University, Lucknow, Uttar Pradesh, India*

[2]*Department of Postharvest Engineering and Technology, Aligarh Muslim University, Aligarh, Uttar Pradesh, India*

Corresponding author. E-mail: osqonline@gmail.com

ABSTRACT

Nanotechnology employs formulating, characterizing, and/or manipulating structures, designs, or materials at nanometer level (0.2–100 nm). It has wide applications and many products are in use currently in food packaging sector. Further, more research work is on-going for its usage likelihoods in food packaging industry. Numerous workers have reported the merits of nanomaterials used in packaging in terms of product quality, processing attributes and prolonged storage duration. No harmful effects have been reported till date. Nevertheless, steps should be taken to ensure its usage keeping in view consumer acceptability and safety. Appropriate labeling and adherence to guidelines (FDA 2014) is required to ensure safety and efficacy. This chapter provides an insight into applications of nanotechnology in various sectors of food science, namely, food processing, food additive delivery systems, nanosensors and fruit and vegetable packaging besides discussing their antimicrobial, barrier and coating properties. Nanotechnology is quickly making its own niche in processing and packaging of food products.

6.1   INTRODUCTION

Packaging is one of the important aspects of food processing. The energy, money, and efforts that are utilized in processing of the food are simply of no use, if not complimented by an effective packaging. The foods obtained from nature are mostly packed in the best way till it is delivered to a processer or consumer, for instance, the peel of fruits. The packaging has evolved all through the civilizations and has been a focus of research in the present scientific era as well. Some remarkable achievements in the field of food packaging include aseptic packaging, intelligent packaging, smart packaging, and others. With all these developments and success, this field still remains one of the broadly researched areas in food science.

"Nano" is the buzzword for recent science. The field provides us the liberty to manipulate atoms, molecules, and chemical reactions at precise scales. The surge of nanotechnology and its application in different fields has inculcated a new hope of revolutionizing the approach of studying the processes and technologies and eventually improve them. Nanotechnology deals with working at tiny scales of nanometer range. Nanotechnology is the science of fabricating, characterizing, and/or manipulating structures, designs, or materials at a nanometer level (0.2–100 nm). Nanotechnology has reshaped our view pertaining to different physical and chemical properties of materials which are significantly different corresponding to their properties at larger scale (Imran et al., 2010).

The research over the last decade has been dominated by nanotechnology and different nanomaterials possessing special properties are already available which find their application in diversified fields including pharmaceuticals, molecular computing, structural materials, and others (Roco et al., 2010). Duncan (2011) predicted that nanotechnology is set to impact about $3 trillion across the global economy by 2020 and manpower of about 6 million workers would be required to fulfill the industry demand. The food industry has been slow in accepting the nanotechnology compared to other industries. The possible justifiable reason could be the consumer perception toward the emergent technologies. Consumers are always reluctant in accepting the new technologies when it comes to food and so is the case with "nanofoods" (Duncan, 2011).

6.2   APPLICATION OF NANOTECHNOLOGY IN FOOD SCIENCE

The applications of nanotechnology are not limited to any field; rather, they are more diversified than we can imagine. Nanotechnology has not only found its place in agriculture and food sector but has influenced to a great deal the productivity, process, and shelf-life of food products. However, these applications in food industry are relatively new as compared to the use of nanotechnology in other fields. Figure 6.1 shows the scope of nanotechnology in food. The field offers us such manipulation capabilities that help us to work for the betterment of postharvest management, in processing of food products, in better packaging of materials to enhance shelf-life of perishable substances, and the nanomaterials can also be used as a food additive to enhance the quality of the food. But before we realize its full potential, we have hurdles to face. The hurdles are in the form of scarce data pertaining to safety of such food products. The field and the technologies are novice and proper safety data is not yet available. This influences the minds of general public and for the same reason such technologies are not readily accepted (as evident from the fate of genetically modified food). The director, Rutgers' Centre for Advanced Food Technology; New Brunswick quotes, "This is one technology that will have profound implications for the food industry, even though they're not very clear to a lot of people" (Gardener, 2002). The market for nanotechnology in food and food processing was valued over \$2 billion in 2004 and has grown above \$20 billion in 2010. With large investments in nanotech R&D the field has the potential to come out of the research labs and move to our dining tables. In fact, knowingly or unknowingly we are already consuming food that has some element of nanotechnology to it. According to Helmut Kaiser Consultancy (Chaudhry and Castle, 2011), more than 200 multinational food companies have started investing in nanotechnology and have commercialized products out of the same. The list includes major giants, such as Heinz, Nestle, Sara Lee, McCain Foods, Kraft Foods, Unilever, Ajinomoto, and many more.

Both food companies and scientists have equally contributed to the application of nanotechnology in food. Scientists at Chiang Mai University have been able to change the color of the grain (Prasanna, 2007), by drilling a nanosize hole and manipulating DNA by inserting nitrogen atom. Syngenta, BASF, and Monsanto have developed nanocapsules and nanopesticides (Kah, 2015) that can be programmed to time release. USDA-funded researchers at Clemson University have developed alternatives to antibiotics for industrial chicken production (Stutzenberger et al., 2010). Bioactive polystyrene nanoparticles were developed and successfully tested. Similar approach was adopted by Clear Spring Trout, United States' largest farmed fish company; for delivering nanoparticle vaccine to trout ponds. Unilever and Nestle are using nanoparticles to improve the texture of spreads and ice creams while Kraft has developed nanocapsules that can change the color and flavor of the food and tried hands on interactive drinks. Kraft and BASF are also involved in developing materials at nanoscale that can extend shelf-life of perishable food. These novel nanomaterials change the color when there is spoilage of food. A range of intelligent packing material is available, by limiting the passage of oxygen and carbon dioxide in and out of the pack the food material is kept fresh for a longer time. The market for intelligent packing material is worth more than $50 billion (Asadi and Mousavi, 2006). Other nanotechnolgy food products such as functional foods that are currently available in the global market include encapsulated cooking oil, "Canola Active" that reduces cholesterol absorption. Nanocapsules to prevent flavor and aroma loss are also available.

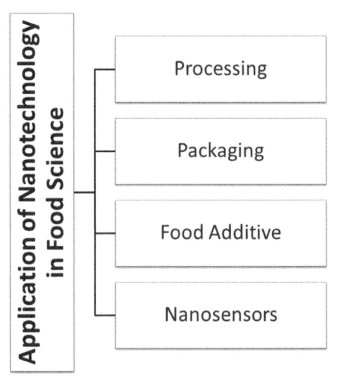

FIGURE 6.1 Applications of nanotechnology in food science.

The specific role of nanotechnology in different divisions of food sector is discussed as:

6.2.1 FOOD PROCESSING

Once the food material is harvested in raw form, it needs to be processed for better consumer compliance and acceptability. Conventional food processing and additives had shortcomings which have been ameliorated by the introduction of nanotechnology in food science. Functional components such as vitamins, antioxidants, flavor, antimicrobials, and preservatives can be used in various forms. The advent of nanotechnology has developed delivery systems that have increased the life of these components in food. The development of various delivery systems fulfills a number of functions, primarily the transport of active ingredient for desired taste. Other functions include providing texture, shelf-life, enhanced taste, and protection from chemical and biological agents of oxidation (Sozer and Kokini, 2009). The nanomaterials that perform the above-stated functions include colloids, nanoemulsions, and nanobiopolymers. Hydrocolloids serve several functions such as compactness, thickening quality, hardness, crispness, gel-forming ability, and mouth feel (Yadollahi et al., 2010; Stephen, 1995). Edible films (1–100 nm/layer) are used to coat a variety of food products such as fruits, vegetables, candies, chocolate, French fries, and baked goods (Yadollahi et al., 2010). These films protect the food from gases, lipids, and moister and may also improve texture of the food. These may also serve as carrier of several flavors, colors, nutrients, antioxidants, and more (Yadollahi et al., 2010; Rhim et al., 2006).

6.2.2 FOOD ADDITIVE DELIVERY SYSTEMS

Nanocapsules incorporating nutraceuticals and nanoencapsulated flavor enhancers are being used in cooking oils. Several already approved additives such as chitosan, gelatin, carrageenan, polyglycolic acid, alginate, and polylactic acid are being used for nanoencapsulation.

6.2.3 NANOSENSORS

Nanotechnology has played a crucial role in the development of technologies that can be used for virus detection, detection of spoilage, detection of food-borne toxins, etc. The available nanosensors and their applications have been summarized in Figure 6.2. Nanosensors can play direct role in packaging material such as electronic tongue or nose, which can detect spoilage by detecting specific chemicals released when food goes foul. Other sensors include microfluidic sensors for the real-time detection of pathogens. Silicon-based systems, so-called lab-on-a-chip technology, have become popular these days. NEMS (nanoelectro-mechanical systems) are being used on a commercial level. Polychromix (USA) has developed NEMS to detect trans-fat content in foods (Ritter, 2005). NEMS can also be used in quality control devices to detect specific biochemical and chemical signals. Nano-cantilevers are another class of smart and highly precise sensors that can be used to detect microbes based on specific interactions such as antigen–antibody, enzyme–substrate, etc. or detection of mass change via production of electromechanical signals (Arora et al., 2011). These devices have already established themselves with success in the market.

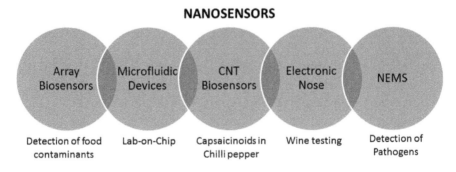

NANOSENSORS

FIGURE 6.2 Use of different nanosensors in food packaging.

6.3 NANOTECHNOLOGY IN FOOD PACKAGING

The demand of new packaging systems has been ever increasing because of many reasons such as:

- The number of food products is continuously increasing and new foods need to be complimented by proper packaging.
- The routine diet of people is being influenced by the modern research because of a growing awareness, and the consumers want better and safe packaging of foods.
- There has been an increase in demand of processed foods in developing countries. Presently, the facilities available in these countries

do not support the chain packaging systems completely, so there is a requirement of new packaging technologies which are self-sufficient and need not to be a part of the chain.

- The distance between producers and consumers is increasing and the food that is to be transported over a long distances requires advanced and sophisticated packaging.

The potential of nanotechnology to revolutionize different processes has provided an opportunity to food scientists to exploit this technology for development and improvement of food packaging. The development of nanotech food packaging and its effect on different aspects of food is already a major focus of research in food industry. Incorporation of nanomaterials in packaging may add variety to already recognized groups of active, smart, and intelligent packaging. The use of nanoparticles may help with better preservation, less wastage, and improved safety of foods. A general comparison of conventional and nanotechnology-based packaging has been illustrated in Figure 6.3 which depicts the advantages of nanotechnology packaging over the conventional packaging.

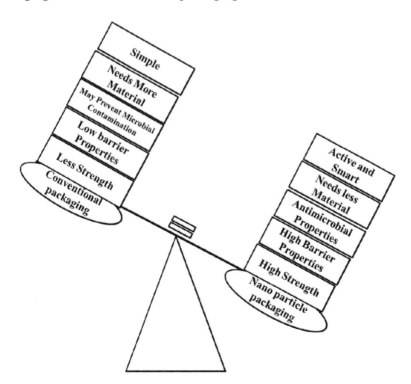

FIGURE 6.3 Advantages of nanopackaging over conventional packaging.

The safety of food products majorly depends on the right selection of packaging materials and technologies. Polymer composites have been used for safety and preservation of food products (Sozer and Kokini, 2009). Nanocomposites promise to expand the horizons of edible and biodegradable films. It helps to reduce the waste associated with packaging. General requirements for food packaging material involve gas barrier, mechanical properties, vapor barrier, aroma barrier, thermal properties, and optical properties (Yadollahi et al., 2010). Table 6.1 presents some of the current and proposed future applications of nanotechnology in food packaging. Nano-TiO_2 (titanium dioxide) with photocatalyses capabilities can oxidize ethylene (hormone responsible for ripening of fruits) into carbon dioxide and water, thus preventing loss due to excessive ripening (Hu and Fu, 2003). Silver nanoparticles also have a photocatalytic along with antibacterial properties. Nano-Ag has also been used for ethylene absorption and degradation and prevention of bacterial contamination (Hu and Fu, 2003). Nanomaterials such as nano-silicon oxide (SiOx)/chitosan have been used for coating Fuji apples. Green tea with nanopacking had a better retention of functional vitamin C, polyphenols, amino acids, and chlorophyll as compared to normal packing (Huang and Hu, 2006). By the addition of suitable nanoparticles, it is possible to produce better packing in terms of mechanical and thermal barrier properties.

TABLE 6.1 Current and Proposed Future Applications of Nanotech in Food Packaging.

Application	Function	Example	References
Polymer-based nanocomposites	Better barrier properties		Bott et al. (2014), Mihindukulasuriya and Lim (2014)
Polymer-based antimicrobial nanocomposites	Antimicrobial	Zinc oxide, magnesium oxide, nanosilver	Kuorwel et al. (2015), Fortunati et al. (2013), Llorens et al. (2012), Motlagh et al. (2012)
UV absorbers	Prevents UV degradation	Nanotitanium dioxide	Reig et al. (2014), Sanchez-Garcia et al. (2010), Siracusa et al. (2008), Smolander and Chaudhry (2010)
Titanium nitride nanoparticles introduced in packaging materials	Provides strength	Titanium nitride	Bott et al. (2012)

TABLE 6.1 *(Continued)*

Application	Function	Example	References
Wheel-shaped alumina platelets	Fillers for plastic materials	Alumina	Smolander and Chaudhry (2010), Bajpai et al. (2012)
Nanoprecipitated calcium carbonate	Improve mechanical properties, heat resistance, and printing quality of polyethylene		Chaudhry et al. (2008)
Nanosensors and biosensors	Detecting food spoilage	BioSilicon™	Chaudhry et al. (2008)

6.4 APPLICATION OF NANOTECHNOLOGY IN FRUIT AND VEGETABLE PACKAGING

Fruits and vegetables are consumed as fresh or in cooked form after their harvest. The surplus fruits and vegetables need to be stored for off-seasons or need to be delivered to the consumers, other than producers, since the shelf-life of fruits and vegetables is very less because of high moisture content and fragile nature, which may result in injuries or bruises. In addition to this, fruits and vegetables still respire after harvesting which causes them to spoil. Extensive care of many factors such as physiological status, diseases, pests, mechanical injuries, and postharvest handling is taken during raising the crop and after harvest; but wrong packaging can spoil all the energy and time spent on the fruit and vegetable cultivation. A right packaging for particular fruit or vegetable is required for the increase of their shelf-life. The conventional packaging used for the fruits and vegetables is wooden boxes, corrugated fiber boards, polyethene, and jute/leno/net bags which provide the bulk packaging and mechanical support during transportation and storage. Although the conventional packaging can increase the shelf-life of fruits and vegetables, still there are huge losses of fruits and vegetables throughout the year which force the food scientists to modify these conventional methods of packaging.

The respiration of most fruits and vegetables continues after harvesting resulting in weight loss and expressed visually as wilting and shriveling of product and may cause monetary loss to the producer. The packed fruits and vegetables may be kept in cold stores, which is one way of reducing the respiration rate. Modified atmosphere packaging is also practiced in which the composition of gases within the package is changed that reduces the rate of respiration, hence increasing the shelf-life of fresh produce. Due to continuous respiration, the gaseous composition of the modified atmosphere packaging changes, limiting its use for the fresh produce. Another effective way that is in practice is controlled atmosphere storage. In controlled atmosphere storage, the composition of gas surrounding the fresh produce is maintained at a predetermined standard composition along with temperature and humidity, which are also controlled. This technology of storage provides a better control over the postharvest physiological changes of fruits and vegetables.

The combined effect of packaging technique and temperature control can help in a better preservation and quality of fruits and vegetables. Temperature has a significant effect on ripening of fruits and vegetables because the rate of respiration has a direct proportionality to temperature. Lower temperatures reduce respiration rate and also inhibit the microbial growth on the fruit and vegetable surfaces. However, all fruits and vegetables are not suitable for low-temperature storage and an advanced and extensive research is a prerequisite before subjecting any fresh commodity to low-temperature storage; otherwise, physiological disorders are likely to occur.

Prevention of the fruit and vegetable losses through good packaging can be beneficial in various ways:

1. Producers get better returns for their crop and consumers get better quality products.
2. Fruits and vegetables remain available throughout the year, especially in off seasons.
3. Packaging preserves the wholesomeness and nutritive value of fruits and vegetables which are directly linked to the health of consumers.
4. Good packaging can ensure the safety of food, both physically and chemically.

To maintain the above standards, new concept called smart and active packaging has emerged. In active packaging, the environment modification within the package during the period of storage is achieved by attaching some additives to the packaging, thus making the packaging active. For instance, ethylene scrubbers, oxygen scavengers, desiccants, carbon dioxide generators, ethanol generators, etc. are available to help maintain the appropriate conditions around the foods within the package. Smart packaging is a sophisticated type of packaging which senses the quality attributes of food during storage period. For example, sensors can be installed in the packaging material which may help consumers to easily detect if there has been a breakdown of the cold chain of frozen peas during any stage of storage and help them in selecting the best quality. There is a continuous demand of a new and advanced means of packaging for better preservation of fruits and vegetables quality because the priority of consumers is shifting from "safe foods" to "nutritious foods." The use of nanotechnology in food packaging represents one such attempt in this direction as has been discussed at length in the above sections. A substantial number of research studies are available that report the use of nanotechnology in fruit and vegetable packaging, given in Table 6.1.

Nanoparticles have been used in the fruit and vegetable packaging by incorporating the nanoparticles into the packaging material or a coating material incorporated with active nanoparticles is applied on the packaging material or it can be applied directly on the surface of fruits and vegetables to enhance their shelf-life. The various functions of nanoparticles in fruit and vegetable packaging are as follows.

6.4.1 BARRIER PROPERTIES

The barrier properties of nanoparticles are directly related to the moisture loss and the rate of respiration. Those fruits and vegetables which are prone to moisture loss (green leafy vegetables, tomatoes, capsicum, reddish, beans, etc.) or belong to the climacteric group of fruits (avocado, breadfruit, papaya, mango, kiwifruit, etc.) need to be packed in a material possessing good barrier properties for oxygen and water. Plastic materials have low barrier properties as compared to glass or metals but these are heavy and expensive, since nanoparticles produce a dense structure as compared to macro-materials, which is associated with reduction in pore size in the packaging material (Syverud and Stenius, 2009). Further, nanoparticles are more efficient as an additive in active packaging as a moisture or oxygen absorbent due to their high surface area. Nanocellulose has been proven to possess good oxygen and water barrier properties and thus used for food packaging purposes (Ferrer et al., 2016). Climacteric behavior and postharvest shelf-life study of apples coated with nanochitosan showed good results of moisture retention and retardation of respiration rate (Sahraei et al., 2016). Nanocomposite-based packaging has retained the quality of cherry such as weight loss during storage due to the nanosilicate beads on polyethylene (Zandi et al., 2014). Similarly, prevention of weight loss during the storage of asparagus was observed due to the silver nanoparticles—polyvinylpyrrolidone (An et al., 2008).

6.4.2 ANTIMICROBIAL PROPERTIES

The surface of fruits and vegetables is in direct contact with the inner surface of packaging material which itself can be a source of contamination. For this reason, aseptic packaging is done for a long shelf-life of some foods. However, aseptic packaging is not practical for every food type. Recently, nanoparticles have been proven to exhibit antimicrobial properties which can be employed in packaging to improve the shelf-life of the packaged fruits and vegetables (Table 6.2). The biocidal effect of metals was long back discovered in 1893 by Karl Wilhelm von Nageli and is known as "oligodynamic" action which is basically the reaction of metals with thiol or amine groups of enzymes of microorganisms. The oligodynamic action of metals can be enhanced by converting them into nanoparticles which have an extremely large surface area, thus enhancing tremendously the surface of contact between the metals and microorganisms. In addition, metals have low volatility and are stable at high temperatures. Among the metal nanoparticles, copper (Cu), zinc (Zn), gold (Au), titanium (Ti), and silver (Ag) have been reported to possess potential antimicrobial activity against pathogenic microorganisms (Toker et al., 2013). Several antimicrobial nanopackaging studies have been carried out on different fruits and vegetables. Storage of barberry and strawberry packed with LDPE–AgNPs (low-density polyethylene–silver nanoparticles) and LDPE–AgNPs–TiO$_2$, respectively, has inhibited the total aerobic bacteria in both cases (Motlagh et al., 2012; Yang et al., 2010). Lettuce and apples were tested by storing in EVOH–AgNPs (ethylene vinyl alcohol–silver nanoparticles) packs and it was found effective against the *Salmonella* spp., and *Listeria monocytogenes* (Martinez-Abad et al., 2012). Similarly, fresh apples and carrots were stored in the nanocomposite packs (polyethylene–Ag–TiO$_2$) and the results depicted inhibitory effect of the packaging against *Staphylococcus aureus, coliforms, Escherichia coli, Listeria* (Metak, 2015). The above examples undoubtedly support the use of antimicrobial nanoparticles in the fruit and vegetable packaging. Since there is a controversy regarding the possible metal migration into the food system during the storage, extensive and elaborate research on the effect of nanoparticles on health is a need of the hour to clear air regarding their use.

TABLE 6.2 Recent Studies of Nanotechnology-based Packaging in Fruits and Vegetables.

Nanoparticles	Method	Fruit/ vegetable	Results	References
Chitosan nanoparticles loaded with *Cinnamomum zeylanicum* essential oil	Coating with refrigerated storage	Cucumber	Extended shelf-life for about 21 days	Mohammadi et al. (2015)
Nanochitosan coatings	Coating with cold storage (1 ± 1°C)	Apple cv. Golab Kohanz	Enhanced the shelf-life and quality of fruit over the period of 9 weeks	Sahraei et al. (2016)
Nano-CaCO$_3$– LDPE	Packaging with cold storage	Fresh-cut sugarcane	Inhibited browning and maintained the quality over 5 days storage	Luo et al. (2014)
Nanochitosan with and without copper	Coating with refrigerated storage	Strawberry (*Fragaria* × *ananassa Duchesne*)	Enhanced shelf-life more than 2.5-fold than control over 12 days storage	Eshghi et al. (2014)
Mixture of polyethylene, nano-Ag, nano-TiO$_2$, and montmorillonite	Nanocomposite-based packaging with cold storage	Kiwifruit (*Actinidia deliciosa*)	Enhanced the shelf-life and quality of fruit over 42 days storage	Hu et al. (2011)
Nanosilver and nano-silicate (polyethene/polypropylene)	Nanocomposite-based packaging	Sweet cherry (*Prunus avium* cv. Syahe Mashhad)	Retard ripening and quality degradation over 45 days storage	Zandi et al. (2014)
Silver nanoparticles-PVP	Coating with refrigerated storage	Green asparagus spears	Increased shelf-life for 25 days with good quality	An et al. (2008)
Polyethylene with nanopowder of (nano-Ag, kaolin, anatase TiO$_2$, and rutile TiO$_2$)	Packaging (room temperature)	Chinese jujube (*Ziziphus jujuba* Mill. var. *inermis* (Bunge) Rehd)	Maintained the quality over 12 days storage	Li et al. (2009)
Silver-montmorillonite nanoparticles	Packaging (Ag-MMT were left on the bottom of polypropylene	Fresh fruit salad	Prolong the shelf-life of about 5 days	Costa et al. (2011)

TABLE 6.2 *(Continued)*

Nanoparticles	Method	Fruit/vegetable	Results	References
nano-CaCO$_3$-LDPE	Packaging with refrigeration	Fresh-cut Chinese yam	Inhibited browning and maintained quality	Luo et al. (2015)
Chitosan/nanosilica	Coating with ambient temperature	Longan fruit (*Dimocarpus longan* Lour. cv. Shijia)	Retained good quality over the period of 10 days	Shi et al. (2013)
Silver nanoparticles with sodium alginate films	Coating stored at 27°C	Carrot and pear	Maintained the shelf-life quality over the 10 days storage	Mohammed-Fayaz et al. (2009)
Nano-Ag–LDPE	Packaging	Dried barberry	Showed antimicrobial activity and maintained the quality during storage	Motlagh et al. (2012)
Nano-TiO$_2$–LDPE	Packaging stored at refrigeration temperature	Strawberry (*Fragaria ananassa* Duch.)	Retained the quality during 14 days storage scavenging of reactive oxygen species	Li et al. (2016)
Ethylene−vinyl alcohol AgNPs	Films	Lettuce, apples	Inhibited the growth of *Salmonella* spp., *L. monocytogenes*	Martinez-Abad et al. (2012)
Nano-Ag, TiO$_2$, kaolin	Polyethene packing combined with controlled atmosphere at 0°C	Gynura (*Gynura bicolor* DC)	Improved shelf-life for 20 days of storage	Jiang et al. (2014)
Nano-Ag, kaolin, anatase TiO$_2$, rutile TiO$_2$	Polyethene with refrigerated storage	Fresh Strawberry (*Fragaria ananassa* Duch. cv. Fengxiang)	Reduced fruit decay rates for 12 days of storage	Yang et al. (2010)
Nanosilver and nanotitanium dioxideon	Packaging container	Fresh apples and carrots	Showed the antimicrobial activity	Metak (2015)
Nanosilver particles	LDPE polyethylene packaging	Dried barberry	Preserved the appearance quality for 2–3 weeks more than control	Motlagh et al. (2012)

6.4.3 COATING

As described earlier, fruits and vegetables respiration continues after harvesting resulting in loss of moisture and other nutrients. Several fruits and vegetables are naturally covered with a coating, which acts as a barrier between the fruit surface and the surrounding environment, reducing the transfer of moisture and gas. Coating of different fresh produce with waxes and other edible polymers to preserve their quality and enhance shelf-life is a common practice. In addition, coating also provides glitter to the products making them attractive. Due to a high surface area and versatile properties of nanoparticles, coating of fruits and vegetables with nanoparticles forms a superior film as compared with macromolecular coating. Such coatings possess enhanced functional property of protecting fruits and vegetables from moisture loss, fat oxidation, and respiration rate reduction, compared to simple coatings of same composition. Further, these films also serve as a carrier for colors, flavors, antioxidants, nutrients, and antimicrobials (Yadollahi et al., 2010). Coating of silver nanoparticles combined with sodium alginate or polyvinylpyrrolidone was able to increase shelf-life and preserve quality of carrot and pear for about 10 and 25 days for green asparagus spears (Mohammed-Fayaz et al., 2009; An et al., 2008). Chitosan/Nanosilica coatings helped in retaining the quality of longan fruit (*Dimocarpus longan* Lour. cv. Shijia) for over a period of 10 days (Shi et al., 2013). Similarly, chitosan nanoparticle coating has been applied on different fruits and vegetables such as cucumber, apple, and strawberry which have improved their shelf-life as compared to control samples (Mohammadi et al., 2015; Sahraei et al., 2016; Eshghi et al., 2014). Recently, nanocoatings have been tried on the fresh-cut fruits and vegetables, for example, fresh-cut Chinese yam and sugarcane were coated with nano-$CaCO_3$–LDPE which enhanced the shelf-life for many days by inhibiting the browning (Luo et al., 2015, 2014).

6.4.4 OTHER PROPERTIES

As fruits and vegetables are delicate, strength of the package is very important. Nanoparticles such as montmorillonite, polylactic acid, polycaprolactone, and polyhydroxy butyrate containing nanoclays were used in packaging to improve the mechanical strength and thermal stability of the delicate foods such as fruits and vegetables (Tharanathan, 2003; Cha and Chinnan, 2004). Use of nanoparticles such as antimicrobial, antioxidants, browning inhibitors, flavor promoters, and enzymes into active packaging of perishable fruits and vegetables has successfully enhanced the shelf-life by absorbing the ethylene gas and inhibiting microbial growth (Weiss et al., 2006; Rashidi and Khosravi-Darani, 2011). From cultivation to processing, fruits and vegetables are subjected to different treatments which may cause health hazards later such as pesticides, a major health concern. Decontamination of foods and packaging equipment with the help of nanoemulsions has also been reported for instance; nanomicelle-based product containing the natural glycerin removes the pesticide residues from the fruits and vegetables (Sekhon, 2010).

Several research studies have been conducted to find out the advantages of nanotechnology in packaging of fruits and vegetables and most of the researchers have reported positive results, but discussing all of them is out of scope of this chapter.

6.5 RISK AND BENEFIT ANALYSIS

Nanotechnology has become increasingly important in the processing and packaging of food products. Innovative devices and processes developed with the help of nanotechnology have influenced a lot in the food sector in a positive sense. But before we start praising and applauding the benefits of nanotechnology, we must discuss about the potential unforeseen hazards. It is due to the large surface area per unit volume that nanoparticles show such varying functions and a great potential. The large surface area might have toxic effects in the body that are not prevalent or visible with their bulk counterparts. Also, migration of nanoparticles into the food or breakdown of nanocomposite polymers and the subsequent transfer of nanoparticles into the food may expose consumers to nanomaterials. So far, no cases of harmful effect of such products have been reported. Bott et al. (2014) in a study concluded that the nanoparticle exposure due to transfer from food contact plastics is not expected. A study on 153 people showed nanotechnology-derived packing material was perceived to be beneficial and better than conventional packing.

Till date, no regulations have been formulated to control or limit the use of nanotechnology in food products, although FDA (2014) has issued some guidelines to help the industry in identification of potential implications for safety, health impact, and effectiveness upon application of nanotechnology in manufacturing products regulated by the administration.

Despite lack of regulations, nanotechnology food products are being manufactured and sold in the market, although nanotechnology-based food additives were not so readily accepted (Sozer and Kokini, 2009). We must take lessons from genetically modified organisms/food and must implement the technology in a more socially acceptable manner. Various laws and regulations must be set to use nanotechnology in our food. The products must be appropriately labeled and other regulations must be set to increase the consumer acceptability.

6.6 CONCLUSION

To conclude, it can be inferred that different nanotechnologies have greatly influenced the food sector. Many products are already available in the market and many more are in research phase. These products have shown a great potential and market. Several investigations support the beneficial aspects of nanomaterials used in packing and have reported many physiochemical and physiological improvements in the products as compared to conventional packing. These also lure us with the advantages of easy processing, feasibility, time-saving, and extended storage life. Although nanotechnology has more advantages than one can think of, the road to the amalgamation of the food sector with nanotechnology needs to be well thought of.

KEYWORDS

- nanotechnology
- packaging
- nanosensor
- nanopolymer
- food
- fruit

REFERENCES

An, J.; Zhang, M.; Wang, S.; Tang, J. Physical, Chemical and Microbiological Changes in Stored Green Asparagus Spears as Affected by Coating of Silver Nanoparticles-PVP. *LWT-Food Sci. Technol.* **2008,** *41,* 1100–1107.

Arora, P.; Sindhu, A.; Dilbaghi, N.; Chaudhury, A. Biosensors as Innovative Tools for the Detection of Food Borne Pathogens. *Biosens. Bioelectron.* **2011**, *28*, 1–12.

Bajpai, S.; Chand, N.; Chaurasia, V. Nano Zinc Oxide-loaded Calcium Alginate Films with Potential Antibacterial Properties. *Food Bioprocess Technol.* **2012**, *5*, 1871–1881.

Bott, J.; Stormer, A.; Franz, R. A Model Study into the Migration Potential of Nanoparticles from Plastics Nanocomposites for Food Contact. *Food Packag. Shelf Life* **2014**, *2*, 73–80.

Bott, J.; Stormer, A.; Wolz, G.; Franz, R. Studies on the Migration of Titanium Nitride Nanoparticles in Polymers. In *Poster Presentation at the 5th International Symposium on Food Packaging*, Berlin, Nov. 14–16, 2012; Fraunhofer Institute for Process Engineering and Packaging: Berlin, 2012.

Cha, D. S.; Chinnan, M. S. Biopolymer-Based Antimicrobial Packaging: A Review. *Crit. Rev. Food Sci. Nutr.* **2004**, *44*, 223–237.

Chaudhry, Q.; Castle, L. Food Applications of Nanotechnologies: An Overview of Opportunities and Challenges for Developing Countries. *Trends Food Sci. Technol.* **2011**, *22*, 595–603.

Chaudhry, Q.; Castle, L.; Bradley, E.; Blackburn, J.; Aitken, R.; Boxall, A. Assessment of Current and Projected Applications of Nanotechnology for Food Contact Materials in Relation to Consumer Safety and Regulatory Implications. *Final Report Project A03063*. Food Standards Agency; 2008. http://www.food.gov.uk/sites/default/files/a03063.pdf.

Costa, C.; Conte, A.; Buonocore, G. G.; Del Nobile, M. A. Antimicrobial Silver-Montmorillonite Nanoparticles to Prolong the Shelf Life of Fresh Fruit Salad. *Int. J. Food Microbiol.* **2011**, *148*, 164–167.

Duncan, T. V. Applications of Nanotechnology in Food Packaging and Food Safety: Barrier Materials, Antimicrobials and Sensors. *J. Colloid Interface Sci.* **2011**, *363*, 1–24.

Eshghi, S.; Hashemi, M.; Mohammadi, A.; Badii, F.; Mohammadhoseini, K. A. Z. Effect of Nanochitosan Based Coating with and without Copper Loaded on Physicochemical and Bioactive Components of Fresh Strawberry Fruit (*Fragaria* × *ananassa* Duchesne) during Storage. *Food Bioprocess Technol.* **2014**, *7*, 2397–2409.

Ferrer, A.; Pal, L.; Hubbe, M. Nanocellulose in packaging: Advances in barrier layer technologies. *Ind. Crops Prod.* **2016**. DOI:10.1016/j.indcrop.2016.11.012.

Fortunati, E.; Peltzer, M.; Armentano, I.; Jimenez, A.; Kenny, J. M. Combined Effects of Cellulose Nanocrystals and Silver Nanoparticles on the Barrier and Migration Properties of PLA Nano-biocomposites. *J. Food Eng.* **2013**, *118*, 117–124.

Gardener, E. Brainy Food: Academia, Industry Sink their Teeth into Edible Nano. *Accessed 22 February 2006 from the website: www.Smalltimes.com.*

Hu, Q.; Fang, Y.; Yang, Y.; Ma, N.; Zhao, L. Effect of Nanocomposite-Based Packaging on Postharvest Quality of Ethylene-Treated Kiwifruit (*Actinidia deliciosa*) during Cold Storage. *Food Res. Int.* **2011**, *44*, 1589–1596.

Hu, A. W.; Fu, Z. H. Nanotechnology and its Application in Packaging and Packaging Machinery. *Packag. Eng.* **2003**, *24*, 22–24.

Huang, Y.; Hu, Q. Effect of a New Fashion Nano-packing on Preservation Quality of Green Tea. *J. Food Sci.* **2006**, *4*, 61.

Imran, M.; Revol-Junelles, A. M.; Martyn, A.; Tehrany, E. A.; Jacquot, M.; Linder, M.; Desobry, S. Active Food Packaging Evolution: Transformation from Micro- to Nanotechnology. *Crit. Rev. Food Sci. Nutr.* **2010**, *50* (9), 799–821.

Jiang, L.; Jiang, J.; Luo, H.; Yu, Z. Effect of Nano-packing Combined Controlled Atmosphere on Postharvest Physiology and Biochemistry of Gynura Bicolor D. C. *J. Food Process Preserv.* **2014**, *38*, 1181–1186.

Kah, M. Nanopesticides and Nanofertilizers: Emerging Contaminants or Opportunities for Risk Mitigation? *Front. Chem.* **2015**, *3*, 64.

Kuorwel, K.; Cran, M. J.; Orbell, J. D.; Buddhadasa, S.; Bigger, S. W. Review of Mechanical Properties, Migration, and Potential Applications in Active Food Packaging Systems Containing Nanoclays and Nanosilver. *Comp. Rev. Food Sci. Food Saf.* **2015**, *14*, 411–430.

Li, H.; Li, F.; Wang, L.; Sheng, J.; Xin, Z.; Zhao, L.; Xiao, H.; Zheng, Y.; Hu, Q. Effect of Nano-packing on Preservation Quality of Chinese Jujube (*Ziziphus jujuba* Mill. var. Inermis (Bunge) Rehd). *Food Chem.* **2009**, *114*, 547–552.

Li, D.; Ye, Q.; Jiang, L.; Luo, Z. Effects of nano-TiO_2 -LDPE Packaging on Postharvest Quality and Antioxidant Capacity of Strawberry. *J. Sci. Food Agric.* **2016**. DOI:10.1002/jsfa.7837.

Llorens, A.; Lloret, E.; Picouet, P. A.; Trbojevich, R.; Fernandez, A. Metallic-Based Micro and Nanocomposites in Food Contact Materials and Active Food Packaging. *Trends Food Sci. Technol.* **2012**, *24*, 19–29.

Luo, Z.; Wang, Y.; Feng, S. Impact of Nano-$CaCO_3$-LDPE Packaging on Quality of Fresh-Cut Sugarcane. *J. Sci. Food Agric.* **2014**, *94* (15), 3273–3280. DOI:10.1002/jsfa.6680.

Luo, Z.; Wang, Y.; Jiang, L.; Xu, X. Effect of Nano-$CaCO_3$-LDPE Packaging on Quality and Browning of Fresh-Cut Yam. *LWT—Food Sci. Technol.* **2015**, *60*, 1155–1161.

Martinez-Abad, A.; Lagaron, J. M.; Ocio, M. J. Development and Characterization of Silver-based Antimicrobial Ethylene-Vinyl Alcohol Copolymer (EVOH) Films for Food-Packaging Applications. *J. Agric. Food Chem.* **2012**, *60*, 5350–5359.

Metak, A. M. Effects of Nanocomposite Based Nano-silver and Nano-titanium Dioxide on Food Packaging Materials. *Int. J. Appl. Sci. Technol.* **2015**, *5* (2), 26–40.

Mihindukulasuriya, S.; Lim, L. T. Nanotechnology Development in Food Packaging: A Review. *Trends Food Sci. Technol.* **2014**, *40*, 149–167.

Mohammadi, A.; Hashemi, M.; Masoud, S. Chitosan Nanoparticles Loaded with *Cinnamomum zeylanicum* Essential Oil Enhance the Shelf Life of Cucumber during Cold Storage. *Postharvest Biol. Technol.* **2015**, *110*, 203–213.

Mohammed-Fayaz, A.; Balaji, K.; Girilal, M.; Kalaichelvan, P. T.; Venkatesan, R. Myco-based Synthesis of Silver Nanoparticles and Their Incorporation into Sodium Alginate Films for Vegetable and Fruit Preservation. *J. Agric. Food Chem.* **2009**, *57*, 6246–6252. DOI:10.1021/jf900337h.

Motlagh, N. V.; Mosavian, M. T. H.; Mortazavi, S. A. Effect of Polyethylene Packaging Modified with Silver Particles on the Microbial, Sensory and Appearance of Dried Barberry. *Packag. Technol. Sci.* **2012**, *23*, 253–266.

Prasanna, B. M. Nanotechnology in Agriculture. In *ICAR National Fellow*; Division of Genetics, Indian Agricultural Research Institute: New Delhi, 2007; pp 111–118.

Rashidi, L.; Khosravi-Darani, K. The Applications of Nanotechnology in Food Industry. *Crit. Rev. Food Sci. Nutr.* **2011**, *51*, 723–730.

Reig, C. S.; Lopez, A. D.; Ramos, M. H.; Ballester, V. A. C. Nanomaterials: A Map for Their Selection in Food Packaging Applications. *Packag. Technol. Sci.* **2014**, *27*, 839–866.

Rhim, J. W.; Hong, S. I.; Park, H. M.; Ng, P. K. W. Preparation and Characterization of Chitosan-Based Nanocomposite Films with Antimicrobial Activity. *J. Agric. Food Chem.* **2006**, *54*, 5814–5822.

Ritter, S. K. An Eye on Food. *Chem. Eng. News* **2005,** *83,* 28–34.

Roco, M. C.; Mirkin, C. A.; Hersam, M. C., Eds. *Nanotechnology Research Directions for Societal Needs in 2020: Retrospective and Outlook, World Technology Evaluation Center (WTEC) and the National Science Foundation (NSF)*; Springer: Berlin, 2010.

Sahraei, A.; Gardesh, K.; Badii, F.; Hashemi, M.; Ardakani, A. Y.; Maftoonazad, N.; Gorji, A. M. Effect of Nanochitosan Based Coating on Climacteric Behavior and Postharvest Shelf-life Extension of Apple cv. Golab Kohanz. *LWT—Food Sci. Technol.* **2016,** *70,* 33–40. DOI:10.1016/j.lwt.2016.02.002.

Sanchez-Garcia, M. D.; Lopez-Rubio, A.; Lagaron, J. M. Natural Micro and Nanobiocomposites with Enhanced Barrier Properties and Novel Functionalities for Food Biopackaging Applications. *Trends Food Sci. Technol.* **2010,** *21* (11), 528–536.

Sekhon, B. S. Food Nanotechnology: An Overview. *Nanotechnol. Sci. Appl.* **2010,** *3,* 1–15.

Shi, S.; Wang, W.; Liu, L.; Wu, S.; Wei, Y.; Li, W. Effect of Chitosan/Nano-silica Coating on the Physicochemical Characteristics of Longan Fruit under Ambient Temperature. *J. Food Eng.* **2013,** *118,* 125–131. DOI:10.1016/j.jfoodeng.2013.03.029.

Siracusa, V.; Rocculi, P.; Romani, S.; Rosa, M. D. Biodegradable Polymers for Food Packaging: A Review. *Trends Food Sci. Technol.* **2008,** *19,* 634–643.

Smolander, M.; Chaudhry, Q. Nanotechnologies in Food Packaging. In *Nanotechnologies in Food*; The Royal Society of Chemistry: London, 2010; pp 86–101.

Sozer, N.; Kokini, J. L. Nanotechnology and Its Applications in the Food Sector. *Trends Biotechnol.* **2009,** *27,* 82–89.

Stephen, A. M. *Food Polysaccharides and Their Applications*, CRC Press: Boca Raton, FL, 1995.

Stutzenberger, F. J.; Latour, Jr., R. A.; Sun, Y. P.; Tzeng, T. R. *Adhesin-Specific Nanoparticles and Process for Using Same*, U.S. Patent 7,682,631, issued March 23, 2010.

Syverud, K.; Stenius, P. Strength and Barrier Properties of MFC Films. *Cellulose* **2009,** *16,* 75–85.

Tharanathan, R. N. Biodegradable Films and Composite Coatings: Past, Present and Future. *Trends Food Sci. Technol.* **2003,** *14,* 71–78. DOI:10.1016/S0924-2244(02)00280-7.

Toker, R. D.; Kayaman-Apohan, N.; Kahraman, M. V. UV-Curable Nano-silver Containing Polyurethane Based Organic–Inorganic Hybrid Coatings. *Prog. Org. Coat.* **2013,** *76,* 1243–1250.

Weiss, J.; Takhistov, P.; McClements, D. J. Functional Materials in Food Nanotechnology. *J. Food Sci.* **2006,** *71,* 107–116.

Yadollahi, A.; Arzani, K.; Khoshghalb, H. The Role of Nanotechnology in Horticultural Crops Postharvest Management. *Acta. Hortic.* **2010,** *875,* 49–56.

Yang, F. M.; Li, H. M.; Li, F.; Xin, Z. H.; Zhao, L. Y.; Zheng, Y. H.; Hu, Q. H. Effect of Nano-packing on Preservation Quality of Fresh Strawberry (*Fragaria ananassa* Duch. cv. Fengxiang) during Storage at 4°C. *J. Food Sci.* **2010.** DOI:10.1111/j.1750-3841.2010.01520.x.

Zandi, K.; Weisany, W.; Naseri, L. Evaluation of Nanocomposite-Based Packaging to Prolong the Shelf-Life of Sweet Cherry (*Prunus avium* cv. Syahe Mashhad) during Storage. *Adv. Biores.* **2014,** *5,* 188–194.

CHAPTER 7

BIOLOGICAL CONTROL OF POSTHARVEST DISEASES

R. R. SHARMA[1]* and S. VIJAY RAKESH REDDY[2]

[1]*Division of Food Science & Postharvest Technology, ICAR-Indian Agricultural Research Institute, New Delhi 110012, India*

[2]*ICAR-Central Institute of Arid Horticulture, Beechwal, Bikaner, Rajasthan, India*

Corresponding author. E-mail: rrs_fht@rediffmail.com

ABSTRACT

Postharvest diseases are one of the major causes of the postharvest losses of fresh fruits and vegetables during transportation as well as storage. Traditionally, synthetic fungicides are primarily used to control postharvest decay loss. However, with the increased public awareness and changing life-style, a shift toward safer and more eco-friendly alternatives for the control of postharvest decays is being observed. Among various biological approaches, the use of antagonistic microorganisms is becoming popular throughout the world. Several postharvest diseases can now be controlled by microbial antagonists. Though the major mechanism(s) of microbial antagonists in suppressing the postharvest diseases is still unknown, competition for nutrients and space could be the most widely accepted mechanism. In addition, production of antibiotics, direct parasitism, and possibly induced resistance in the harvested commodity are other modes of their actions by which they suppress the activity of postharvest pathogens in fruits and vegetables. Microbial antagonists are applied either before or after harvest, but postharvest applications are more effective than pre-harvest applications. Microbial antagonists used in combinations were found to provide better control of postharvest diseases over individual cultures or strains. Similarly, the efficacy of the microbial antagonist(s) can be enhanced if they are used with low doses of fungicides, salt additives, and physical treatments like hot water dips, irradiation with ultraviolet light, etc. Although the results of this technology are encouraging, we need to continue to explore the potential uses on the commercial scale in different corners of the world. Nonetheless, it is necessary to continue identifying new potential microorganisms, better understanding their mode of action, and pathogen, antagonist and host interactions, to increase the potential of bio-control agents in becoming a real alternative to synthetic postharvest fungicides.

7.1 INTRODUCTION

Postharvest losses might occur at any stage during postharvest handling, from harvest till consumption. Among all the causes, postharvest diseases contribute to a greater share for the losses in perishable commodities, especially horticultural crops, namely, fruits, vegetables, and flowers. In addition to the economic losses caused by the postharvest diseases, the diseased produce poses potential health risk. A large number of microbes, namely, fungi, bacteria, and yeasts, are known to produce mycotoxins/spores under certain conditions which pose great threat when used for processing and value addition. Postharvest diseases may start before or after harvesting. Plants or fruits infected in the field may not develop symptoms until stored. Once in storage, infections continue to develop on the fruits and vegetables. Wounds, cuts, or bruises caused during harvesting are common entry points for bacteria and fungi. Penetration can also occur during storage through natural openings, such as lenticels, or directly through the cuticle and epidermis. The losses due to postharvest diseases are influenced by a great number of factors such as type of commodity, cultivar susceptibility, storage environment (temperature and relative humidity), produce maturity and stage of ripening, harvesting and handling methods, pre- or postharvest treatments, etc.

Traditionally, fungicides are used extensively for postharvest disease control in perishable commodities. For postharvest pathogens which infect produce before harvest, field application of fungicides is often necessary. In the postharvest situation, fungicides are often applied to control infections already established in the surface tissues of produce or to protect against infections which may occur during storage and handling. In the case of quiescent field infections present at the time of harvest, fungicides must be able to penetrate to the site of infection to be effective (Coates and Johnson, 1997). Along with the use of fungicides, controlled atmosphere storage, heat treatments, hygienic practices for prevention of injury, and ionizing radiations are under vogue for controlling postharvest diseases. With the consumer's health awareness regarding the harmful effects of pesticide residues, the scientists are in search of new eco-friendly approaches for controlling the postharvest diseases, namely, biocontrol strategies using microbial antagonists, constitutive/induced host resistance, and development of natural fungicides (Montesinos and Bonaterra, 2009; Junaid et al., 2013). In recent years, there has been considerable interest in the use of antagonistic microorganisms, namely, yeasts, fungi, and bacteria for the control of postharvest diseases (Wisniewski and Wilson, 1992; Droby, 2006; Korsten, 2006; Dalal and Kulkarni, 2013). This chapter deals with the use of microbial antagonists for controlling postharvest diseases in fruits and vegetables.

7.2   BIOLOGICAL CONTROL AND ITS IMPORTANCE

Biological control is defined as the reduction of inoculum density or disease-producing activities of a pathogen or parasite in its active or dormant state, by one or more organisms accomplished naturally or through manipulation of the environment or host or antagonist or by mass introduction of one or more antagonists (Baker and Cook, 1974). It appears to have a significant role in addressing both environmental and economic issues raised through conventional approaches. Various biocontrol agents such as fungi and bacteria have been identified for the control of postharvest diseases which play an important role in sustainable horticulture and management of plant pathogens (Wisniewski and Wilson, 1992; Ragsdale and Sisler, 1994; Montesinos, 2003; Sobowale et al., 2008; Montesinos and Bonaterra, 2009; Junaid et al., 2013). The effectiveness of microbial antagonists depends on their ability to colonize fruit surfaces and adapt to various environmental conditions (Wilson and Wisniewski, 1989; Droby et al., 2002; Sharma, 2014). Microbial antagonists for biological control of postharvest diseases can be used in two different ways, namely, (1) using microbes already present on the produce itself, which can be promoted and managed or (2) those that can be artificially introduced against postharvest pathogens.

7.2.1   NATURAL MICROBIAL ANTAGONISTS

These microbes are present naturally on the surface of fruits and vegetables, which were identified, isolated, and used for the control of postharvest diseases (Janisiewicz, 1987; Sobiczewski et al., 1996). Chalutz and Wilson (1990) found that when concentrated washings from the surface of citrus fruit were plated out on agar medium, only bacteria and yeast appeared while after dilution of these washings, several rot fungi appeared on the agar, suggesting that yeast and bacteria may be suppressing fungal growth. Thus, it indicates that when fruits and vegetables are washed, they are more susceptible to decay than those which are not washed at all.

7.2.2 ARTIFICIALLY INTRODUCED MICROBIAL ANTAGONISTS

Artificial introduction of microbial antagonists is more effective in controlling postharvest diseases of perishable commodities than other means of biological control. The first report on the use of a microbial antagonist was made in strawberry (*Fragaria* × *ananassa* Duch.) with *Trichoderma* spp. against *Botrytis* rot (Tronsmo and Denis, 1977), while the first classical work was done for the control of brown rot in stone fruits using *Bacillus subtilis* (Pusey and Wilson, 1984). Several microbial antagonists have been identified and artificially introduced on a variety of harvested commodities including citrus, pome, and stone fruits and vegetables for control of postharvest diseases (Table 7.1). For instance, effective control of fruit rot decay of citrus was observed with yeasts such as *Pichia guilliermondii* Wiskerham, *Candida oleophila* Montrocher, *Candida sake* Saito and Ota, *Candida formata* Meyer & Yarrow, *Candida saitona* Nakase & Suzuki, *Debaryomyces hansenii* Lodder & Kre-Van Rij, *Aureobasidium pullulans* (de Bary) Arnaud, *Pantoea agglomerans* (Ewing & Fife), *Saccharomyces cerevisiae* Hansen and *Metschnikowia fructicola* Kurtzman & Droby, and *Metschnikowia pulcherrima* Pittes & Miller (Wilson and Chalutz, 1989; Chalutz and Wilson, 1990; Ippolito et al., 2000; Nunes et al., 2001a,b; Teixido et al., 2001; Karabulut et al., 2003; Spadaro et al., 2004; Lahlali et al., 2005; Droby, 2006; Long et al., 2006, 2007; Torres et al., 2007; Morales et al., 2008). Control of decay of citrus fruit caused by *Penicillium digitatum* (Pers.: Fr.) Sacc. and *Penicillium italicum* Wehmer was also reported with bacterial antagonists such as *B. subtilis* (Ehrenberg) Cohn, *Burkholderia* (*Pseudomonas*) *cepacia* Palleroni & Holmes, and *Pesudomonas syringae* van Hall (Singh, 2002; Long et al., 2007). Fungal antagonists including *Myrothecium roridum* Tode.: Fries (Appel et al., 1988) and *Trichoderma viride* Persoon.: Fries (Kota et al., 2006) were also shown to reduce decay of citrus fruit. *Trichoderma harzianum* Rifai has been effective in controlling anthracnose in banana (Devi and Arumugam, 2005) and rambutan (*Nephelium lappaceum* L.) (Sivakumar et al., 2000) and gray mold in grapes, kiwifuits, and pears (Batta, 2007).

TABLE 7.1 Microbial Antagonists Used for Controlling Postharvest Diseases of Horticultural Produce.

Microbial antagonist	Disease (pathogen)	Commodity	Reference
Acremonium brevae (Sukapure and Thirumulachar) Gams	Gray mold (*Botrytis cinerea*)	Apple	Janisiewicz (1988)
Aureobasidium pullulans	Monilinia rot (*Monilinia laxa*)	Banana	Wittig et al. (1997)
	Penicillium rots (*Penicillium* spp.)	Citrus	Wilson and Chalutz (1989)
	Botrytis rot (*Botrytis cinerea*)	Grape	Schena et al. (2003)
	Soft rot (*Monilinia laxa*)	Grape	Barkai-Golan (2001)
Bacillus subtilis	Brown rot (*Lasiodiplodia theobromae*)	Apricot	Pusey and Wilson (1984)
	Stem end rot (*Botryodiplodia theobromae* Pat.)	Avocado	Demoz and Korsten (2006)
	Botrytis rot (*Botrytis cinerea*)	Cherry	Utkhede and Sholberg (1986)
	Green mold (*Penicillium digitatum*)	Citrus	Singh and Deverall (1984)
	Sour rot (*Geotrichum candidum* Link)	Citrus	Singh and Deverall (1984)
	Stem end rot (*Botryodiplodia theobromae, Phomopsis citri* Fawc., *Alternaria citri* Ell. & Pierce)	Citrus	Singh and Deverall (1984)
	Alternaria rot (*Alternaria alternata* (Fr.) Keissler)	Litchi	Jiang et al. (1997, 2001)
	Brown rot (*Lasiodiplodia theobromae*)	Nectarine	Pusey and Wilson (1984)
	Brown rot (*Lasiodiplodia theobromae*)	Peach	Pusey and Wilson (1984)
	Brown rot (*Lasiodiplodia theobromae*)	Plum	Pusey and Wilson (1984)

TABLE 7.1 *(Continued)*

Microbial antagonist	Disease (pathogen)	Commodity	Reference
	Gray mold (*Botrytis cinerea*)	Strawberry	Zhao et al. (2007)
	Alternaria rot (*Alternaria alternata*)	Muskmelon	Yang et al. (2006), Wang et al. (2010)
Bacillus licheniformis (Weigmann) Verhoeven	Anthracnose (*Colletotrichum gloeosporioides*) and stem end rot (*Dothiorella gregaria* Sacc.)	Mango	Govender et al. (2005)
Bacillus pumilus	Gray mold (*Botrytis cinerea*)	Pear	Mari et al. (1996)
Burkholderia cepacia	Anthracnose (*Colletotrichum musae*)	Banana	Costa and Erabadupitiya (2005)
	Blossom end rot (*Colletotrichum musae*)	Banana	Costa and Erabadupitiya (2005)
Burkholderia gladioli	Phytopathogenic fungi (*Botrytis cinerea, Penicillium expansum, Penicillium digitatum, Aspergillus flavus, Aspergillus niger, Phytophthora cactorum, Sclerotinia sclerotiorum*)	In vitro	Elshafie et al. (2012)
Brevundimonas diminuta (Leifson & Hugh) Segers	Anthracnose (*Colletotrichum gloeosporioides*)	Mango	Kefialew and Ayalew (2008)
Candida guilliermondii	Gray mold (*Botrytis cinerea*)	Nectarine	Tian et al. (2002)
	Gray mold (*Botrytis cinerea*)	Peach	Tian et al. (2002)
	Gray mold (*Botrytis cinerea*)	Tomato	Saligkarias et al. (2002)
Candida membranifaciens Hansen	Anthracnose (*Colletotrichum gloeosporioides*)	Mango	Kefialew and Ayalew (2008)
Candida oleophila	Penicillium rot (*Penicillium expansum*)	Apple	El-Neshawy and Wilson (1997)
	Penicillium rots (*Penicillium digitatum* and *Penicillium italicum*)	Citrus	El-Neshawy and El-Sheikh (1998), Lahlali et al. (2004, 2005)
	Crown rot (*Colletotrichum musae*)	Banana	Lassois et al. (2008)
	Anthracnose (*Colletotrichum gloeosporioides*)	Papaya	Gamagae et al. (2003)
	Gray mold (*Botrytis cinerea*)	Peach	Karabulut and Baykal (2004)

TABLE 7.1 *(Continued)*

Microbial antagonist	Disease (pathogen)	Commodity	Reference
	Gray mold (*Botrytis cinerea*)	Tomato	Saligkarias et al. (2002)
Candida sake (CPA-1)	Penicillium rot (*Penicillium expansum*)	Apple	Vinas et al. (1996), Usall et al. (2001), Torres et al. (2006), Morales et al. (2008)
	Gray mold (*Botrytis cinerea*)	Apple	Vinas et al. (1998)
	Rhizopus rot (*Rhizopus nigricans* Ehrenberg)	Apple	Vinas et al. (1998)
	Blue mold (*Penicillium expansum*)	Pear	Torres et al. (2006)
	Botrytis bunch rot (*Botrytis cinerea*)	Wine grapes	Calvo-Garrido et al. (2013)
Clonostachys rosea	Fusarium dry rot (*Fusarium avenaceum, Fusarium caeruleum*)	Potato	Jima (2013)
	Gray mold (*Botrytis cinerea*)	Wild blueberry	Reeh (2012)
Cryptococcus laurentii	Bitter rot (*Glomerella cingulata*)	Apple	Blum et al. (2004)
	Brown rot (*Monilinia fructicola*)	Cherry	Karabulut and Baykal (2003), Tian et al. (2004), Qin et al. (2006)
	Alternaria rot (*Alternata alternata*) and Penicillium rot (*Penicillium expansum*)	Jujube	Qin and Tian (2004), Tian et al. (2005)
	Rhizopus rot (*Rhizopus stolonifer*)	Peach	Zhang et al. (2007c)
	Gray mold (*Botrytis cinerea*)	Peach	Zhang et al. (2007c)
	Brown rot (*Monilinia fructicola*)	Peach	Yao and Tian (2005)
	Blue mold (*Penicillium expansum*)	Peach	Zhang et al. (2007c)
	Mucor rot (*Mucor piriformis* Fischer)	Pear	Roberts (1990b)
	Gray mold (*Botrytis cinerea*)	Pear	Zhang et al. (2005)
	Blue mold (*Penicillium expansum*)	Pear	Zhang et al. (2003)

TABLE 7.1   *(Continued)*

Microbial antagonist	Disease (pathogen)	Commodity	Reference
	Rhizopus rot (*Rhizopus stolonifer*)	Strawberry	Zhang et al. (2007b)
	Gray mold (*Botrytis cinerea*)	Tomato	Xi and Tian (2005)
Cryptococcus flavus	Mucor rot (*Mucor piriformis*)	Pear	Roberts (1990b)
Cryptococcus albidus (Saito) Skinner	Mucor rot (*Mucor piriformis*)	Pear	Roberts (1990b)
	Gray mold (*Botrytis cinerea*)	Apple	Fan and Tian (2001)
	Blue mold (*Penicillium expansum*)	Apple	Fan and Tian (2001)
Cryptococcus spp.	Blue mold (*Penicillium expansum*)	Apple	Chand-Goyal and Spotts (1997)
Cystofilobasidium infirmominiatum	Penicillium rot of apple (*Penicillium expansum*)	Apple	Liu et al. (2011b)
Debaryomyces hansenii	Green and blue mold (*Penicillium digitatum* and *Penicillium italicum*)	Citrus	Singh (2002)
	Blue mold (*Penicillium italicum*)	Citrus	Chalutz and Wilson (1990)
	Sour rot (*Geotrichum candidum*)	Citrus	Chalutz and Wilson (1990)
	Rhizopus rot (*Rhizopus stolonifer*)	Peach	Mandal et al. (2007), Singh (2004, 2005)
Enterobacter aerogenes Hormaeche & Edwards	Alternaria rot (*Alternaria alternata*)	Cherry	Utkhede and Sholberg (1986)
Enterobacter cloacae	Rhizopus rot (*Rhizopus stolonifer*)	Peach	Wilson et al. (1987)
	Fusarium dry rot (*Fusarium sambucinum*)	Potato	Al-Mughrabi (2010)
Kloeckera apiculate (Rees) Janke	Botrytis rot (*Botrytis cinerea*)	Cherry	Karabulut et al. (2005)
	Penicillium rots (*Penicillium* spp.)	Citrus	Long et al. (2006, 2007)
	Green (*Penicillium digitatum*) and blue mold (*Penicillium italicum*)	Citrus	Long et al. (2006, 2007)

TABLE 7.1   *(Continued)*

Microbial antagonist	Disease (pathogen)	Commodity	Reference
Leucosporidium scottii	Blue mold of apple (*Penicillium expansum*)	Apple	Vero et al. (2013)
	Gray mold of apple (*Botrytis cinerea*)	Apple	Vero et al. (2013)
Metschnikowia fructicola	Botrytis rot (*Botrytis cinerea*)	Grape	Karabulut et al. (2003)
	Apple rot (*Penicillium expansum*)	Apple	Liu et al. (2011a)
Metschnikowia pulcherrima	Blue mold (*Penicillium expansum*) and Gray mold (*Botrytis cinerea*)	Apple	Spadaro et al. (2002, 2004)
Pantoea agglomerans	Penicillium rot (*Penicillium expansum*)	Apple	Nunes et al. (2002a), Morales et al. (2008)
	Green (*Penicillium digitatum*) and blue mold (*Penicillium italicum*)	Citrus	Teixido et al. (2001), Torres et al. (2007)
	Penicillium rots (*Penicillium* spp.)	Citrus	Plaza et al. (2001)
	Rhizopus rot (*Rhizopus stolonifer*)	Pear	Nunes et al. (2001a,b)
Penicillium sp. (Attenuated strains)	Penicillium rot (*Penicillium* spp.)	Pineapple	Tong and Rohrbock (1980)
Penicillium frequentans Westling	Brown rot (*Monilinia* spp.)	Peach	Guijarro et al. (2007)
Penicillium roqueforti and *Penicillium viridicatum*	Black rot disease (*Aspergillus niger*)	Onion	Khokhar et al. (2013)
Pestalotiopsis neglecta (Thuemen) Steyaert	Anthracnose (*Colletotrichum gloeosporioides*)	Apricot	Adikaram and Karunaratne (1998)
Pichia anomala (Hansen) Kurtzman	Penicillium rots (*Penicillium* spp.)	Citrus	Lahlali et al. (2004)
	Crown rot (*Colletotrichum musae*)	Banana	Lassois et al. (2008)
Pichia caribbica	Rhizopus rot (*Rhizopus stolonifer*)	Peach	Xu et al. (2013)
Pichia fermentans	Fruit decay (*Monilinia fructicola* and *Botrytis cinerea*)	Apple, peach	Fiori et al. (2012)

TABLE 7.1   *(Continued)*

Microbial antagonist	Disease (pathogen)	Commodity	Reference
Pichia guilliermondii	Blue mold (*Penicillium expansum*)	Apple	McLaughlin et al. (1990)
	Gray mold (*Botrytis cinerea*)	Apple	Janisiewicz et al. (1998)
	Green mold (*Penicillium digitatum*)	Citrus	Chalutz and Wilson (1990), Wilson and Chalutz (1989)
	Rhizopus rot (*Rhizopus stolonifer*)	Grape	Chalutz et al. (1988)
	Gray mold (*Botrytis cinerea*)	Grape	Chalutz et al. (1988)
	Anthracnose (*Colletotrichum capsici* (Syd.) Butler & Bisby	Chillies	Chanchaichaovivat et al. (2007)
	Gray mold (*Botrytis cinerea*)	Tomato	Chalutz et al. (1988)
	Alternaria rot (*Alternata alternata*)	Tomato	Chalutz et al. (1988)
	Rhizopus rot (*Rhizopus nigricans*)	Tomato	Zhao et al. (2008)
Pseudomonas aeruginosa (Schroter) Migula	Bacterial soft rot (*Erwinia carotovora* subsp. *carotovora*)	Cabbage	Adeline and Sijam (1999)
Pseudomonas cepacia	Blue mold (*P. expansum*)	Apple	Janisiewicz and Roitman (1988)
	Mucor rot (*Mucor piriformis*)		
	Gray mold (*Botrytis cinerea*)	Pear	Janisiewicz and Roitman (1988)
	Blue mold (*Penicillium expansum*)		
	Green mold (*Penicillium digitatum*)	Orange	Huang et al. (1993)
	Brown rot (*Monilinia fructicola*)	Nectarine	Smilanik et al. (1993)
	Brown rot (*Monilinia fructicola*)	Peach	Smilanik et al. (1993)
Pseudomonas corrugata Roberts & Scarlett	Brown rot (*Monilinia fructicola*)	Peach	Smilanik et al. (1993)
	Brown rot (*Monilinia fructicola*)	Nectarine	Smilanik et al. (1993)
Pseudomonas fluorescens Migula	Gray mold (*Botrytis mali* Ruehle)	Apple	Mikani et al. (2008)

TABLE 7.1   *(Continued)*

Microbial antagonist	Disease (pathogen)	Commodity	Reference
Pseudomonas glathei	Green mold (*Penicillium digitatum*)	Citrus	Huang et al. (1995)
Pseudomonas putida (Trevisan) Migula	Soft rot (*Erwinia carotovora* subsp. *carotovora*)	Potato	Colyer and Mount (1984)
Pseudomonas syringae	Blue mold (*Penicillium expansum*)	Apple	Janisiewicz (1987), Zhou et al. (2002)
	Green and blue mold (*Penicillium digitatum* and *P. italicum*)	Citrus	Wilson and Chalutz (1989)
	Gray mold (*Botrytis cinerea*)	Apple	Zhou et al. (2001)
	Brown rot (*Monilinia laxa*)	Peach	Zhou et al. (1999)
Pseudomonas sp.	Crown rot (*Colletotrichum musae*)	Banana	Costa and Subasinghe (1998)
Rahuella aquatilis	Gray mold (*Botrytis cinerea*)	Apple	Calvo et al. (2003, 2007)
	Blue mold (*Penicillium expansum*)	Apple	Calvo et al. (2007)
Rhodotorula glutinis	Blue mold (*Penicillium expansum*)	Apple	Zhang et al. (2009)
	Gray mold (*Botrytis cinerea*)	Apple	Zhang et al. (2009)
	Alternaria rot (*Alternata alternata*)	Jujube	Tian et al. (2005)
	Penicillium rot (*Penicillium expansum*)	Jujube	Tian et al. (2005)
	Blue rot (*Penicillium expansum*)	Pear	Zhang et al. (2008b)
	Gray mold (*Botrytis cinerea*)	Pear	Zhang et al. (2008b)
	Gray mold (*Botrytis cinerea*)	Strawberry	Zhang et al. (2007a)
Trichoderma atroviride	*Phomopsis vexans*	Brinjal	Das et al. (2014)
Trichoderma hamatum	Fungal diseases (*Phytophthora palmivora, Rhizoctonia solani, Fusarium* spp., *Sclerotium rolfsii, Pythium* sp.)	Chilly	Ha (2010), Ngullie et al. (2010)
Trichoderma harzianum	Anthracnose (*Colletotrichum musae*)	Banana	Devi and Arumuga (2005)
	Gray mold (*Botrytis cinerea*)	Grape	Batta (2007)

TABLE 7.1   *(Continued)*

Microbial antagonist	Disease (pathogen)	Commodity	Reference
	Gray mold (*Botrytis cinerea*)	Kiwifruit	Batta (2007)
	Gray mold (*Botrytis cinerea*)	Pear	Batta (2007)
	Anthracnose (*Colletotrichum gloeosporioides*)	Rambutan	Sivakumar et al. (2000)
	Brown spot (*Gliocephalotrichum microchlamydosporum* (Mey) Wiley & Simmons	Rambutan	Sivakumar et al. (2002a,b)
	Stem end rot (*Botryodiplodia theobromae*)	Rambutan	Sivakumar et al. (2001)
	Gray mold (*Botrytis cinerea*)	Strawberry	Batta (2007)
Trichoderma koningii	Alternaria diseases (*Alternaria alternata*)	In vitro	Odebode (2006), Shaikh and Nasreen (2013)
Trichoderma viride	Green mold (*Penicillium digitatum*)	Citrus	De-Matos (1983)
	Stem-end rot (*Botryodiplodia theobromae*)	Mango	Kota et al. (2006)
	Gray mold (*Botrytis cinerea*)	Strawberry	Tronsmo and Denis (1977)
	Fruit rot (*Colletotrichum gloeosporioides*)	Chilli	Ngullie et al. (2010)
Trichoderma spp.	Sour rot (*Geotrichum candidum*)	Citrus	De-Matos (1983)
	Fruit rots caused by *Lasiobasidium theobromae*, *Phomopsis psdi*, and *Rhizopus* spp.	Guava	Majumdar and Pathak (1995)
	Fruit rots (*Lasiobasidium theobromae* and *Rhizopus* spp.)	Mango	Pathak (1997)
Trichosporon pullulans (Lindner) Didlens & Lodder	Alternaria rot (*Alternata alternata*)	Cherry	Qin et al. (2004)
	Gray mold (*Botrytis cinerea*)		

7.3   CHARACTERISTICS OF AN IDEAL BIOAGENT

For biocontrol to be successful, the conditions that favor a potential antagonist should be the similar to those that favor the pathogen. The best antagonists perform well over the full spectrum of conditions conducive to pathogen development. Strains of an antagonist species can be compared for effectiveness in controlling fruit decay and for phenotypic characteristics that are useful in determining their commercial potential (Janisiewicz and Korsten, 2002). A potential microbial antagonist should have certain desirable characteristics to make it an ideal bioagent (Wilson and Wisniewski, 1989; Barkai-Golan, 2001). The antagonist should be (1) genetically stable, (2) effective at low concentrations, (3) not fastidious in its nutritional requirements, (4) capable of surviving under adverse environmental conditions, (5) effective against a wide range of the pathogens and different harvested commodities, (6) resistant to pesticides, (7) a nonproducer of metabolites harmful to human, (8) nonpathogenic to the host, (9) preparable in a form that can be effectively stored and dispensed, and (10) compatible with other chemical and physical treatments. In addition, a microbial antagonist should have an adaptive advantage over specific pathogen (Wilson and Wisniewski, 1989). For example, *Rhizopus stolonifer* is more sensitive to low temperature than many other pathogens. Thus, for its effective control, a microbial antagonist should have the ability to grow, multiply, and suppress the pathogen at low temperature. Similarly, *C. oleophila* was effective along with dicloran to reduce the incidence of *Penicillium expansum* and Rhizopus rot in nectarine even under controlled atmosphere storage conditions (Lurie et al., 1995). Most of the pome fruits are stored in cold storage. Thus, for controlling their postharvest diseases to a satisfactory level, a microbial antagonist should have the ability to survive under cold store conditions as well. Considering these factors, research work on the use of microbial antagonists for the control of postharvest diseases of fruits and vegetables has been reoriented in many countries. Accordingly, a new strain of *C. sake* Saito & Ota was isolated, which controlled *P. expansum*, *Botrytis cinerea*, and *R. stolonifer* even under various storage conditions (Vinas et al., 1996). Biocontrol agents are more acceptable if they can be applied together with current practices, and information on the compatibility of the biocontrol agents with current chemicals used in the postharvest system is generated. However, even if an antagonist has all the desirable characteristics, economic factor decides whether it has to be commercialized or not. If there is no potential market for the product, then it cannot be commercialized.

7.4   SCREENING AND ISOLATION OF BIOAGENTS

Screening and isolation of bioagents is one of the important steps involved in the development of biological control agents for managing postharvest decay of fruits and vegetables. In general, the fruit surface acts as a primary source for the naturally occurring antagonists against postharvest fruit decay. For example, searching for antagonists on healthy apples in the orchard and storage, "a place where a disease can be expected but it does not occur" (Baker and Cook, 1974), resulted in the isolation of many ecologically fit bacterial and yeast antagonists effective against postharvest decays of apple (Janisiewicz, 1987, 1991). Similar isolations from other fruits also yielded effective antagonists against postharvest fruit decay pathogens (Adikaram and Karunaratne, 1998; Arras, 1993; Borras and Aguilar, 1990; Guinebretiere et al., 2000; Huang et al., 1992; Kanapathipillai and Jantan, 1986; Zahavi et al., 2000). There are greater chances for isolation of potential antagonists from unmanaged orchards, where natural populations have not been disturbed by chemical usage (Falconi and Mendgen, 1994; Janisiewicz, 1996). A variety of enrichment procedures have been used that favor isolation of microorganisms growing efficiently on the substrate, which occurs at the infection site (wound) that must be protected. These include isolation from natural cracks on the fruit surface (W. J. Janisiewicz, unpublished), agar plates containing apple juice which were seeded with fruit washings (Janisiewicz, 1991), fruit wounds treated with fruit washings and incubated for several days (Wilson et al., 1993), freshly made wounds on apples in the orchard which were exposed to colonization by fruit-associated microbiota from 1 to 4 weeks before harvest (Janisiewicz, 1996), and from an apple juice culture resulting from seeding-diluted apple juice with the orchard colonized wounds and repeated re-inoculation to fresh apple juice.

7.5   METHODS OF APPLICATION OF BIOAGENTS

For effectiveness of any microbial antagonist, the method of application is quite important and needs to be standardized before commercial recommendation. In general, microbial antagonists are applied by two different ways, that is, preharvest application and postharvest application.

7.5.1 PREHARVEST APPLICATION

Generally, pathogen infestation occurs in the field, and these latent infections become major factor for decay of fruits and vegetables during transportation or storage. Hence, preharvest application(s) of microbial antagonistic culture are often effective to control postharvest decay of fruits and vegetables (Ippolito and Nigro, 2000; Janisiewicz and Korsten, 2002; Ippolito et al., 2004; Irtwange, 2006). Major purpose of this treatment is to precolonize the fruit surface with an antagonist immediately before harvest so that wounds inflicted during harvesting can be colonized by the antagonist before pathogen attack (Ippolito and Nigro, 2000). This approach could not become commercially viable, due to poor survivability of the microbial antagonists in the harsh field conditions; however, it has been quite successful in certain cases. For instance, the antagonists *Cryptococcus infirmominiatus* (Okanuki) Phaff & Fell, *Cryptococcus laurentii*, and *Rhodopholus glutinis* (Fresenius) Harrison, applied to "d Anjou" and "Bosc" pears in the field 3 weeks before harvest, reduced gray mold on "Bosc" pears from 13% to 4% and on "d Anjou" pear from 7% to nearly 1% (Benbow and Sugar, 1999). *C. sake* CPA-1 reduced blue mold by nearly 50% on wounded apples if inoculated with antagonist 2 days prior to harvest and inoculation with *P. expansum* and cold storage for 4 months (Teixido et al., 1999). Although it is difficult to control postharvest diseases of strawberry even with preharvest application of fungicides, some success has been achieved with field application(s) of various microbial antagonists like *Gliocladium roseum* Bainier (Sutton et al., 1997), *T. harzianum* (Kovach et al., 2000), and *Epicoccum nigrum* Link (Larena et al., 2005). The highest levels of control, however, were obtained with application of pyrrolnitrin, a secondary metabolite produced by *Pseudomonas cepacia* (Janisiewicz and Korsten, 2002). Near-harvest application of *M. fructicola* Kurtzman & Droby alone or in combination with ethanol or sodium bicarbonate controlled postharvest diseases of grapes significantly over control (Karabulut et al., 2003). Preharvest spray of *M. fructicola* Kurtzman & Droby was also effective in controlling preharvest and postharvest fruit rots in strawberry (Karabulut et al., 2004). Similarly, preharvest application of *A. pullulans* reduced storage rots in strawberry (Lima et al., 1997), grapes (Schena et al., 1999, 2003), cherries (Schena et al., 2003), and apples significantly (Leibinger et al., 1997). The incidence of green mold

(*P. digitatum*) on grapefruit was reduced by preharvest spray of *P. guillier-mondii* (Droby et al., 1992).

In citrus, preharvest application of the biocontrol yeast *P. agglomerans* CPA-2 effectively controlled postharvest rots under laboratory conditions. Similarly, preharvest application(s) of *C. laurentii* and *C. oleophila* reduced storage rots in pear (Benbow and Sugar, 1999). Field application of *E. nigrum* was reported to be effective for controlling postharvest brown rot (*Monilinia* spp.) in peaches. Canamas et al. (2008) have very recently reported that preharvest application of different concentrations of *P. agglomerans* was effective for protecting oranges [*Citrus sinensis* (L.) Obseck.] against *P. digitatum* during storage. However, it appears that this approach has still many limitations, and in commercial practice, it is used only in avocado.

7.5.2 POSTHARVEST APPLICATION

Postharvest application of microbial antagonists appears to be more practicable, useful, and better approach for controlling postharvest diseases of fruits and vegetables. In this method, microbial cultures are applied either as postharvest dips or sprays of antagonist's solution (Barkai-Golan, 2001; Irtwange, 2006). This approach was found to be more effective than preharvest application of microbial antagonists and has several successes. For example, postharvest application of *T. harzianum*, *T. viride*, *G. roseum*, and *Paecilomyces variotii* Bainier resulted in better control of *Botrytis rot* in strawberries and *Alternaria rot* in lemons than preharvest application(s) (Pratella and Mari, 1993). In lemons, postharvest application of *Pseudomonas variotii* was more effective in controlling *Aspergillus* rot than iprodion treatment, and in potatoes (*Solanum tuberosum* L.), postharvest application of *T. harzianum* controlled *Fusarium* rot effectively than benomyl dip treatment. A significant reduction in storage decay was achieved by bringing several yeast species in direct contact with wounds in the peel of harvested fruits. For instance, direct contact of microbial antagonist and infested fruit peel has been quite useful for the suppression of pathogens like *P. digitatum*, *P. italicum* in citrus (Chalutz and Wilson, 1990); *B. cinerea* in apples (Gullino et al., 1992; Wisniewski et al., 1988); *B. cinerea* and *P. expansum* in pears (Chand-Goyal and Spotts, 1997; Sugar and Spotts, 1999); and *B. cinerea*, *R. stolonifer*, and *Alternaria alternata* in tomatoes (Chalutz et al., 1988). However, all the pathogens do not react in a similar fashion to a given antagonist.

7.6 ENHANCEMENT OF THE BIO-EFFICACY OF BIOAGENTS

The efficacy of the microbial antagonists can be improved by various kinds of approaches. Various techniques such as manipulation of storage environment, use of mixed cultures, addition of salt additives and low-dose fungicides, nutrients and plant products in microbial cultures, etc., are in vogue for improving the bio-efficacy of the microbial antagonists.

7.6.1 MANIPULATION OF STORAGE ENVIRONMENT

Fruits and vegetables are usually stored at predetermined temperature, relative humidity, and gas combinations for varying periods with the primary objective of maintaining the quality to meet the market demands. Microbial antagonists should be selected only after proper screening for their ability to develop rapidly under the required storage conditions. However, manipulation of storage environment can also be a useful strategy for enhancing the efficacy of microbial antagonists, as it is possible to manipulate the physical and chemical environment to the advantage of microbial antagonists in storage (Janisiewicz and Korsten, 2002). These manipulations should, however, be such that they should not affect the quality of the produce and should be well suited for the establishment of the microbial antagonist (Dock et al., 1998; Usall et al., 2000). Fruits and vegetables are often treated and/or handled in water before, during, and after the storage which provides an excellent opportunity to modify the environment. Nitrogen is likely to be a limiting nutrient in the carbon-rich environment of apple and pear wounds, which can be increased by the addition of L-asparagine and L-proline to enhance the population of microbial antagonist *Pseudomonas syringae*. This treatment prevented blue mold decay completely as against 50% decay in control (Janisiewicz et al., 1992). The bio-efficacy of *C. sake* against *P. expansum* on apples was enhanced significantly with the addition of L-serine and L-aspartic nitrogenous compounds. In cold storage, addition of ammonium molybdate to *C. sake* entirely eliminated the incidence of blue mold on pears and reduced the severity and incidence of the disease by more than 80% on apples (Nunes et al., 2001b). The efficiency of *Rhodotorula glutinis* against *P. expansum* was enhanced by the addition of siderophores. The addition of siderophores reduces decay by sequestering iron required for germination of some postharvest pathogens (Calvente et al., 1999). The bio-efficacy of *P. syringae* for the control of crown rot and anthracnose was considerably enhanced by the addition of low doses of thiobendazole (TBZ) or imazalil (250 μg/mL), which brought control similar to higher doses of fungicides (Williamson et al., 2008).

7.6.2 MIXED CULTURES

Application of mixed cultures of microbial antagonists widens the spectrum of activity against the major postharvest pathogens. A caution of compatibility is also a must for effective control of the postharvest diseases (Barkai-Golan, 2001; El-Ghaouth et al., 2004; Singh and Sharma, 2007). Application of mixtures of microbial antagonists has certain special advantages, namely, (1) widening the spectrum of antagonists resulting in control of more than one disease; (2) increasing the effectiveness under different situations such as cultivars, maturity stages, and locations; (3) enhancing the efficiency and reliability of biocontrol as the components of the mixtures act through different mechanisms like antagonism, parasitism, and induction of resistance in the host; and (4) combination of different biocontrol traits without the transfer of alien genes through genetic transformation.

The enhancement of bio-efficacy of microbial antagonists may be due to better utilization of substrate, resulting in acceleration of the growth rate; removal of substances inhibitory to one organism by the other microbial agent; production of nutrients by one microbe that may be used by another; and formation of more stable microbial community that may exclude other microbes, including pathogens (Janisiewicz, 1998). Use of mixed strains of microbial antagonists is a challenging work, as microorganisms have different growth habits and requirements for nutrition and cultural conditions. However, some success has been achieved in this area as well. For instance, a combination of the bacteria *P. syringae* and the yeast *Sporobolomyces roseus* proved to have a marked advantage over each of the antagonists in controlling *P. expansum* in apple, both in reducing the incidence of wound infections and in limiting rot diameter (Janisiewicz and Bors, 1995). The advantage of antagonistic pairs over a single antagonist was described by Schisler et al. (1997) in the control of *Fusarium* dry rot (*Gibberella pulcaris* Hohn & Desjardins) in stored potatoes. The black rot of pineapple [*Ananas comosus* (L.) Merrrill], caused by *Ceratomyces paradoxa* (Dade) Moreau, could be controlled by the yeast *P. guilliermondii*; its combination with five yeast isolates was still more effective and the level of control was comparable to current industry practice of holding fruit at a low temperature (8–10°C) (Reyes et al., 2004).

The efficiency of an antagonist is affected both by the concentration of the yeast cells in the wound and by the number of pathogen spores used for inoculation. In apple, a broader spectrum of pathogens was controlled when microbial antagonists were applied in mixtures than individual microbial strains (Calvo et al., 2003; Conway et al., 2005). In potato, antagonist pairs effectively controlled *Fusarium* dry rot over their single use (Schisler et al., 1997). Mixed cultures of *C. sake* and *P. agglomerans* gave better control on blue and gray mold both in apple and pears than their individual use (Nunes et al., 2002b). Janisiewicz et al. (2008) reported that mixed cultures of *M. pulcherrima* and *C. laurentii* exhibited greater biocontrol activity on blue mold (*P. expansum*) than either yeast applied alone, or in combination with sodium carbonate or bicarbonate in a pilot test conducted on citrus in controlled atmosphere. Although the use of antagonistic mixtures offers more effective control, the economic viability of this approach appears to be a major obstacle for its adoption, as registration of two microbial antagonists will cause additional burden for the industry.

7.6.3 USE OF ADDITIVES

Various kinds of additives, namely, salts, nutrients, plant products, and low-dose fungicides, are known to enhance the bio-efficacy of the microbial antagonists in controlling the postharvest diseases of fruits and vegetables compared to use as sole antagonists. Among various salt additives, calcium chloride, calcium propionate, sodium carbonate, sodium bicarbonate, potassium metabisulphite, ethanol, ammonium molybdate, etc., have been found to be very successful when used with microbial antagonists for controlling postharvest diseases of fruits and vegetables (Plaza et al., 2001; Teixido et al., 2001; Tian et al., 2002; Xi and Tian, 2005; Zhang et al., 2005; Qin et al., 2006; Torres et al., 2007; Cao et al., 2008) (Table 7.2). However, the effectiveness of microbial antagonists depends upon the concentration of the antagonist, concentration of salt additive(s), their mutual compatibility, and duration and time of application. Usually, the cultures should be applied well before the initiation of infection process (Barkai-Golan, 2001).

TABLE 7.2   Salt Additives for Improving Bio-efficacy of Microbial Antagonists (Sharma et al., 2009).

Fruit	Salt additive	Microbial agent	Disease	References
Apple	Calcium chloride	*Candida* spp.	Gray and blue molds	Wisniewski et al. (1995)
	Sodium carbonate	*Metschnikowia pulcherrima* and *Cryptococcus laurentii*	Blue mold	Conway et al. (2007), Janisiewicz et al. (2008)
	Calcium propionate	Aspire (*Candida oleophila*)	Blue mold	Droby et al. (2003)
	Sodium bicarbonate	Aspire (*Candida oleophila*)	Blue mold	Droby et al. (2003)
Pear	Sodium carbonate	*Cryptococcus laurentii*	Blue mold and Alternaria rot	Yao et al. (2004)
	Sodium carbonate	*Trichosporon pullulans*	Blue mold and Alternaria rot	Yao et al. (2004)
	Calcium chloride	*Candida saitona*	Gray and blue molds	McLaughlin et al. (1990), Wisniewski et al. (1995)
	Calcium chloride	*Cryptococcus laurentii*	Gray mold rot	Zhang et al. (2005)
	Ammonium molybdate	*Rhodotorula glutinis*	Blue mold	Wan and Tian (2005)
	Ammonium molybdate	*Trochosporon* spp.	Alternaria rot	Wan and Tian (2005)
Peach	Calcium chloride	*Debaryomyces hansenii*	Rhizopus rot	Singh (2004, 2005)
	Calcium propionate	Aspire	Brown rot	Droby et al. (2003)
	Sodium bicarbonate	Aspire	Rhizopus rot	Droby et al. (2003)

TABLE 7.2 *(Continued)*

Fruit	Salt additive	Microbial agent	Disease	References
Cherry	Ammonium molybdate	*Pichia membranaefaciens*	Brown rot	Qin et al. (2006)
		Crptococcus laurentii	Brown rot	Qin et al. (2006)
	Calcium chloride	*Aureobasidium pullulans*	Brown rot	Ippolito et al. (2005)
	Sodium bicarbonate	*Aureobasidium pullulans*	Brown rot	Karabulut et al. (2005)
	Potassium sorbate	*Candida oleophila*	Postharvest decay	Karabulut et al. (2001)
Grapefruit	Calcium chloride	*Pichia guilliermondii*	Green mold	Droby et al. (1997)
Oranges	Calcium chloride	*Pseudomonas syringae*	Blue mold	Janisiewicz et al. (1998)
	Calcium chloride	*Candida oleophila*	Penicillium rots	El-Neshawy and El-Sheikh (1998)
	Sodium carbonate	*Pseudomonas syringae*	Green mold	Smilanick et al. (1999)
	Sodium bicarbonate	*Pseudomonas syringae*	Green mold	Plaza et al. (2001)
Grape	Sodium bicarbonate	*Metschnikowia fruticola*	Botrytis rot	Karabulut et al. (2003)
Citrus	Sodium carbonate	*Cryptococcus laurentii*	Green mold	Zhang et al. (2004), Usall et al. (2008)
	Sodium bicarbonate	*Bacillus subtilis*	Green and blue molds	Obagwu and Korsten (2003)
	Sodium bicarbonate	*Pantoea agglomerans*	Penicillium rots	Plaza et al. (2001), Torres et al. (2007), Usall et al. (2008)
Papaya	Sodium bicarbonate	*Candida oleophila*	Anthracnose	Gamagae et al. (2003)
Loquat	Calcium chloride	*Pichia membranifaciens*	Anthracnose	Cao et al. (2008)
Rambutan	Potassium metabisulphite	*Trichoderma* spp.	Postharvest rots	Sivakumar et al. (2002a,b)
Tomato	Sodium bicarbonate	*Cryptococcus laurentii*	Botrytis rot	Xi and Tian (2005)

Source: Reprinted/adapted with permission from Sharma, R. R.; Singh, D.; Singh, R. Biological Control of Postharvest Diseases of Fruits and Vegetables by Microbial Antagonists: A Review. *Biol. Cont.* **2009**, *50*, 205–221. © 2009 Elsevier.

Addition of certain nutrient compounds or natural plant extracts is known to enhance the bio-efficacy of microbial antagonists. Addition of nitrogenous compounds like L-aspargine and L-proline, and 2-deoxy-D-glucose, a sugar analog, helped in enhancing the bio-efficacy of microbial antagonists in controlling the postharvest decay rots in some fruits and vegetables (El-Ghaouth et al., 2000a,b). When applied in fruit wounds, the combination of *C. saitona* and 2-deoxy-D-glucose (0.2%) controlled fruit decay on apples, oranges, and lemons caused by *B. cinerea*, *P. expansum*, and *P. digitatum* (El-Ghaouth et al., 2000a,b) than when either *C. saitona* or 2-deoxy-D-glucose was applied alone. The treatment of peaches with *C. laurentii* (1 × 10^8 CFU/mL) alone or in combination with methyl jasmonate (200 µM/L) inhibited the lesion diameter of brown rot and blue mold rots caused by *Monilinia fructicola* and *P. expansum*, respectively (Yao and Tian, 2005).

Addition of low doses of synthetic fungicides to the microbial antagonists is found to be highly effective providing nearly 100% control of postharvest diseases (Wisniewski et al., 2001). Applying *P. guilliermondii* to citrus fruit in combination with substantially reduced concentration of TBZ reduced *P. digitatum* decay to a level similar to that achieved by currently recommended concentration of TBZ application alone (Droby et al., 1993), which helps in maintaining very low level of chemical residue in the fruit (Hofstein et al., 1994). Mixing *P. syringae* with low doses of cypronidil brought effective control in decay caused by *P. expansum* on apples, and pear decay in storage was reduced significantly by combining low doses of fungicides with biocontrol agent (Errampalli and Brubacher, 2006). Chand-Goyal and Spotts (1997) also reported control of blue mold on apple and brown rot on pear when yeasts were used with a low dose of a fungicide. Similarly, fruit decay in citrus was controlled effectively with *C. oleophila* + TBZ as comparable to commercial fungicide treatment (Droby et al., 1998). Zhou et al. (2002) achieved over 90% control in blue and gray mold rots on apples by treating the fruit with cypronidil (20 ppm) and *P. syringae* (3 × 10^7 CFU/mL). Similarly, *C. laurentii* + imazalil (25 ppm) treatment was highly effective in controlling storage rots of jujube than applying *C. laurentii* or imazalil alone (Qin and Tian, 2004).

7.7   MODE OF ACTION OF BIOCONTROL ANTAGONISTS

The mechanism(s) by which microbial antagonists exert their influence on the pathogens has not yet been fully understood. It is important to understand the mode of action of the microbial antagonists because it will help in developing some additional means and procedures for better results from the known antagonists and it will also help in selecting more effective and desirable antagonists or strains of antagonists (Wisniewski and Wilson, 1992). Several modes of action have been suggested to explain the biocontrol activity of microbial antagonists (Table 7.3). Still, competition for nutrient and space between the pathogen and the antagonist is considered as the major modes of action by which microbial agents control pathogens causing postharvest decay (Droby et al., 1992; Wilson et al., 1993; Ippolito et al., 2000). In addition, production of antibiotics (antibiosis), direct parasitism, and possibly induced resistance are other modes of action of the microbial antagonists by which they suppress the activity of postharvest pathogens on fruits and vegetables (Barkai-Golan, 2001; El-Ghaouth et al., 2004).

TABLE 7.3 Mode of Action of Microbial Antagonists Used in Controlling Postharvest Diseases of Perishable Commodities (Sharma et al., 2009).

Commodity	Disease	Antagonist	Reference
Antibiotic production			
Apple	Blue mold	*Pseudomonas cepacia*	Janisiewicz and Roitman (1988)
	Mucor rot	*Pseudomonas cepacia*	Janisiewicz and Roitman (1988)
	Gray mold	*Pseudomonas syringae*	Bull et al. (1998)
Apricot	Brown rot	*Bacillus subtilis*	Pusey et al. (1988)
Cherry	Brown rot	*Bacillus subtilis*	Utkhede and Sholberg (1986)
	Alternaria rot	*Enterobacter aerogenes*	Utkhede and Sholberg (1986)
Citrus	Sour rot	*Bacillus subtilis*	Singh and Deverall (1984)
	Green mold	*Bacillus subtilis*	Singh and Deverall (1984)
	Green mold	*Pseudomonas syringae*	Bull et al. (1998)
	Stem-end rot		Singh and Deverall (1984)
	Sour rot	*Trichoderma* spp.	De-Matos (1983)
Nectarine	Brown rot	*Bacillus subtilis*	Pusey et al. (1988)
	Brown rot	*Pseudomonas corrupta*	Smilanik et al. (1993)
Peach	Brown rot	*Bacillus subtilis*	Pusey et al. (1988)
	Brown rot	*Pseudomonas cepacia*	Smilanik et al. (1993)
Pear	Blue mold	*Pseudomonas cepacia*	Janisiewicz and Roitman (1988)
	Gray mold	*Pseudomonas cepacia*	Janisiewicz and Roitman (1988)
Plum	Brown rot	*Bacillus subtilis*	Pusey et al. (1988)

TABLE 7.3 *(Continued)*

Commodity	Disease	Antagonist	Reference
Nutritional competition (N) and/or induced host resistance (HR)			
Apple	Blue mold	*Pseudomonas cepacia* (HR)	Janisiewicz (1987)
	Gray mold	*Acremonium brevae* (HR)	Janisiewicz (1988)
	Gray mold	*Debaryomyes hansenii* (N + HR)	Wisniewski et al. (1988), Roberts (1990a)
	Gray mold	*Cryptococcus humicola* Fricke (N)	Filonow et al. (1996)
	Gray mold	*Aureobasidium pullulans* (N + HR)	Ippolito et al. (2000), Castoria et al. (2001)
	Blue mold	*Aureobasidium pullulans* (N + HR)	Ippolito et al. (2000), Castoria et al. (2001)
	Blue mold	*Aureobasidium pullulans* (N)	Bencheqroun et al. (2007)
Citrus	Green mold	*Debaryomyces hansenii* (N + HR)	Chalutz and Wilson (1990), Droby et al. (1992)
	Blue mold	*Debaryomyces hansenii* (N + HR)	Chalutz and Wilson (1990)
	Sour rot	*Debaryomyces hansenii* (N + HR)	Chalutz and Wilson (1990)
Grapes	Gray mold	*Debaryomyces hansenii* (N)	Chalutz et al. (1988)
	Rhizopus rot	*Debaryomyces hansenii* (N)	Chalutz et al. (1988)
	Gray mold	*Aureobasidium pullulans* (N + HR)	Castoria et al. (2001)
	Blue mold	*Aureobasidium pullulans* (N + HR)	Castoria et al. (2001)
	Rhizopus rot	*Aureobasidium pullulans* (N + HR)	Castoria et al. (2001)
Peach	Rhizopus rot	*Enterobacter cloacae* (N)	Wisniewski et al. (1988)
Strawberry	Gray mold	*Cryptococcus laurentii* (N)	Castoria et al. (1997)
	Gray mold	*Rhodotorula glutinis* (N)	Castoria et al. (1997)
Tomato	Rhizopus rot	*Debaryomyces hansenii* (N)	Chalutz et al. (1988)
	Gray mold	*Debaryomyces hansenii* (N)	Chalutz et al. (1988)
	Alternaria rot	*Debaryomyces hansenii* (N)	Chalutz et al. (1988)

7.7.1 COMPETITION FOR NUTRIENTS AND SPACE

Competition for nutrition and space between the microbial antagonist and the pathogen is considered as the major mode of action by which microbial antagonists suppress pathogens causing decay in harvested fruits and vegetables (Wilson and Wisniewski, 1989; Liu et al., 2012). To compete successfully with pathogen at the wound site, the microbial antagonist should be better adapted to various environmental and nutritional conditions than the pathogen (Barkai-Golan, 2001; El-Ghaouth et al., 2004).

The biocontrol activity of microbial antagonists with most harvested commodities increased with the increasing concentrations of antagonists and decreasing concentrations of pathogen. For example, *C. saitona* was effective at a concentration of 10^7 CFU/mL for controlling *P. expansum* on apples (McLaughlin et al., 1990). Attachment by microbial antagonist to the pathogen hyphae appears to be an important factor necessary for competition for nutrients as shown by the interactions of *Enterobacter cloacae* (Jordon) Hormaeche & Edwards and *R. stolonifer* (Ehrenberg: Fries) Lind (Wisniewski et al., 1989), and *P. guilliermondii* Wickerham and *P. italicum* Wehmer (Arras et al., 1998). In vitro studies conducted on such interactions revealed that due to direct attachment, antagonistic yeasts and bacteria take nutrients more rapidly than target pathogens and thereby prevent spore germination and growth of the pathogens (Wisniewski et al., 1989).

Biocontrol of gray mold (*B. cinerea*) on apple by *M. pulcherrima* was reduced or totally suppressed by the addition of several nutrients suggesting that competition for nutrients plays a role in the biocontrol capability of *M. pulcherrima* against *B. cinerea* (Piano et al., 1997). *M. pulcherrima* out competes pathogens like *B. cinerea* and *P. expansum* in apple through iron depletion (Saravanakumar et al., 2008). As a result of its ability for suppressing postharvest diseases, Kurtzman and Droby (2001) and Grebenisan et al. (2008) have recommended it as potential yeast for controlling fruit rots. The level of control provided by the microbial antagonists is also highly dependent on the initial concentration of the antagonists applied on the wound site and the ability of the antagonist to rapidly colonize the wound site (Wisniewski et al., 1989; McLaughlin et al., 1990). In general, microbial antagonists are most effective in controlling postharvest decay on fruits and vegetables when applied at a concentration of 10^7–10^8 CFU/mL (El-Ghaouth et al., 2004), and rarely, higher concentrations are required.

7.7.2 ANTIBIOTIC PRODUCTION

Production of antibiotics is one of the major mechanisms by which microbial antagonists suppress the pathogens of harvested fruits and vegetables. For instance, bacterial antagonists like *B. subtilis* and *P. cepacia* Burkh are known to kill pathogens by producing the antibiotic iturin (Gueldner et al., 1988; Pusey, 1989). The antagonism so produced by *B. subtilis* was effective in controlling fungal rot in citrus (Singh and Deverall, 1984) and *M. fructicola* (Winter) Honey in peaches and cherries (Utkhede and Sholberg, 1986). Similarly, the bacterial antagonist, *P. syringae* van Hall, controlled green mold of citrus and gray mold of apple, by producing an antibiotic syringomycin (Bull et al., 1998).

Although antibiosis might be an effective tool for controlling postharvest diseases in a few fruits and vegetables, at present, emphasis is being given for the development of nonantibiotic producing microbial antagonists for the control of postharvest diseases of fruits and vegetables (El-Ghaouth et al., 2004; Singh and Sharma, 2007). Researchers are aiming to isolate, evaluate, and/or develop those antagonistic microorganisms that control postharvest diseases of harvested commodities by the mechanism of competition for space and nutrient, direct parasitism, or induced resistance (Droby, 2006).

7.7.3 DIRECT PARASITISM

Meagre information is available on direct parasitism of microbial antagonists for controlling postharvest diseases. Wisniewski et al. (1991) observed that *P. guilliermondii* cells had the ability to attach to the hyphae of *B. cinerea* and *Penicillium*. After yeast cells were dislodged from the hyphae, the hyphal surface appeared to be concave and there was partial degradation of the cell wall of *B. cinerea* at the attachment sites. Microbial antagonists also produce lytic enzymes such as gluconase, chitinase, and proteinases that help in the cell-wall degradation of the pathogenic fungi (Castoria et al., 1997, 2001; Mortuza and Ilag, 1999; Chernin and Chet, 2002). Bonaterra et al. (2003) reported that direct parasitism was a major factor that permitted *P. agglomerans* (Ewing & Fife) to control *M. laxa* (Aderh. & Ruhl.) Honey or *R. stolonifer* decay on stone fruits. Thus, strong attachment of microbial antagonist with enhanced activity of cell-wall degradation enzymes may be responsible for enhancing the efficacy of microbial agents in controlling the postharvest diseases of fruits and vegetables (Wisniewski et al., 1991). And, attachment of the microbial antagonists to a site enhances their potential activity for the utilization of nutrients at the invasion site; it partly affects the access of the pathogen to nutrients as well (El-Ghaouth et al., 2004).

7.7.4 INDUCED RESISTANCE

Induction of defense responses in the harvested fruits and vegetables by the microbial antagonists has been suggested and shown as another mode of action of microbial antagonists for controlling postharvest decay in them (Ippolito et al., 2000). For example, *Cryptococcus saitona* induced chitinase activity and formed structural barrier (papillae) on host cell walls in apple against *P. expansum* (El-Ghaouth et al., 1998). Similarly, *A. pullulans* caused a transient increase in the activity of 1,3-gluconase, peroxidase, and chitinase enzymes in apple wounds which stimulated wound healing processes and induced defense mechanisms against *P. expansum* (Ippolito et al., 2000). Microbial antagonists induced disease resistance in the harvested commodities by the production of antifungal compounds, as in avocado (*Persea Americana* Mill) fruit (Yakoby et al., 2001), and accumulation of phytoalexins, like scoparone and scopoletin in citrus fruit (Rodov et al., 1994; Arras, 1996). Production of such antifungal compounds by microbial antagonists in the host cells helps in inducing defense mechanism and hence provides biocontrol on the harvested commodities.

7.8 POTENTIAL FOR COMMERCIALIZATION

Though a large number of microbial antagonists are reported to control the postharvest diseases under laboratory conditions, only few were commercialized subjecting to various limitations in the field level, namely, (1) their relative ineffectiveness compared to the synthetic chemicals and (2) lack of economic incentives (Wilson and Wisniewski, 1989). However, there are quite a good number of developed and approved commercial biocontrol products like Trichodermil, Bio-tricho, Supresivit, Eco-77, Trichodex (*T. harzianum*), Trichdermax EC, Ecohope, Quality WG, Trichotech (*Trichoderma asperellum*), Trichospray, Trichopel, Trichodry, Vinevax (*T. atroviride*), Remedier WP (*Trichoderma gamsii*), Biocure F, Bio-shield, Binab T (*Trichoderma viride*), BW240 G, BW240WP, G-41 technical (*Trichoderma virens*), and Floragard (*Trichoderma hamatum*) (Kabaluk et al., 2010; Bettiol et al., 2012; Woo et al., 2014; Mishra et al., 2013). Products, namely, Aspire™ and BioSave™, lead the way in commercial application of biocontrol agents to fruits. Other products such as YieldPlus™ based on *Cryptococcus albidus*, Avogreen™ based on *B. subtilis*, and Shemer™ based on *M. fructicola* are also on the market in various countries (Liu et al., 2011a). Aspire™ has been registered in the United States for postharvest application to citrus and pome fruits. This product was taken off the market 3 years after its large-scale commercial introduction. BioSave™ (two strains of *P. syringae*) was originally registered for postharvest application to pome and citrus fruits, and this was later extended to cherries, potatoes, and more recently to sweet potatoes. YieldPlus™ was developed in South Africa for postharvest application to pome fruits but the success of this product is largely unknown and there is no published literature or information available to determine extent of its use. Avogreen™ has been used for control of postharvest disease of avocado but its usage has been limited due to inconsistent results. More recently, Shemer™ was registered in Israel for both pre- and postharvest application on various fruits and vegetables including apricot, citrus, grapes, peach, pepper, strawberry and sweet potato. There are three more products coming to the market: Candifruit™ based on *C. sake*, developed in Spain; Boni-Protect®, based on *A. pullulans*, developed in Germany; and NEXY, based on *C. oleophila*, developed in Belgium. Most of these products have been registered for control of postharvest diseases of pome fruits (Table 7.4).

TABLE 7.4 Commercialized Biocontrol Products Available for Controlling Postharvest Diseases (Sharma et al., 2009).

Product	Microbial antagonist	Commodity	Disease controlled	Manufacturer/ distributer
AQ-10 bio-fungicide	*Ampelomyces quisqualis* Cesati ex. Schlechtendahl	Apples, grapes, strawberries, tomatoes, and cucurbits	Powdery mildew	Ecogen, Inc., USA
Aspire	*Candida oleophila* strain 1–182	Apple, pear and citrus	Blue, gray, and green molds	Ecogen, Inc., USA
Biosave 10LP, 110	*Pseudomonas syringae* strain 10 LP, 110	Apple, pear, citrus, cherries, and potatoes	Blue and gray mold, mucor, and sour rot	Eco Science Corporation, USA
Blight Ban A 506	*Pseudomonas fluorescence* A 506	Apple, pear, strawberries and potatoes	Fire blight and soft rots	Nu Farm, Inc., USA

TABLE 7.4 (*Continued*)

Product	Microbial antagonist	Commodity	Disease controlled	Manufacturer/ distributer
Contains WG, Intercept WG	*Coniothyrium minitans* Campbell	Onion	Basal and neck rots	Prohyta Biologischer, Germany
Messenger	*Erwinia amylovora* (Burrill)	Vegetables	Fire blight	EDEN Bioscience Corporation, USA
Rhio-plus	*Bacillus subtilis* FZB 24	Potatoes and other vegetables	Powdery mildew and root rots	KFZB Biotechnick, Germany
Serenade	*Bacillus subtilis*	Apple, pear, grapes, and vegetables	Powdery mildew, late blight, brown rot, and fire blight	Agro Quess Inc., USA

Source: Reprinted/adapted with permission from Sharma, R. R.; Singh, D.; Singh, R. Biological Control of Postharvest Diseases of Fruits and Vegetables by Microbial Antagonists: A Review. *Biol. Cont.* **2009,** *50*, 205–221. © 2009 Elsevier.

7.9 CONCLUSION

Traditionally, postharvest diseases are being controlled through application of manmade fungicides, and with the increasing awareness of the end user toward the health hazards and environment pollution, there is a great need for developing alternate strategies for controlling postharvest diseases of perishable horticultural commodities such as fruits and vegetables. Among various alternative approaches available, application of microbial antagonists was found to be the most suitable method considering its eco-friendly nature. This method has certain limitations under definite circumstances, which can be effectively addressed, as this method is amenable to manipulations such as modification of the environment, use of mixed cultures, manipulation of formulations, use of additives, etc.

The postharvest system has a well-defined environment, giving a unique opportunity for using microbial antagonists in the delivery system. In future, scientists could try developing strains adapted to specific postharvest conditions and introduce genes of biocontrol activity as per the specific requirements. However, this strategy of using biocontrol agents is still in infancy compared to the chemical treatment methods, but the progress made in this area during the last two decades is noteworthy. Some biocontrol antagonists can be lethal and/or cause environmental pollution, and therefore, the key usage of this technology is directly linked to the biosafety and environmental impact of these biocontrol agents. Hence, it is vital to carry out more research on lesser known aspects of biological control including development of novel formulations from them, etc., which needs substantial increase in funding and cooperation from the government bodies.

KEYWORDS

- **postharvest handling**
- **storage environment**
- **biological control**
- **fungicides**
- **microbial antagonist**

REFERENCES

Adeline, T. S. Y.; Sijam, K. Biological Control of Bacterial Soft Rot of Cabbage. In *Biological Control in the Tropics: Towards Efficient Biodiversity and Bioresource Management for Effective Biological Control: Proceedings of the Symposium on Biological Control in the Tropics*; Hong, L. W., Sastroutomo, S. S., Caunter, I. G., Ali, J., Yeang, L. K., Vijaysegaran, S., Sen, Y. H., Eds.; CABI Publishing, Wallingford, UK, 1999; pp 133–134.

Adikaram, N. K. B.; Karunaratne, A. Suppression of Avocado Anthracnose and Stem-End Rot Pathogens by Endogenous Antifungal Substances and a Surface Inhabiting *Pestalotiopsis* sp. *ACIAR Proc. Ser.* **1998,** *80,* 72–77.

Al-Mughrabi, K. I. Biological Control of *Fusarium* Dry Rot and Other Potato Tuber Diseases Using *Pseudomonas fluorescens* and *Enterobacter cloacae. Biol. Cont.* **2010,** *53,* 280–284.

Appel, D. J.; Gees, R.; Coffey, M. D. Biological Control of the Postharvest Pathogen *Penicillium digitatum* on Eureka Lemons. *Phytopathology* **1988,** *12,* 1595–1599.

Arras, G. Inhibition of Postharvest Fungal Pathogens by *Bacillus subtilis* Strains Isolated from Citrus Fruit. *Adv. Hortic. Sci.* **1993,** *7,* 123–127.

Arras, G. Mode of Action of an Isolate of *Candida famata* in Biological Control of *Penicillium digitatum* in Orange Fruit. *Postharvest Biol. Technol.* **1996,** *8,* 191–198.

Arras, G.; de-Cicco, V.; Arru, S.; Lima, G. Biocontrol by Yeasts of Blue Mold of Citrus Fruits and the Mode of Action of an Isolate of *Pichia guilliermondii. J. Hortic. Sci. Biotechnol.* **1998,** *73,* 413–418.

Baker, K. F.; Cook, R. J. *Biological Control of Plant Pathogens.* W. H. Freeman and Company, San Francisco, 1974; p 433.

Barkai-Golan, R. *Postharvest Diseases of Fruit and Vegetables: Development and Control.* Elsevier Sciences: Amasterdam, The Netherlands, 2001.

Batta, Y. A. Control of Postharvest Diseases of Fruit with an Invert Emulsion Formulation of *Trichoderma harzianum* Rifai. *Postharvest Biol. Technol.* **2007**, *43* (1), 143–150.

Benbow, J. M.; Sugar, D. Fruit Surface Colonization and Biological Control of Postharvest Diseases of Pear by Preharvest Yeast Applications. *Plant Dis.* **1999**, *83*, 839–844.

Bencheqroun, S. M.; Bajji, M.; Massart, S.; Labhilili, M.; El-Jaafari, S.; Jijakli, M. H. *In Vitro* and *In Situ* Study of Postharvest Apple Blue Mold Biocontrol by *Aureobasidium pullulans*: Evidence for the Involvement of Competition for Nutrients. *Postharvest Biol. Technol.* **2007**, *46* (2), 128–135.

Bettiol, W.; Morandi, M. A. B.; Pinto, Z. V.; De Paula, T. J.; Corrêa, E. B.; Moura, A. B.; Lucon, C. M. M.; Costa, J.; Bezerra, J. L. Produtos comerciais f base de agentes de biocontrole de doençasdeplantas. *1 Aediçãoeletrônica 2012.* 2012. http://ainfo.cnptia.embrapa.br/digital/bitstream/item/66628/1/Doc-88-1.pdf [accessed 10 August 2015].

Blum, L. E. B.; Amarante, C. V. T.; Valdebenito-Sanhueza, R. M.; Guimaraes, L. S.; Dezanet, A.; Hack-Neto, P. Postharvest Application of *Cryptococcus laurentii* Reduces Apple Fruit Rots. *Fitopatol. Bras.* **2004**, *29*.

Bonaterra, A.; Mari, M.; Casalini, L.; Montesinos, E. Biological Control of *Monilinia laxa* and *Rhizopus stolonifer* in Postharvest of Stone Fruit by *Pantoea agglomerans* EPS125 and Putative Mechanisms of Antagonism. *Int. J. Food Microbiol.* **2003**, *84* (1), 93–104.

Borras, A. D.; Aguilar, R. V. Biological Control of *Penicillium digitatum* by *Trichiderma viride* on Postharvest Citrus Fruits. *Int. J. Food Microbiol.* **1990**, *11*, 179–184.

Bull, C. T.; Wadsworth, M. L. K.; Sorenson, K. N.; Takemoto, J.; Austin, R.; Smilanick, J. L. Syringomycin E Produced by Biological Agents Controls Green Mold on Lemons. *Biol. Cont.* **1999**, *12*, 89–95.

Calvente, V.; Benuzzi, D.; de Tosetti, M. I. D. Antagonistic Action of Siderophores from *Rhodotorula glutinis* upon the Postharvest Pathogen *Penicillium expansum.* *Int. Biodeter. Biodegrad.* **1999**, *43*, 167–172.

Calvo, J.; Calvente, V.; de Orellano, M. E.; Benuzzi, D.; de Tosetti, M. I. S. Biological Control of Postharvest Spoilage Caused by *Penicillium expansum* and *Botrytis cinerea* in Apple by Using the Bacterium *Rahnella aquatilis.* *Int. J. Food Microbiol.* **2007**, *113* (3), 251–257.

Calvo, J.; Calvente, V.; Orellano, M.; Benuzzi, D.; Sanz-de-Tosetti, M. I. Improvement in the Biocontrol of Postharvest Diseases of Apples with the Use of Yeast Mixtures. *Biol. Cont.* **2003**, *48* (5), 579–593.

Calvo-Garrido, C.; Elmer, P. A. G.; Vinas, I.; Usall, J.; Bartra, E.; Teixido, N. Biological Control of Botrytis Bunch Rot in Organic Wine Grapes with the Yeast Antagonist *Candida sake* CPA-1. *Plant Pathol.* **2013**, *62* (3), 510–519.

Canamas, T. P.; Vinas, I.; Usall, J.; Torres, R.; Anguera, M.; Teixido, N. Control of Postharvest Diseases on Citrus Fruit by Preharvest Applications of Biocontrol Agent *Pantoea agglomerans* CPA-2: Part II. Effectiveness of Different Cell Formulations. *Postharvest Biol. Technol.* **2008**, *49* (1), 96–106.

Cao, S.; Zheng, Y.; Tang, S.; Wang, K. Improved Control of Anthracnose Rot in Loquat Fruit by a Combination Treatment of *Pichia membranifaciens* with $CaCl_2$. *Int. J. Food Microbiol.* **2008**, *126* (1–2), 216–220.

Castoria, R.; Curtis, F.; Lima, G.; Cicco, V. β-1,3-Glucanase Activity of Two Saprophytic Yeasts and Possible Mode of Action as Biocontrol Agents against Postharvest Diseases. *Postharvest Biol. Technol.* **1997**, *12* (3), 293–300.

Castoria, R.; de Curtis, F.; Lima, G.; Caputo, L.; Pacifico, S.; de Cicco, V. *Aureobasidium pullulans* (LS-30), an Antagonist of Postharvest Pathogens of Fruits: Study on Its Mode of Action. *Postharvest Biol. Technol.* **2001**, *32*, 717–724.

Chalutz, E.; Ben-Arie, R.; Droby, S.; Cohen, L.; Weiss, B.; Wilson, C. L. Yeasts as Biocontrol Agents of Postharvest Diseases of Fruit. *Phytoparasitica* **1988**, *16*, 69–75.

Chalutz, E.; Wilson, C. L. Postharvest Biocontrol of Green and Blue Mold and Sour Rot of Citrus Fruit by *Debaryomyces hansenii*. *Plant Dis.* **1990**, *74*, 134–137.

Chanchaichaovivat, A.; Ruenwongsa, P.; Panijpan, B. Screening and Identification of Yeast Strains from Fruit and Vegetables: Potential for Biological Control of Postharvest Chilli Anthracnose (*Colletotrichum capscii*). *Biol. Cont.* **2007**, *42* (3), 326–335.

Chand-Goyal, T.; Spotts, R. A. Biological Control of Postharvest Diseases of Apple and Pear under Semi-commercial and Commercial Conditions Using Three Saprophytic Yeasts. *Biol. Cont.* **1997**, *10* (3), 199–206.

Chernin, L.; Chet, I. *Microbial Enzymes in the Biocontrol of Plant Pathogens and Pests.* In *Enzymes in the Environment: Activity, Ecology, and Applications*; Burns, R. G.; Dick, R. P., Eds.; Marcel Dekker Inc.: New York, NY, 2002.

Coates, L.; Johnson, G. Postharvest Diseases of Fruit and Vegetables. *Plant Pathol. Plant Dis.* **1997**, *33*, 533–548.

Colyer, P. D.; Mount, M. D. Bacterization of Potatoes with *Pseudomonas putida* and Its Influence on Postharvest Soft Rot Diseases. *Plant Dis.* **1984**, *68*, 703–706.

Conway, W. S.; Janisiewicz, W. J.; Leverentz, B.; Saftner, R. A.; Camp, M. J. Control of Blue Mold of Apple by Combining Controlled Atmosphere, an Antagonist Mixture, and Sodium Bicarbonate. *Postharvest Biol. Technol.* **2007**, *45* (3), 326–332.

Conway, W. S.; Leverentz, B.; Janisiewicz, W. J.; Blodgett, A. B.; Saftner, R. A.; Camp, M. J. Improving Biocontrol Using Antagonist Mixtures with Heat and/or Sodium Bicarbonate to Control Postharvest Decay of Apple Fruit. *Postharvest Biol. Technol.* **2005**, *36* (3), 235–244.

Costa, D. M.; Erabadupitiya, H. R. U. T. An Integrated Method to Control Postharvest Diseases of Banana Using a Member of the *Burkholderia cepacia* Complex. *Postharvest Biol. Technol.* **2005**, *36* (1), 31–39.

Costa, D. M.; Subasinghe, S. S. N. S. Antagonistic Bacteria Associated with the Fruit Skin of Banana in Controlling its Postharvest Diseases. *Trop. Sci.* **1998**, *38* (4), 206–212.

Dalal, J.; Kulkarni, N. Antagonistic and Plant Growth Promoting Potentials of Indigenous Endophytic Bacteria of Soybean (*Glycine max* (L) Merril). *Curr. Res. Microbiol. Biotechnol.* **2013**, *1* (2), 62–69.

Das, S. N.; Sarma, T. C.; Tapadar, S. A. In Vitro Evaluation of Fungicides and Two Species of *Trichoderma* against *Phomopsis vexans* Causing Fruit Rot of Brinjal (*Solanum melongena* L.). *Int. J. Scientific Res. Publ.* **2014**, *4* (9), 1–2.

De-Matos, A. P. Chemical and Microbiological Factors Influencing the Infection of Lemons by *Geotrichum candidum* and *Penicillium digitatum*. *Ph.D. Thesis*, University of California: Riverside, CA, 1983; p 106.

Demoz, B. T.; Korsten, L. *Bacillus subtilis* Attachment, Colonization, and Survival on Avocado Flowers and Its Mode of Action on Stem-End Rot Pathogens. *Biol. Cont.* **2006**, *37* (1), 68–74.

Devi, A. N.; Arumugam, T. Studies on the Shelf Life and Quality of Rasthali Banana as Affected by Postharvest Treatments. *Orissa J. Hortic.* **2005**, *33* (2), 3–6.

Dock, L. L.; Nielsen, P. V.; Floros, J. D. Biological Control of *Botrytis cinerea* Growth on Apples Stored under Modified Atmospheres. *J. Food Protec.* **1998**, *61*, 1661–1665.

Droby, S. Biological Control of Postharvest Diseases of Fruits and Vegetables: Difficulties and Challenges. *Phytopathology* **2006**, *39*, 105–117.

Droby, S.; Vinokur, V.; Weiss, B.; Cohen, L.; Daus, A.; Goldschmidt, E. E.; Poral, R. Induction of resistance to *Penicillium digitatum* in grapefruit by the yeast biocontrol agent *Candida oleophila*. *Phytopathology* **2002**, *92*, 393–399.

Droby, S. Improving Quality and Safety of Fresh Fruit and Vegetables after Harvest by the Use of Biocontrol Agents and Natural Materials. *Acta Hortic.* **2006**, *709*, 45–51.

Droby, S.; Chalutz, E.; Wilson, C. L.; Wisniewski, M. E. Biological Control of Postharvest Diseases: A Promising Alternative to the Use of Synthetic Fungicides. *Phytoparasitica* **1992**, *20*, 1495–1503.

Droby, S.; Cohen, A.; Weiss, B.; Horev, B.; Chalutz, E.; Katz, H.; Keren-Tzur, M.; Shachnai, A. Commercial Testing of Aspire: A Yeast Preparation for the Biological Control of Postharvest Decay of Citrus. *Biol. Cont.* **1998**, *12*, 97–100.

Droby, S.; Hofstein, R.; Wilson, C. L.; Wisniewski, M.; Fridlender, B.; Cohen, L.; Weiss, B.; Daus, A.; Timar, D.; Chalutz, E. Pilot Testing of *Pichia guilliermondii*: A Biocontrol Agent for Postharvest Diseases of Citrus Fruit. *Biol. Cont.* **1993**, *3*, 47–52.

Droby, S.; Wisniewski, M.; El-Ghaouth, A.; Wilson, C. Influence of Food Additives on the Control of Postharvest Rots of Apple and Peach and Efficacy of the Yeast-Based Biocontrol Product Aspire. *Postharvest Biol. Technol.* **2003**, *27* (2), 127–135.

Droby, S.; Wisniewski, M. E.; Cohen, L.; Weisis, B.; Toutou, D.; Eilam, Y.; Chalutz, E. Influence of CaCl$_2$ on *Penicillium digitatum* Grapefruit Peel Tissue and Biocontrol Activity of *Pichia guilliermondii*. *Phytopathology* **1997**, *87*, 310–315.

El-Ghaouth, A.; Smilanick, J. L.; Brown, G. E.; Ippolito, A.; Wisniewski, M. E.; Wilson, C. L. Application of *Candida saitona* and Glycochitosan for the Control of Postharvest Diseases of Apple and Citrus Fruit under Semi-commercial Conditions. *Plant Dis.* **2000a**, *84*, 243–248.

El-Ghaouth, A.; Smilanick, J. L.; Wilson, C. L. Enhancement of the Performance of *Candida saitona* by the Addition of Glycochitosan for Control of Postharvest Decay of Apple and Citrus Fruit. *Postharvest Biol. Technol.* **2000b**, *19*, 249–253.

El-Ghaouth, A.; Wilson, C. L.; Wisniewski, M. E. Ultrastructural and Cytochemical Aspects of Biocontrol Activity of *Candida saitona* in Apple Fruit. *Phytopathology* **1998**, *88*, 282–291.

El-Ghaouth, A.; Wilson, C. L.; Wisniewski, M. E. Biologically Based Alternatives to Synthetic Fungicides for the Postharvest Diseases of Fruit and Vegetables. In *Diseases of Fruit and Vegetables*; Naqvi, S. A. M. H.; Ed.; Kluwer Academic Publishers: Dordrecht, The Netherlands, 2004; vol 2, pp 511–535.

El-Neshawy, S. M.; El-Sheikh, M. M. Control of Green Mold on Oranges by *Candida oleophila* and Calcium Treatments. *Ann. Agri Sci. Cairo* **1998**, *3*, 881–890.

El-Neshawy, S. M.; Wilson, C. L. Nisin Enhancement of Biocontrol of Postharvest Diseases of Apple with *Candida oleophila*. *Postharvest Biol. Technol.* **1997**, *10* (1), 9–14.

Elshafie, S. H.; Camele, I.; Racioppi, R.; Scrano, L.; Iacobellis, N. S.; Bufo, S. A. *In vitro* Antifungal Activity of *Burkholderia gladioli* pv. *agaricicola* against Some Phytopathogenic Fungi. *Int. J. Mol. Sci.* **2012**, *13*, 16291–16302.

Errampalli, D.; Brubacher, N. R. Biological and Integrated Control of Postharvest Blue Mold (*Penicillium expansum*) of Apples by *Pseudomonas syringae* and Cyprodinil. *Biol. Cont.* **2006,** *36* (1), 49–56.

Falconi, C. J.; Mendgen, K. Epiphytic Fungi on Apple Leaves and Their Value for Control of the Postharvest Pathogens *Botrytis cinerea, Monilinia fructigena* and *Penicillium expansum. J. Plant Pathol.* **1994,** *101,* 38–47.

Fan, Q.; Tian, S. P. Postharvest Biological Control of Grey Mold and Blue Mold on Apple by *Cryptococcus albidus* (Saito) Skinner. *Postharvest Biol. Technol.* **2001,** *21* (3), 341–350.

Filonow, A. B.; Vishniac, H. S.; Anderson, J. A.; Janisiewicz, W. J. Biological Control of *Botrytis cinerea* in Apple by Yeasts from Various Habitats and their Putative Mechanism of Antagonism. *Biol. Cont.* **1996,** *7* (2), 212–220.

Fiori, S.; Scherm, B.; Liu, J.; Farrell, R.; Mannazzu, I.; Budroni, M.; Maserti, B. E.; Wisniewski, M. E.; Migheli, Q. Identification of Differentially Expressed Genes Associated with Changes in the Morphology of *Pichia fermentans* on Apple and Peach Fruit. *FEMS Yeast Res.* **2012,** *12,* 785–795.

Gamagae, S. U.; Sivakumar, D.; Wilson-Wijeratnam, R. S.; Wijesundra, R. L. C. Use of Sodium Bicarbonate and *Candida oleophila* to Control Anthracnose in Papaya During Storage. *Crop Prot.* **2003,** *22* (5), 775–779.

Govender, V.; Korsten, L.; Sivakumar, D. Semi-commercial Evaluation of *Bacillus licheniformis* to Control Mango Postharvest Diseases in South Africa. *Postharvest Biol. Technol.* **2005,** *38* (1), 57–65.

Grebenisan, I.; Cornea, P.; Mateesu, R.; Cimpeanu, C.; Olteanu, V.; Canpenn, G. H.; Stefan, L. A.; Oancea, F.; Lupa, C. *Metschnikowia pulcherrima,* a New Yeast with Potential for Biocontrol of Postharvest Fruit Rots. *Acta Hortic.* **2008,** *767,* 355–360.

Gueldner, R. C.; Reilly, C. C.; Pussey, P. L.; Costello, C. E.; Arrendale, R. F.; Cox, R. H.; Himmelsbach, D. S.; Crumley, F. G.; Culter, H. G. Isolation and Identification of Iturins as Antifungal Peptides in Biological Control of Peach Brown Rot with *Bacillus subtilis. J. Agric. Food Chem.* **1988,** *36,* 366–370.

Guijarro, B.; Melgarejo, P.; Torres, R.; Lamarca, N.; Usall, J.; de Cal, A. Effects of Different Biological Formulations of *Penicillium frequentans* on Brown Rot of Peaches. *Biol. Cont.* **2007,** *42* (1), 86–96.

Guinebretiere, M. H.; Nguyen-The, C.; Morrison, N.; Reich, M.; Nicot, P. Isolation and Characterization of Antagonists for the Biocontrol of the Postharvest Wound Pathogen *Botrytis cinerea* on Strawberry Fruits. *J. Food Prot.* **2000,** *63,* 386–394.

Gullino, M. L.; Benzi, D.; Aloi, C.; Testoni, A.; Garibaldi, A. Biological Control of Botrytis Rot of Apple. In *Recent Advances in Botrytis Research. Proceedings of the 10th International Botrytis Symposium,* Heraklion, Crete, 1992; pp 197–200.

Ha, T. N. Using *Trichoderma* Species as Biological Control of Plant Pathogens in Vietnam City. *J. ISSASS* **2010,** *16* (1), 17–21.

Hofstein, R.; Fridlender, B.; Chalutz, E.; Droby, S. Large Scale Production and Pilot Testing of Biological Control Agents for Postharvest Diseases. In *Biological Control of Postharvest Diseases—Theory and Practice*; Wilson, C. L., Wisniewski, M. E., Eds.; CRC Press: Boca Raton, FL; pp 89–100.

Huang, Y.; Wild, B. L.; Morris, C. Postharvest Biological Control of *Penicilium digitatum* Decay on Citrus Fruit by *Bacillus pumilus. Ann. Appl. Biol.* **1992,** *120,* 367–372.

Huang, Y.; Daverall, B. J.; Morris, S. C.; Wild, B. L. Biocontrol of Postharvest Orange Diseases by a Strain of *Pseudomonas cepacia* under Semi-Commercial Conditions. *Postharvest Biol. Technol.* **1993**, *3* (4), 293–304.

Huang, Y.; Deverall, B. J.; Morris, S. C. Postharvest Control of Green Mold on Oranges by a Strain of *Pseudomonas glathei* and Enhancement of Its Biocontrol by Heat Treatment. *Postharvest Biol. Technol.* **1995**, *13*, 129–137.

Ippolito, A.; El-Ghaouth, A.; Wilson, C. L.; Wisniewski, M. A. Control of Postharvest Decay of Apple Fruit by *Aureobasidium pullulans* and Induction of Defense Responses. *Postharvest Biol. Technol.* **2000**, *19*, 265–272.

Ippolito, A.; Leonardo, S.; Isabella, P.; Franco, N. Control of Postharvest Rots of Sweet Cherries by Pre- and Postharvest Applications of *Aureobasidium pullulans* in Combination with Calcium Chloride or Sodium Bicarbonate. *Postharvest Biol. Technol.* **2005**, *36*, 245–252.

Ippolito, A.; Nigro, F. Impact of Preharvest Application of Biological Control Agents on Postharvest Diseases of Fresh Fruit and Vegetables. *Crop Prot.* **2000**, *19* (8/10), 715–723.

Ippolito, A.; Nigro, F.; Schena, L. Control of Postharvest Diseases of Fresh Fruit and Vegetables by Preharvest Application of Antagonistic Microorganisms. In *Crop Management and Postharvest Handling of Horticultural Products: Diseases and Disorders of Fruit and Vegetables*; Niskanen, D. R., Jain, R., Eds.; SM Science Publishers, Inc.: Enfield, USA, 2004; Vol IV, pp 1–30.

Irtwange, S. Application of Biological Control Agents in Pre and Post-Harvest Operations. *Agric. Eng. Int.* 8, *Invited Overview* 3; A & M University Press: Texas, 2006.

Janisiewicz, W. J. Postharvest Biological Control of Blue Mold on Apples. *Phytopathology* **1987**, *77*, 481–485.

Janisiewicz, W. J. Control of Post Harvest Diseases of Fruits with Biocontrol Agents. In *The Biological Control of Plant Diseases*; Bay-Petersen, J., Ed.; Food Fertil. Technol. Cent. Asian Pac. Reg.: Taipei, Taiwan, 1991; pp 56–68.

Janisiewicz, W. J. Ecological Diversity, Niche Overlap, and Coexistence of Antagonists Used in Developing Mixtures for Biocontrol of Postharvest Diseases of Apples. *Phytopathology* **1996**, *86*, 473–479.

Janisiewicz, W. J. Biological Control of Postharvest Diseases of Apple with Antagonists' Mixtures. *Phytopathology* **1988**, *78*, 194–198.

Janisiewicz, W. J. Biocontrol of Postharvest Diseases of Temperate Fruit: Challenges and Opportunities. In *Plant Microbe Interactions and Biological Control*; Boland, J., Kuykendall, L. D., Eds.; Marcel Dekker: New York, NY, 1998; pp 171–198.

Janisiewicz, W. J.; Conway, W. S.; Glenn, D. M.; Sams, C. E. Integrating Biological Control and Calcium Treatment for Controlling Postharvest Decay of Apple. *HortSci.* **1998**, *33*, 105–109.

Janisiewicz, W. J.; Roitman, J. Biological Control of Blue Mold and Gray Mold on Apple and Pear with *Pseudomonas cepacia*. *Phytopathology* **1988**, *78*, 1697–1700.

Janisiewicz, W. J.; Saftner, R. A.; Conway, W. S.; Yoder, K. S. Control of Blue Mold Decay of Apple During Commercial Controlled Atmosphere Storage with Yeast Antagonists and Sodium Bicarbonate. *Postharvest Biol. Technol.* **2008**, *49* (3), 374–378.

Janisiewicz, W. J.; Usall, J.; Bors, B. Nutritional Enhancement of Biocontrol of Blue Mold of Apples. *Phytopathology* **1992**, *82*, 1364–1370.

Janisiewicz, W. J.; Bors, B. Development of a Microbial Community of Bacterial and Yeast Antagonists to Control Wound-Invading Postharvest Pathogens of Fruits. *Appl. Environ. Microbiol.* **1995**, *61*, 3261–3267.

Janisiewicz, W. J.; Korsten, L. Biological Control of Postharvest Diseases of Fruit. *Ann. Rev. Phytopathology* **2002**, *40*, 411–441.

Jiang, Y. M.; Chen, F.; Li, Y. B.; Liu, S. X. A Preliminary Study on the Biological Control of Postharvest Diseases of Litchi Fruit. *J. Fruit Sci.* **1997**, *14* (3), 185–186.

Jiang, Y. M.; Zhu, X. R.; Li, Y. B. Postharvest Control of Litchi Fruit Rot by *Bacillus subtilis*. *Lebensmittel Wissenschaft Technol.* **2001**, *34* (7), 430–436.

Jima, T. A. *Postharvest Biological Control of Fusarium Dry-Rot Disease in Potato Tubers Using Clonostachys rosea Strain IK726*; Swedish University of Agricultural Sciences (SLU), Department of Forest Mycology and Plant Pathology, Uppsala, 2013; p 42.

Junaid, J. M.; Dar, N. A.; Bhat, T. A.; Bhat, A. H.; Bhat, M. A. Commercial Biocontrol Agents and Their Mechanism of Action in the Management of Plant Pathogens. *Int. J. Mod. Plant Animal Sci.* **2013**, *1* (2), 39–57.

Kabaluk, J. T.; Svircev, A. M.; Goettel, M. S.; Woo, S. G., Eds. *The Use and Regulation of Microbial Pesticides in Representative Jurisdictions Worldwide*; IOBC Global, 2010; p 99. http://www.iobcglobal.org/publications_iobc_use_and_regulation_of_microbial_pesticides.html.

Kanapathipillai, V. S.; Jantan, R. Approach to Biological Control of Anthracnose Fruit Rot of Bananas. In *First Reg. Symp. Biol. Control. Tropics*; Hussein, M. Y., Ibrahim, A. G., Ed.; 1986; pp 387–398.

Karabulut, O. A.; Arslan, U.; Kadir, I.; Gul, K. Integrated Control of Post Harvest Diseases of Sweet Cherry with Yeast Antagonist and Sodium Bicarbonate Applications within a Hydro-cooler. *Postharvest Biol. Technol.* **2005**, *37*, 135–141.

Karabulut, O. A.; Baykal, N. Biological Control of Postharvest Diseases of Peaches and Nectarines by Yeasts. *J. Phytopathol.* **2003**, *151* (3), 130–134.

Karabulut, O. A.; Baykal, N. Integrated Control of Postharvest Diseases of Peaches with a Yeast Antagonist, Hot Water and Modified Atmosphere Packaging. *Crop Prot.* **2004**, *23* (5), 431–435.

Karabulut, O. A.; Lurie, S.; Droby, S. Evaluation of the Use of Sodium Bicarbonate, Potassium Sorbate and Yeast Antagonists for Decreasing Postharvest Decay of Sweet Cherries. *Postharvest Biol. Technol.* **2001**, *23* (3), 233–236.

Karabulut, O. A.; Smilanick, J. L.; Gabler, F. M.; Mansour, M.; Droby, S. Near-harvest Applications of *Metschnikowia fructicola*, Ethanol, and Sodium Bicarbonate to Control Postharvest Diseases of Grape in Central California. *Plant Dis.* **2003**, *87* (11), 1384–1389.

Karabulut, O. A.; Tezean, H.; Daus, A.; Cohen, L.; Wiess, B.; Droby, S. Control of Preharvest and Postharvest Fruit Rot in Strawberry by *Metschnikowia fructicola*. *Biocontr Sci. Technol.* **2004**, *14*, 513–521.

Kefialew, Y.; Ayalew, A. Postharvest Biological Control of Anthracnose (*Colletotrichum gloeosporioides*) on Mango (*Mangifera indica*). *Postharvest Biol. Technol.* **2008**, *50* (1), 8–11.

Khokhar, I.; Haider, M. S.; Mukhtar, I.; Mushtaq, S. Biological Control of *Aspergillus niger*, the Cause of Black Rot Disease of *Allium cepa* L. (Onion), by *Penicillium* Species. *J. Agrobiol.* **2013**, *29* (1), 23–28.

Korsten, L. Advances in Control of Postharvest Diseases in Tropical Fresh Produce. *Int. J. Postharvest Technol. Innov.* **2006**, *1* (1), 48–61.

Kota, V. R.; Kulkarni, S.; Hegde, Y. R. Postharvest Diseases of Mango and their Biological Management. *J. Plant Dis. Sci.* **2006**, *1* (2), 186–188.

Kovach, J.; Petzoldi, R.; Harman, G. E. Use of Honey Bees and Bumble Bees to Dissemi-nate *Trichoderma harzianum* to Strawberries for *Botrytis* Control. *Biol. Cont.* **2006,** *18,* 235–242.

Kurtzman, C. P.; Droby, S. *Metschnikowia fructicola,* a New Ascosporic Yeast with Potential for Biocontrol of Postharvest Fruit Rots. *Syst. Appl. Microbiol.* **2001,** *24* (3), 395–399.

Lahlali, R.; Serrhini, M. N.; Jijakli, M. H. Efficacy Assessment of *Candida oleophila* (Strain O) and *Pichia anomala* (Strain K) against Major Postharvest Diseases of Citrus Fruit in Morocco. *Comm. Agric. Appl. Biol. Sci.* **2004,** *69* (4), 601–609.

Lahlali, R.; Serrhini, M. N.; Jijakli, M. H. Development of a Biological Control Method against Postharvest Diseases of Citrus Fruit. *Commun. Agric. Appl. Biol. Sci.* **2005,** *70* (3), 47–58.

Larena, I.; Torres, R.; de Cal, A.; Linan, M.; Melgarejo, P.; Domenichini, P.; Bellini, A.; Mandrin, J. F.; Lichou, J.; Ochoa de Eribe, X.; Usall, J. Biological Control of Postharvest Brown Rot (*Monilinia* spp.) of Peaches by Field Applications of *Epicoccum nigrum.* *Biol. Cont.* **2005,** *32* (2), 305–310.

Lassois, L.; de Bellaire, L.; Jijakli, M. H. Biological Control of Crown Rot of Bananas with *Pichia anomala* Strain K and *Candida oleophila* Strain O. *Biol. Cont.* **2008,** *45* (3), 410–418.

Leibinger, W.; Breuker, B.; Hahn, M.; Mendgen, K. Control of Postharvest Pathogens and Colonization of the Apple Surface by Antagonistic Microorganisms in the Field. *Phytopa-thology* **1997,** *87,* 1103–1110.

Lima, G.; Ippolito, A.; Nigro, F.; Salerno, M. Effectiveness of *Aureobasidium pullulans* and *Candida oleophila* against Postharvest Strawberry Rots. *Postharvest Biol. Technol.* **1997,** *10,* 169–178.

Liu, J.; Wisniewski, M.; Droby, S.; Norelli, J.; Hershkovitz, V.; Tian, S.; Farrell, R. Increase in Antioxidant Gene Transcripts, Stress Tolerance and Biocontrol Efficacy of *Candida oleophila* Following Sublethal Oxidative Stress Exposure. *FEMS Microbiol. Ecol.* **2012,** *80,* 578–590.

Liu, J.; Wisniewski, M.; Droby, S.; Tian, S.; Hershkovitz, V.; Tworkoski, T. Effect of Heat Shock Treatment on Stress Tolerance and Biocontrol Efficacy of *Metschnikowia fructicola.* *FEMS Microbiol. Ecol.* **2011a,** *76,* 145–155.

Liu, J.; Wisniewski, M.; Droby, S.; Vero, S.; Tian, S.; Hershkovitz, V. Glycine Betaine Improves Oxidative Stress Tolerance and Biocontrol Efficacy of the Antagonistic Yeast *Cystofilobasidium infirmominiatum. Int. J. Food Microbiol.* **2011b,** *146,* 76–83.

Long, C. A.; Deng, B. X.; Deng, X. X. Pilot Testing of *Kloeckera apiculata* for the Biological Control of Postharvest Diseases of Citrus. *Ann. Microbiol.* **2006,** *56* (1), 13–17.

Long, C. A.; Deng, B. X.; Deng, X. X. Commercial Testing of *Kloeckera apiculata,* Isolate 34-9, for Biological Control of Postharvest Diseases of Citrus Fruit. *Ann. Microbiol.* **2007,** *57* (2), 203–207.

Lurie, S.; Droby, S.; Chalupowicz, L.; Chalutz, E. Efficacy of *Candida oleophila* Strain 182 in Preventing *Penicillium expansum* Infection of Nectarine Fruit. *Phytoparasitica* **1995,** *23,* 231–234.

Majumdar, V. L.; Pathak, V. N. Biological Control of Postharvest Diseases of Guava Fruit by *Trichoderma* spp. *Acta Bot. Indica* **1995,** *23,* 263–267.

Mandal, G.; Singh, D.; Sharma, R. R. Effect of Hot Water Treatment and Biocontrol Agent (*Debaryomyces hansenii*) on Shelf Life of Peach. *Indian J. Hortic.* **2007,** *64* (1), 25–28.

Mari, M.; Guizzardi, M.; Pratella, G. C. Biological Control of Gray Mold in Pears by Antagonistic Bacteria. *Biol. Cont.* **1996,** *7* (1), 30–37.

McLaughlin, R. J.; Wilson, C. L.; Chalutz, E.; Kurtzman, W. F.; Osman, S. F. Characterization and Reclassification of Yeasts Used for Biological Control of Postharvest Diseases of
Fruit and Vegetables. *Appl. Environ. Microbiol.* **1993,** *56,* 3583–3586.

Mikani, A.; Etebarian, H. R.; Sholberg, P. L.; Gorman, D. T.; Stokes, S.; Alizadeh, A. Biological Control of Apple Gray Mold Caused by *Botrytis mali* with *Pseudomonas fluorescens*
Strains. *Postharvest Biol. Technol.* **2008,** *48* (1), 107–112.

Mishra, D. S.; Kumar, A.; Prajapati, C. R.; Singh, A. K.; Sharma, S. D. Identification of
Compatible Bacterial and Fungal Isolate and Their Effectiveness against Plant Disease. *J.
Environ. Biol.* **2013,** *34,* 183–189.

Montesinos, E. Development, Registration and Commercialization of Microbial Pesticides
for Plant Protection. *Int. Microbiol.* **2003,** *6,* 245–252.

Montesinos, E.; Bonaterra, A. Microbial Pesticides. In *Encyclopedia of Microbiology*, 3rd
ed.; Schaechter, M., Ed.; Elsevier: New York, NY, 2009; pp 110–120.

Morales, H.; Sanchis, V.; Usall, J.; Ramos, A. J.; Marin, S. Effect of Biocontrol Agents
Candida sake and *Pantoea agglomerans* on *Penicillium expansum* Growth and Patulin
Accumulation in Apples. *Int. J. Food Microbiol.* **2008,** *122* (1–2), 61–67.

Mortuza, M. G.; Ilag, L. L. Potential for Biocontrol of *Lasiodiplodia theobromae* in Banana
Fruit by *Trichoderma* species. *Biol. Cont.* **1999,** *15,* 235–240.

Ngullie, M.; Daiho, L.; Upadhyay, D. N. Biological Management of Fruit Rot in the World's
Hottest Chilli (*Capsicum chinense* Jacq.). *J. Plant Prot. Res.* **2010,** *50* (3), 269–273.

Nunes, C.; Teixido, N.; Usall, J.; Vinas, I. Biological Control of Major Postharvest Diseases
on Pear Fruit with Antagonistic Bacterium *Pantoea agglomerans* (CPA-2). *Acta Hortic.*
2001a, *553* (2), 403–404.

Nunes, C.; Usall, J.; Teixido, N.; Miro, M.; Vinas, I. Nutritional Enhancement of Biocontrol
Activity of *Candida sake* (CPA-1) against *Penicillium expansum* on Apples and Pears. *Eur.
J. Plant Pathol.* **2001b,** *107,* 543–551.

Nunes, C.; Usall, J.; Teixido, N.; Fons, E.; Vinas, I. Postharvest Biological Control by
Pantoea agglomerans (CPA-2) on Golden Delicious Apples. *J. Appl. Microbiol.* **2002a,**
92 (2), 247–255.

Nunes, C.; Usall, J.; Teixido, N.; Torres, R.; Vinas, I. Control of *Penicillium expansum* and
Botrytis cinerea on Apples and Pears with a Combination of *Candida sake* (CPA-1) and
Pantoea agglomerans. *J. Food Prot.* **2002b,** *65,* 178–184.

Obagwu, J.; Korsten, L. Integrated Control of Citrus Green and Blue Molds Using *Bacillus
subtilis* in Combination with Sodium Bicarbonate or Hot Water. *Postharvest Biol. Technol.*
2003, *28* (1), 187–194.

Odebode, A. C. Control of Postharvest Pathogens of Fruits by Culture Filtrate from Antagonistic Fungi. *Int. J. Plant Prot. Res.* **2006,** *46* (1), 1–6.

Pathak, V. N. Postharvest Fruit Pathology: Present Status and Future Possibilities. *Indian
Phytopathol.* **1997,** *50,* 161–185.

Piano, S.; Neyrotti, V.; Migheli, Q.; Gullino, M. I. Biocontrol Capability of *Metschnikowia
pulcherrima* against *Botrytis* Postharvest Rot of Apple. *Postharvest Biol. Technol.* **1997,**
11, 131–140.

Plaza, P.; Torres, R.; Teixido, N.; Usall, J.; Abadias, M.; Vinas, I. Control of Green Mold by
the Combination of *Pantoea agglomerans* (CPA-2) and Sodium Bicarbonate on Oranges.
Bull. OILB/SROP **2001,** *24* (3), 167–170.

Pratella, G. C.; Mari, M. Effectiveness of *Trichoderma*, *Gliocladium* and *Paecilomyces* in Postharvest Fruit Protection. *Postharvest Biol. Technol.* **1993**, *3*, 49–56.

Pusey, P. L. Use of *Bacillus subtilis* and Related Organisms as Biofungicides. *Pesticide Sci.* **1989**, *27*, 133–140.

Pusey, P. L.; Hotchkiss, M. W.; Dulmage, H. T.; Banumgardner, R. A.; Zehr, E. I. Pilot Tests for Commercial Production and Application of *Bacillus subtilis* (B-3) for Postharvest Control of Peach Brown Rot. *Plant Dis.* **1988**, *72*, 622–626.

Pusey, P. L.; Wilson, C. L. Postharvest Biological Control of Stone Fruit Brown Rot by *Bacillus subtilis*. *Plant Dis.* **1984**, *68*, 753–756.

Qin, G. Z.; Tian, S. P. Biocontrol of Postharvest Diseases of Jujube Fruit by *Cryptococcus laurentii* Combined with a Low Doses of Fungicides under Different Storage Conditions. *Plant Dis.* **2004**, *88* (5), 497–501.

Qin, G. Z.; Tian, S. P.; Xu, Y. Biocontrol of Postharvest Diseases on Sweet Cherries by Four Antagonistic Yeasts in Different Storage Conditions. *Postharvest Biol. Technol.* **2004**, *31* (1), 51–58.

Qin, G. Z.; Tian, S. P.; Xu, Y.; Chan, Z. L.; Li, B. Q. Combination of Antagonistic Yeasts with Two Food Additives for Control of Brown Rot Caused by *Monilinia fructicola* on Sweet Cherry Fruit. *J. Appl. Microbiol.* **2006**, *100* (3), 508–515.

Ragsdale, N. N.; Sisler, H. D. Social and Political Implications of Managing Plant Diseases with Decreased Availability of Fungicides in the United States. *Ann. Rev. Phytopathol.* **1994**, *32*, 545–557.

Reeh, K. W. Commercial Bumble Bees as Vectors of the Microbial Antagonist *Clonostachys rosea* for Management of *Botrytis* Blight in Wild Blueberry (*Vaccinium angustifolium*). *M.Sc. Dissertation, Depon.* Department of Environmental Science, Dalhousie University: Halifax, 2012; 99 pp.

Reyes, M. E. Q.; Rohrbach, K. G.; Paull, R. E. Microbial Antagonists Control Postharvest Black Rot of Pineapple Fruit. *Postharvest Biol. Technol.* **2004**, *33* (2), 193–203.

Roberts, R. G. Postharvest Biological Control of Gray Mold of Apple by *Cryptococcus laurentii*. *Phytopathology* **1990a**, *80*, 526–530.

Roberts, R. G. Biological Control of Mucor Rot of Pear by *Cryptococcus laurentii*, *C. Flavus*, and *C. albidus*. *Phytopathology* **1990b**, *80*, 1051–1154.

Rodov, V.; Ben-Yehoshua, S.; Albaglis, R.; Fang, D. Accumulation of Phytoalexins, Scoparone and Scopoletin in Citrus Fruit Subjected to Various Postharvest Treatments. *Acta Hortic.* **1994**, *381*, 517–523.

Saligkarias, I. D.; Gravanis, F. T.; Epton, H. A. S. Biological Control of *Botrytis cinerea* on Tomato Plants by the Use of Epiphytic Yeasts *Candida guilliermondii* Strains 101 and US 7 and *Candida oleophila* Strain I-182: *In Vivo* Studies. *Biol. Cont.* **2002**, *25* (2), 143–150.

Saravanakumar, D.; Ciavorella, A.; Spadaro, D.; Garibaldi, A.; Gullino, M. L. *Metschnikowia pulcherrima* Strain MACH-1 Outcompetes *Botrytis cinerea*, *Alternaria alternata* and *Penicillium expansum* in Apples through Iron Depletion. *Postharvest Biol. Technol.* **2008**, *49* (1), 121–128.

Schena, L.; Ippolito, A.; Zehavi, T.; Cohen, L.; Nigro, F.; Droby, S. Genetic Diversity and Biocontrol Activity of *Aureobasidium pullulans* Isolates against Postharvest Rots. *Postharvest Biol. Technol.* **1999**, *17*, 189–199.

Schena, L.; Nigro, F.; Pentimone, I. A.; Ippolito, A. Control of Postharvest Rots of Sweet Cherries and Table Grapes with Endophytic Isolates of *Aureobasidium pullulans*. *Postharvest Biol. Technol.* **2003**, *30* (3), 209–220.

Schisler, D. A.; Slininger, P. J.; Bothast, R. J. Effects of Antagonist Cell Concentration and Two Strain Mixtures of Biological Control of *Fusarium* Dry Rot of Potatoes. *Phytopathology* **1997**, *87*, 177–183.

Shaikh, F. T.; Nasreen, S. Biocontrol Efficacy of *Trichoderma koningii* against Some Plant Pathogenic Fungi. *Indian J. Res.* **2013**, *2* (3), 9–10.

Sharma, N. Biological Control for Preventing Food Deterioration. In *Strategies for Pre- and Postharvest Management*; Wiley-Blackwell: Hoboken, NJ, 2014, 464 p.

Sharma, R. R.; Singh, D.; Singh, R. Biological Control of Postharvest Diseases of Fruits and Vegetables by Microbial Antagonists: A Review. *Biol. Cont.* **2009**, *50*, 205–221.

Singh, D. Bioefficacy of *Debaryomyces hansenii* on the Incidence and Growth of *Penicillium italicum* on Kinnow Fruit in Combination with Oil and Wax Emulsions. *Ann. Plant Prot. Sci.* **2002**, *10* (2), 272–276.

Singh, D. Effect of *Debaryomyces hansenii* and Calcium Salt on Fruit Rot of Peach (*Rhizopus macrosporus*). *Ann. Plant Prot. Sci.* **2004**, *12* (2), 310–313.

Singh, D. Interactive Effect of *Debaryomyces hansenii* and Calcium Chloride to Reduce *Rhizopus* Rot of Peaches. *J. Mycol. Plant Pathol.* **2005**, *35* (1), 118–121.

Singh, D.; Sharma, R. R. Postharvest Diseases of Fruit and Vegetables and their Management. In *Sustainable Pest Management*; Prasad, D., Ed., 2007; Daya Publishing House: New Delhi, India.

Singh, V.; Deverall, B. J. *Bacillus subtilis* as a Control Agent against Fungal Pathogens of Citrus Fruit. *Trans. Br. Mycol. Soc.* **1984**, *83*, 487–490.

Sivakumar, D.; Wijeratnam, R. S. W.; Marikar, F. M. M. T.; Abeyesekere, M.; Wijesundera, R. L. C. Antagonistic Effect of *Trichoderma harzianum* on Post Harvest Pathogens of Rambutans. *Acta Hortic.* **2001**, *553* (2), 389–392.

Sivakumar, D.; Wijeratnam, R. S. W.; Wijesundera, R. L. C.; Abeyesekere, M. Combined Effect of Generally Regarded as Safe (GRAS) Compounds and *Trichoderma harzianum* on the Control of Postharvest Diseases of Rambutan. *Phytoparasitica* **2002a**, *30* (1), 43–51.

Sivakumar, D.; Wijeratnam, R. S. W.; Wijesundera, R. L. C.; Abeyesekere, M. Control of Postharvest Diseases of Rambutan Using Controlled Atmosphere Storage and Potassium Metabisulphite or *Trichoderma harzianum*. *Phytoparasitica* **2002b**, *30* (4), 403–409.

Sivakumar, D.; Wijeratnam, R. S. W.; Wijesundera, R. L. C.; Marikar, F. M. T.; Abeyesekere, M. Antagonistic Effect of *Trichoderma harzianum* on Postharvest Pathogens of Rambutan (*Nephelium lappaceum*). *Phytoparasitica* **2000**, *28* (3), 240–247.

Smilanick, J. L.; Margosan, D. A.; Milkota, F.; Usall, J.; Michael, I. F. Control of Citrus Green Mold by Carbonate and Bicarbonate Salts and the Influence of Commercial Postharvest Practices on their Efficacy. *Plant Dis.* **1999**, *83*, 139–145.

Smilanik, J. L.; Denis-Arrue, R.; Bosch, J. R.; Gonjalez, A. R.; Henson, D.; Janisiwicz, W. J. 1993. Control of Postharvest Brown Rot of Nectarines and Peaches by *Pseudomonas* species. *Crop Prot.* **1993**, *12* (7), 513–520.

Sobiczewski, P.; Bryk, H.; Berezynski, S. Evaluation of Epiphytic Bacteria Isolated from Apple Leaves in the Control of Postharvest Diseases. *J. Fruit Ornam. Plant Res.* **1996**, *4*, 35–45.

Sobowale, A. A.; Cardwell, K. F.; Odebode, A. C.; Bandyopadhyay, R.; Jonathan, S. G. Antagonistic Potential of *Trichoderma longibrachiatum* and *T. hamatum* Resident on Maize (*Zea mays*) Plant against *Fusarium verticillioides* (Nirenberg) Isolated from Rotting Maize Stem. *Arch. Phytopathol. Plant Prot.* **2008**, *43* (8), 744–753.

Spadaro, D.; Garibaldi, A.; Gullino, M. L. Control of *Penicillium expansum* and *Botrytis cinerea* on Apple Combining a Biocontrol Agent with Hot Water Dipping and Acibenzolar-*S*-Methyl, Baking Soda, or Ethanol Application. *Postharvest Biol. Technol.* **2004**, *33* (2), 141–151.

Spadaro, D.; Vola, R.; Piano, S.; Gullino, M. L. Mechanisms of Action and Efficacy of Four Isolates of the Yeast *Metschnikowia pulcherrima* Active against Postharvest Pathogens on Apples. *Postharvest Biol. Technol.* **2002**, *24* (2), 123–134.

Sugar, D.; Spotts, R. A. Control of Postharvest Decay of Pear by Four Laboratory Grown Yeasts and Two Registered Biocontrol Products. *Plant Dis.* **1999**, *83*, 155–158.

Sutton, J. C.; Li, D.; Peng, G.; Yu, H.; Zhang, P.; Valdebenito-Sanhueza, R. M. *Gliocladium roseum*, a Versatile Adversary of *Botrytis cinerea* in Crops. *Plant Dis.* **1997**, *31*, 316–328.

Teixido, N.; Usall, J.; Palou, L.; Asensio, A.; Nunes, C.; Vinas, I. Improving Control of Green and Blue Molds of Oranges by Combining *Pantoea agglomerans* (CPA-2) and Sodium Bicarbonate. *Eur. J. Plant Pathol.* **2001**, *107* (7), 685–694.

Teixido, N.; Usall, J.; Vinas, I. Efficacy of Preharvest and Postharvest *Candida sake* Biocontrol Treatments to Prevent Blue Mold on Apples During Storage. *Int. J. Food Microbiol.* **1999**, *50*, 203–210.

Tian, S. P.; Fan, Q.; Xu, Y.; Qin, G. Z.; Liu, H. B. Effect of Biocontrol Antagonists Applied in Combination with Calcium on the Control of Postharvest Diseases in Different Fruit. *Bull. OILB/SROP* **2002**, *25* (10), 193–196.

Tian, S. P.; Qin, G. Z.; Xu, Y. Survival of Antagonistic Yeasts under Field Conditions and their Biocontrol Ability against Postharvest Diseases of Sweet Cherry. *Postharvest Biol. Technol.* **2004**, *33* (3), 327–331.

Tian, S. P.; Qin, G. Z.; Xu, Y. Synergistic Effects of Combining Biocontrol Agents with Silicon against Postharvest Diseases of Jujube Fruit. *J. Food Prot.* **2005**, *68* (3), 544–550.

Tong, K.; Rohrbock, K. G. Role of *Penicillium funiculosum* Strains in the Development of Pineapple Fruit Diseases. *Phytopathology* **1980**, *70*, 663–665.

Torres, R.; Nunes, C.; Garcia, J. M.; Abadias, M.; Vinas, I.; Manso, T.; Olmo, M.; Usall, J. Application of *Pantoea agglomerans* CPA-2 in Combination with Heated Sodium Bicarbonate Solutions to Control the Major Postharvest Diseases Affecting Citrus Fruit at Several Mediterranean Locations. *Eur. J. Plant Pathol.* **2007**, *118* (1), 73–83.

Torres, R.; Teixido, N.; Vinas, I.; Mari, M.; Casalini, L.; Giraud, M.; Usall, J. Efficacy of Candida Sake CPA-1 Formulation for Controlling *Penicillium expansum* Decay on Pome Fruit from Different Mediterranean Regions. *J. Food Prot.* **2006**, *69* (11), 2703–2711.

Tronsmo, A.; Denis, C. The use of *Trichoderma* Species to Control Strawberry Fruit Rots. *Netherlands J. Plant Pathol.* **1977**, *83*, 449–455.

Usall, J.; Smilanick, J.; Palou, L.; Denis-Arrue, N.; Teixido, N.; Torres, R.; Vinas, I. Preventive and Curative Activity of Combined Treatments of Sodium Carbonates and *Pantoea agglomerans* CPA-2 to Control Postharvest Green Mold of Citrus Fruit. *Postharvest Biol. Technol.* **2008**, *50* (1), 1–7.

Usall, J.; Teixido, N.; Fons, E.; Vinas, I. Biological Control of Blue Mold on Apple by a Strain of *Candida sake* under Several Controlled Atmosphere Conditions. *Int. J. Food Microbiol.* **2000**, *58*, 83–92.

Usall, J.; Teixido, N.; Torres, R.; Ochoa de Eribe, X.; Vinas, I. Pilot Tests of *Candida sake* (CPA-1) Applications to Control Postharvest Blue Mold on Apple Fruit. *Postharvest Biol. Technol.* **2001**, *21* (2), 147–156.

Utkhede, R. S.; Sholberg, P. L. *In vitro* Inhibition of Plant Pathogens: *Bacillus subtilis* and *Enterobacter aerogenes In Vivo* Control of Two Postharvest Cherry Diseases. *Can. J. Microbiol.* **1986**, *32*, 963–967.

Vero, S.; Garmendia, G.; Gonzalez, M. B.; Bentancur, O.; Wisniewski, M. Evaluation of Yeasts Obtained from Antarctic Soil Samples as Biocontrol Agents for the Management of Postharvest Diseases of Apple (*Malus domestica*). *FEMS Yeast Res.* **2013**, *13*, 189–199.

Vinas, I.; Usall, J.; Teixido, N.; Fons, E.; Ochoa-de-Eribe, J. Successful Biological Control of the Major Postharvest Diseases on Apples and Pears with a New Strain of *Candida sake*. *Proc. Br. Crop Prot. Conf. Pests Dis.* **1996**, *6C*, 603–608.

Vinas, I.; Usall, J.; Teixido, N.; Sanchis, V. Biological Control of Major Postharvest Pathogens on Apple with *Candida sake*. *Int. J. Food Microbiol.* **1998**, *40* (1–2), 9–16.

Wan, Y. K.; Tian, S. P. Integrated Control of Postharvest Diseases of Pear Fruit Using Antagonistic Yeasts in Combination with Ammonium Molybdate. *J. Sci. Food Agric.* **2005**, *85* (15), 2605–2610.

Wang, Y.; Wang, P.; Xia, J.; Yu, T.; Luo, B.; Wang, J.; Zheng, X. Effect of Water Activity on Stress Tolerance and Biocontrol Activity in Antagonistic Yeast *Rhodosporidium paludigenum*. *Int. J. Food Microbiol.* **2010**, *143*, 103–108.

Williamson, S. M.; Guzman, M.; Marin, D. H.; Anas, O.; Jin, X.; Sutton, T. B. Evaluation of *Pseudomonas syringae* Strain ESS-11 for Biocontrol of Crown Rot and Anthracnose of Bananas. *Biol. Cont.* **2008**, *46* (3), 279–286.

Wilson, C. L.; Wisniewski, M. E. Biological Control of Postharvest Diseases of Fruits and Vegetables: An Emerging Technology. *Ann. Rev. Phytopathol.* **1989**, *27*, 425–441.

Wilson, C. L.; Wisniewski, M. E.; Droby, E.; Chalutz, E. A Selection Strategy for Microbial Antagonists to Control Postharvest Diseases of Fruits and Vegetables. *Sci. Hortic.* **1993**, *53*, 183–189.

Wilson, C. L.; Chalutz, E. Postharvest Biocontrol of Penicillium Rots of Citrus with Antagonistic Yeasts and Bacteria. *Sci. Hortic.* **1989**, *40*, 105–112.

Wilson, C. L.; Franklin, J. D.; Pusey, P. L. Biological Control of *Rhizopus* Rot of Peach with *Enterobacter cloacae*. *Phytopathology* **1987**, *77*, 303–305.

Wisniewski, M. E.; Wilson, C. L. Biological Control of Postharvest Diseases of Fruits and Vegetables: Recent Advances. *HortSci.* **1992**, *27*, 94–98.

Wisniewski, M.; Biles, C.; Droby, S. The Use of Yeast *Pichia guilliermondii* as a Biocontrol Agent: Characterization of Attachment to *Botrytis cinerea*. In *Biological Control of Postharvest Diseases of Fruit and Vegetables*; Wilson, C. L., Chalutz, E., Eds.; *Proc. Workshop, US Department of Agriculture*, ARS-92, 1991; pp 167–183.

Wisniewski, M.; Wilson, C. L.; Hershberger, W. Characterization of Inhibition of *Rhizopus stolonifer* Germination and Growth by *Enterobacter cloacae*. *Plant Dis.* **1989**, *81*, 204–210.

Wisniewski, M.; Wilson, C. L.; Chalutz, E.; Hershberger, W. Biological Control of Postharvest Diseases of Fruit: Inhibition of *Botrytis* Rot on Apples by an Antagonistic Yeast. *Proc. Electron Microsc. Soc. Am.* **1988**, *46*, 290–291.

Wisniewski, M. E.; Droby, S.; Chalutz, E.; Eilam, Y. Effect of Ca^{2+} and Mg^{2+} on *Botrytis cinerea* and *Penicillium expansum* In Vitro and on the Biocontrol Activity of *Candida oleophilla*. *Plant Pathol.* **1995**, *44*, 1016–1024.

Wisniewski, M. E.; Droby, S.; El-Ghaouth, A.; Droby, S. Non-chemical Approaches to Postharvest Disease Control. *Acta Hortic.* **2001**, *553*, 407–411.

Wittig, H. P. P.; Johnson, K. B.; Pscheidt, J. W. Effect of Epiphytic Fungi on Brown Rot, Blossom Blight and Latent Infections in Sweet Cherry. *Plant Dis.* **1997**, *81*, 383–387.

Woo, S. L.; Ruocco, M.; Vinale, F.; Nigro, M.; Marra, R.; Lombardi, N.; Pascale, A.; Lanzuise, S.; Manganiello, G.; Lorito, M. *Trichoderma*-Based Products and their Widespread Use in Agriculture. *Open Mycol. J.* **2014,** *8* (Suppl. 1, M4), 71–126.

Xi, L.; Tian, S. P. Control of Postharvest Diseases of Tomato Fruit by Combining Antagonistic Yeast with Sodium Bicarbonate. *Sci. Agric. Sin.* **2005,** *38* (5), 950–955.

Xu, B.; Zhang, H.; Chen, K.; Xu, Q.; Yao, Y.; Gao, H. Biocontrol of Postharvest *Rhizopus* Decay of Peaches with *Pichia caribbica*. *Curr. Microbiol.* **2013,** *67,* 255–261.

Yakoby, N.; Zhou, R.; Koblier, I.; Dinoor, A.; Prusky, D. Development of *Colletotrichum gloeosporioides* Restriction Enzyme-Mediated Integration Mutants as Biocontrol Agents against Anthracnose in Avocado Fruit. *Phytopathology* **2001,** *91,* 143–148.

Yang, D. M.; Bi, Y.; Chen, X. R.; Ge, Y. H.; Zhao, J. Biological Control of Postharvest Diseases with *Bacillus subtilis* (B1 Strain) on Muskmelons (*Cucumis melo* L. cv. Yindi). *Acta Hortic.* **2006,** *712* (2), 735–739.

Yao, H.; Tian, S.; Wang, Y. Sodium Bicarbonate Enhances Biocontrol Efficacy of Yeasts on Fungal Spoilage of Pears. *Int. J. Food Microbiol.* **2004,** *93* (3), 297–304.

Yao, H. J.; Tian, S. P. Effects of a Biocontrol Agent and Methyl Jasmonate on Postharvest Diseases of Peach Fruit and the Possible Mechanisms Involved. *J. Appl. Microbiol.* **2005,** *98* (4), 941–950.

Zahavi, T.; Cohen, L.; Weiss, B.; Schena, L.; Daus, A.; et al. Biological Control of *Botrytis*, *Aspergillus* and *Rhizopus* Rots on Table and Wine Grapes in Israel. *Postharvest Biol. Technol.* **2000,** *20,* 115–124.

Zhang, H.; Wang, L.; Dong, Y.; Jiang, S.; Zhang, H.; Zheng, X. Control of Postharvest Pear Diseases Using *Rhodotorula glutinis* and Its Effects on Postharvest Quality Parameters. *Int. J. Food Microbiol.* **2008b,** *126* (1–2), 167–171.

Zhang, H.; Wang, L.; Ma, L.; Dong, Y.; Jiang, S.; Xu, B.; Zheng, X. Biocontrol of Major Postharvest Pathogens on Apple Using *Rhodotorula glutinis* and Its Effects on Postharvest Quality Parameters. *Biol. Cont.* **2009,** *48* (1), 79–83.

Zhang, H.; Zheng, X.; Fu, C.; Xi, Y. Biological Control of Blue Mold Rot of Pear by *Cryptococcus laurentii*. *J. Hortic. Sci. Biotechnol.* **2003,** *78,* 888–893.

Zhang, H.; Zheng, X.; Fu, C.; Xi, Y. Postharvest Biological Control of Gray Mold Rot of Pear with *Cryptococcus laurentii*. *Postharvest Biol. Technol.* **2005,** *35* (1), 79–86.

Zhang, H.; Zheng, X.; Wang, L.; Li, S.; Liu, R. Effect of Antagonist in Combination with Hot Water Dips on Postharvest *Rhizopus* Rot of Strawberries. *J. Food Engg.* **2007b,** *78,* 281–287.

Zhang, H.; Zheng, X. D.; Yu, T. Biological Control of Postharvest Diseases of Peach with *Cryptococcus laurentii*. *Food Cont.* **2007c,** *18* (4), 287–291.

Zhang, H.; Wang, L.; Dong, Y.; Jiang, S.; Cao, J.; Meng, R. Postharvest Biological Control of Gray Mold Decay of Strawberry with *Rhodotorula glutinis*. *Biol. Cont.* **2007a,** *40* (2), 287–292.

Zhang, H. Y.; Fu, C. X.; Zheng, X. D.; Shan, L. J.; Jhon, X. Effect of *Cryptococcus laurentii* (Kufferath) Skinner in Combination with Sodium Bicarbonate on Biocontrol of Postharvest Green Mold Decay of Citrus Fruit. *Bot. Bull. Acad. Sin.* **2004,** *45,* 159–164.

Zhao, Y.; Shao, X. F.; Tu, K.; Chen, J. K. Inhibitory Effect of *Bacillus subtilis* B10 on the Diseases of Postharvest Strawberry. *J. Fruit Sci.* **2007,** *24* (3), 339–343.

Zhao, Y.; Tu, K.; Shao, X.; Jing, W.; Su, Z. Effects of the Yeast *Pichia guilliermondii* against *Rhizopus nigricans* on Tomato Fruit. *Postharvest Biol. Technol.* **2008,** *49* (1), 113–120.

Zhou, T.; Chu, C. L.; Liu, W. T.; Schneider, K. E. Postharvest Control of Blue Mold and Gray Mold on Apples Using Isolates of *Pseudomonas syringae*. *Can. J. Plant Pathol.* **2001,** *23,* 246–252.

Zhou, T.; Northover, J.; Schneider, K. E. Biological Control of Postharvest Diseases of Peach with Phyllosphere Isolates of *Pseudomonas syringae*. *Can. J. Plant Pathol.* **1999,** *21,* 375–381.

Zhou, T.; Northover, J.; Schneider, K. E.; Lu, X. W. Interactions between *Pseudomonas syringae* MA-4 and Cyprodinil in the Control of Blue Mold and Gray Mold of Apples. *Can. J. Plant Pathol.* **2002,** *24* (2), 154–161.

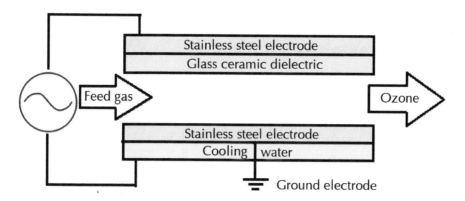

FIGURE 8.1   Corona discharge cell.

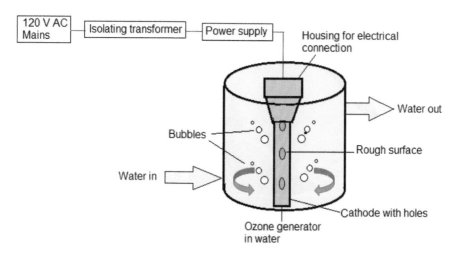

FIGURE 8.2   Schematic housing of ozone generation probe in water (Salama and Salama, 2016).

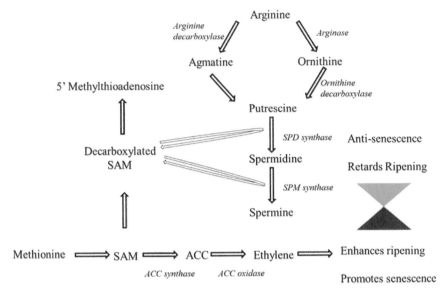

FIGURE 12.1 Simplified schematic pathway of polyamines biosynthesis (Adapted from Pandey et al., 2000).

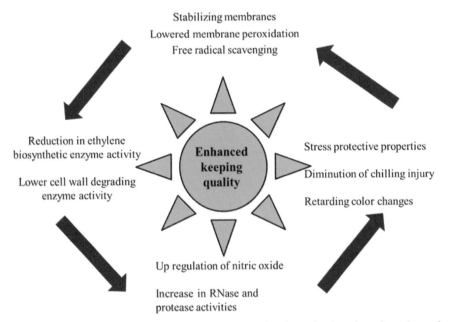

FIGURE 12.2 Schematic diagram of mechanism of action of polyamines for enhanced keeping quality of horticultural produce.

CHAPTER 8

APPLICATION OF OZONE AS A POSTHARVEST TREATMENT

SERDAR ÖZTEKIN[*]

Farm Machinery and Technologies Engineering Department, Faculty of Agriculture, Cukurova University, Adana, Turkey

[*]E-mail: oztekin@cu.edu.tr

ABSTRACT

The quality and quantity losses in fruits and vegetables due to insect and microorganisms in their ecosystem increase in parallel with growing mass production. These problems are becoming more serious due to inappropriate cultivation and storage practices. Ozone is an effective and environmental friendly natural tool for both fresh and stored agricultural crops. It could be artificially generated and applied in gaseous or aqueous form. Unlike chlorine-based sanitizing agents, ozone decomposes easily and fast after the processing and does not leave any remaining substances, odor, or taste. Application of gaseous or aqueous ozone in fresh fruits and vegetables has been deeply studied in previous works as a way of enhancing quality and safety of produces and prolonging the shelf-life. The ozone usage in fruits and vegetables are mainly focused on its effect on microbial population, visual appearance, enzyme activities, pH, color, weight loss, texture, firmness, respiration, ethylene production, pesticide residues, and nutritional value, quality and safety properties, soluble solids content and titratable acidity. The studies show that the efficiency of ozone treatment depends usually on the contacting conditions (e.g., exposure time, concentration, pH, etc.), and characteristics of vegetable and fruit surfaces. A survey on previous works showed that although the results on ozonation of fresh produces are promising on a laboratory scale, data on large-scale ozonation plants in processing of fresh fruits and vegetables stay confidential business information of manufacturer. In order for ozone processing to become commercially viable alternative to other methods, it is necessary to build a balance between overall cost of an ozonation system and its benefits.

8.1   INTRODUCTION

Spoilage microorganisms can cause produce tissue breakdown, thereby shortening the shelf-life of foods and produce illnesses or illness-causing toxins by occurring visible or invisible changes in color, texture, flavor, and/or smell. The consumption of fresh vegetables has resulted in several outbreaks of human gastroenteritis. Fruits and salads that contain raw vegetables have also been identified as a cause of serious illnesses such as diarrhea, etc. Although there are no reliable data that cover the worldwide spoilage losses occurring at farm, retail, and consumer level; national reports indicate the seriousness of the problem in any case. According to a USDA-Economic Research Service study in 2010, 18.9 billion pounds of fresh fruits and vegetables were lost annually due to spoilage, which was 19.6% of all US losses of edible foods that year (Buzby et al., 2014). It is also estimated that about 20% vegetable production is lost each year due to spoilage in India (Saranraj et al., 2012). To reduce the spoilage losses, the complicated relationship between perishable produces and their ecosystem needs to be understood. Normally, microorganisms don't exist in the internal tissues of healthy crops and they are considered as sterile. But different kinds of microorganisms are available on the surface of plants. Most microorganisms that are initially observed on whole fruit or vegetable surfaces are soil inhabitants and members of a very large and diverse community of microbes that collectively are responsible for maintaining a dynamic ecological balance within most agricultural systems (Barth et al., 2009). Most bacteria and fungi that arrive on the developing plant either are completely benign to the crop's health or, in many instances, provide a natural biological barrier to infestation by the subset of microorganisms responsible for crop damage (Janisiewicz and Korsten, 2002; Andrews and Harris, 2000). From the standpoint of nutrient content, fruits and vegetables are capable of supporting the growth of molds, yeasts, and bacteria (Jay et al., 2005). Impact injuries due to mechanical harvesting, long duration from field to farm, improper transport, and prestorage conditions cause the expansion of contamination. Spoilage is further accelerated due to deterioration caused by enzymatic activities in fresh produce. Additionally, microorganisms release their own enzymes as they grow, speeding up the spoiling process. Washing produce in water can effectively remove sand, soil, and other impurities from fresh fruits and vegetables but should not be relied upon to completely remove microorganisms (Baratharaj, n.d.). The methods currently available for disinfection and sterilization are through the use of oxidative and nonoxidative biocides. The disadvantages of nonoxidative biocides are that the process of disinfection often may not be complete (in the concentrations used) or some bacteria may be resistant and the preferred procedure is the use of oxidative biocides. Ozone can be used for prestorage or storage treatment in air or water, or as a continuous or intermittent atmosphere throughout the storage period. Both procedures have recently attracted considerable commercial interest, especially because of the lack of residue on the produce and the possibilities opened by the new regulations.

8.2 PROPERTIES OF OZONE IN TERMS OF POSTHARVEST TREATMENT OF FRUIT AND VEGETABLES

Ozone is the most efficient sanitizer with strong, fast, and broad-spectrum antimicrobial effect against bacteria, bacterial spores, viruses, fungi, fungal spores, and protozoa (White, 1999). Unlike other sanitizing agents, ozone is easy and fast to remove after the process and does not leave any remaining chemicals, odor, or taste. The efficacy of ozone depends on the higher oxidation potential which is measured in volts of electrical energy (Shammas and Wang, 2005). The disinfection process of ozone is governed by Chick's law.

Ozone can oxidize many organic compounds, particularly those with phenolic rings or unsaturated bonds in their structure such as pesticides and mycotoxins (Razumovski and Zaikov, 1984; McKenzie et al., 1997). The influence mechanism of ozone in water is quite complicated. Both molecular ozone (O_3) and its decomposition products hydroxyl radicals (OH˙) react with organic compounds in water (Staehelin and Hoigné, 1985). One common theory is that ozone inactivates bacteria by targeting cell membrane, thereby disrupting permeability functions (Jay et al., 2005; Kim et al., 1999). As ozone molecules make contact with the cell wall of bacteria, a tiny hole occurs that injures the bacterium. This reaction is called an oxidative burst. After thousands of ozone collisions over only a few seconds, the bacterial wall can no longer maintain its shape and the cell dies. Ozone also kills spores and viruses as it oxidizes DNA and proteins in the spore as well as in viruses. Within a pH range between 6 and 8.5, the disinfection potential of ozone remains constant and generally exceeds that of chlorine. The disinfection velocity is even 3000 times higher than that of chlorine-based sanitizers which is the most widely used chemical in washing of whole and fresh-cut produces (Heim and Glas, 2011). Safety concerns about the reaction of chlorine with organic residues in the formation of potentially mutagenic or carcinogenic reaction products, such as trihalomethanes and their impact on human and environmental safety, have been raised in recent years (Chang et al., 1988). For this reason, its use for washing fresh-cut products is banned in several countries. Significant advantages of ozone in water are that it decomposes quickly to oxygen, leaving no residue of it and few disinfection by-products. Ozonated process water in crop-washing facilities could be properly recycled that is another important aspect from the water-saving point of view. Ozone with these features would be a good option for many applications that fulfill the legal requirements, consumer acceptance, and industrial expectations.

8.2.1 PROCEDURE FOR OZONE GENERATION

Ozone is a highly reactive, odorous gas composed of three atoms of oxygen. In public, the terms atmospheric and artificially produced ozone are often confused. Ozone is quite readily available in the nature. Natural ozone occurs in the Earth's upper (the stratosphere) and lower level (the troposphere). Stratospheric ozone is formed naturally through the interaction of solar ultraviolet (UV) radiation with molecular oxygen (O_2). The popular term "ozone layer," approximately 6–30 miles above the Earth's surface, reduces the amount of harmful UV radiation reaching the Earth's surface. Tropospheric or ground-level ozone is formed primarily from photochemical reactions between two major classes of air pollutants, volatile organic compounds and nitrogen oxides (NO_x) (Anonymous I, n.d.).

Artificially generated ozone can be traced back to 1785 when van Marum, a Dutch physicist, found electric discharge in the air results in a characteristic ozone odor (Rideal, 1920 cited by Da Silva et al., 2003). In 1801, the same odor was observed during water electrolysis. However, ozone discovery was only officially announced by Schönbein, in 1840. The name "ozone" was derived by Schönbein from the Greek word "ozein," meaning "to smell." By 1867, the identity and structure of ozone were confirmed with scientists looking toward its usefulness.

The first large-scale application of ozone was the tap-water treatment in the Netherlands in 1880s. Ozonation has been in continuous use in Nice, France since 1906, to ensure disinfection of mountain stream water. Ozone is utilized for a number of specific water treatment applications, including disinfection, taste and odor control, color removal, iron and manganese oxidation, H_2S removal, nitrite and cyanide destruction, oxidation of many organics (e.g., phenols, some pesticides, and some detergents), algae destruction and removal, and as a coagulant aid. The number of ozone-based water treatment installations has been reported as more than 3000 all over the world (Rice et al., 2000).

Gaseous ozone is continuously injected into a water supply system to maintain effective concentrations of dissolved ozone for microbial disinfection. In water, molecular ozone may react directly with dissolved substances, or it may decompose to form secondary oxidants such as *hydroperoxyl* (HO_2) and *hydroxyl* (OH) radicals (Pi et al., 2005). The hydroxyl radical is an important transient species and chain-propagating radical. *Hydroxyl radicals* (OH) are extremely powerful and nonspecific oxidants that react with many organic and some inorganic compounds (Haag and Yao, 1992). Staehelin and Hoignè (1985) observed that organic solutes which can convert radicals into *superoxide radical ion* (O_2^-) always accelerate the decomposition of ozone unless an ample amount of radical scavengers is present. Decomposition of the ozone becomes accelerated by increasing the pH in pure water (Staehelin and Hoignè, 1982). In aqueous solutions, the decomposition of ozone at a given pH is often accelerated by a radical type chain reaction which can be initiated, promoted, or inhibited by various solutes.

Today, artificially generating and applying ozone occurs in three steps (Anonymous II, n.d.). In the first step, ozone is produced by ozone unit (ozonator) and infused into water. Second step is the elimination (off gassing or venting) of the ozone and other gases/odors, such as sulfur, occur by an ozone-stripping action. This step is called aeration. The final step for removing the oxidized material is filtration. A water ozonation system consists of four main sections: feed gas supply, ozone generator, ozone injection, and ozone destruct units.

8.2.2 FEED GAS SUPPLY

One of the first questions to implement a proper ozonation system is whether the application requires concentrated oxygen or simply clean, dry air. The answer to the air versus oxygen question will depend on the required ozone for specific application, type of the ozone generator, and the project budget. Depending on the requirements of the application and the capabilities of the ozone system, the feed gas requirement may be met with clean, dry, compressed air. Nearly 21% of surrounding air comprises diatomic oxygen. But ambient air contains moisture, particulates, and other contaminants that can damage ozone generators. Whether ambient air or concentrated oxygen, care must be taken to remove compressor oil, water, and other contaminants from the feed gas. Dirty feed gas will ultimately reduce feed gas flow, damage the ozone reactor cell, and reduce the effectiveness of the ozone system or shut it down entirely (Smith-McCollum, 2008). Moisture from atmospheric air condenses when the air is compressed and must be removed.

Oxygen concentrators remove nitrogen and residual water vapor by the cyclical processes of adsorbing and desorbing nitrogen on a synthetic molecular sieve material. Commercial oxygen concentrators sequentially pressurize and depressurize multiple beds of adsorbent to produce a continuous stream of concentrated diatomic oxygen (up to 95% O_2). Water vapor, having a larger molecular size than oxygen, is adsorbed and desorbed much like nitrogen and is not passed to the ozone generator. As a result, oxygen concentrators can tolerate a higher level of water vapor in the input air. Continuously, supplying poor quality feed gas to an oxygen concentrator can contaminate the adsorbent, decreasing the oxygen output of the concentrator and the ozone output from the ozone generator. Oil droplets and hydrocarbon vapor can foul the adsorbent beds, blocking the binding of nitrogen during the pressurization cycle of the concentrator. Excess water vapor is adsorbed during pressurization and exceeds the amount that can be desorbed during the depressurization cycle. Over time, the adsorbent beds become increasingly saturated with excess water vapor and the oxygen output of the concentrator decreases. Two other air preparation parameters that must be considered are the feed gas flow rate and feed gas pressure. The performance specifications of the feed gas system for these parameters must match those of the ozone generator. Inadequate feed gas flow rate and pressure will rob the generator of oxygen and reduce total ozone output (Smith-McCollum, 2008). Energy consumption for ozone production is 5–8 kW/kg in oxygen use and 15–18 kW/kg in air use (Smith, n.d.).

8.2.3 OZONE GENERATORS

Main components of ozone production system are power supply unit, generation vessel, and cooling system (Fig. 8.1). Most widely known ozone production methods are as follows:

- Electrical gas discharge (corona discharge),
- Generation ozone from water itself,
- Photochemical production by UV light,

- Electrochemical process, and
- Radiochemical process

The procedure of *electrical gas discharge*, also named *corona discharge method*, was developed by Siemens (1857). Corona discharge process could be basically defined as the condition created when a high voltage passes through an air gap (Fig. 8.1). This process is based on the application of an alternating voltage between two electrodes with dry air or oxygen passing in between. During this process, electrons are accelerated across an air gap so as to give them sufficient energy to split the oxygen double bond, thereby producing atomic oxygen. These oxygen atoms then react with other diatomic oxygen molecules to form ozone. Fundamentals of corona discharge have been extensively discussed by Chang et al. (1991) and Da Silva et al. (2003). The experimental factors affecting the corona efficiency for O_3 production are temperature of the entering gas, oxygen content, contaminants in the gas, ozone concentration, electric power, and the gas flow (Da Silva et al., 2003). At present, corona discharge is the most widely used method for commercial ozone production (Smith, n.d.).

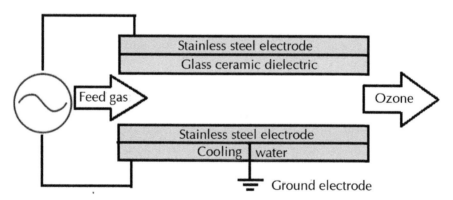

FIGURE 8.1   (**See color insert.**) Corona discharge cell.

The idea to generate ozone from drinking water is an innovative approach carried out in recent years. So-called in situ generation of ozone from water is a system which is capable of producing ozone directly in water by electrocatalytic dissociation; so, there is no need of compressor, Venturi, air dryer, and/or oxygen concentrator as in gaseous ozone production by corona discharge method (Salama and Salama, 2016). Schematically, settlement of in situ ozone generator in water tank is shown in Figure 8.2. To use in situ ozone generator, water has to have a minimum electrical conductivity which is equivalent to 100 mg/L of total dissolved solids. Typical ozone concentration depending on water conductivity is given in Table 8.1. To increase conductivity sodium chloride, sodium sulfate or potassium salts could be added in water. But organic salts use is not proper because it consumes O_3 and other oxidizing radicals. From the energetic point of view, the energy requirement of aqueous generation of ozone by in situ system is lower than gaseous ozone generation by corona discharge method. Higher conductivity will help produce more mixed oxidants and reduce power consumption. Generation of ozone in water is very efficient method not only to produce water in drinking quality but also very convenient for crop washing and equipment disinfecting.

Ozone generation by *UV radiation* was discovered for the first time by Lenard in 1900 and described more exactly by Goldstein in 1903. UV radiation method depends on the absorbing the energy of UV light by oxygen molecules (O_2) and dissociating them to oxygen atoms (O). The oxygen atoms then react with other oxygen molecules to form ozone. For effective ozone production, it is necessary to utilize a short wavelength of 185 nm. The energy for the decomposition of oxygen molecules into its atoms is supplied by photons at wavelengths below 193 nm. If longer wavelengths (>200 nm) are used, dissociation reactions take place, which leads to ozone destruction (Heim and Glas, 2011). In practice, low-pressure mercury lamps used to produce UV light produce not only the 185-nm radiation but also the 254-nm radiation that is the peak wavelength for ozone destruction. In theory, the yield of O_3 from 185 nm UV light is 130 g/kW h of light (Smith, n.d.). The method has been reviewed thoroughly by Langlais et al. (1990). Although it is an advantage to use air without any pretreatment, the low concentration of UV-generated ozone limits its industrial usage for water treatment applications.

The low ozone concentration available using electric discharge in the gaseous phase (corona process) or UV-light absorption (photochemical process) technologies restricts ozone application in processes where a higher O_3 concentration is necessary (e.g., decomposition of resistant organic pollutants). To overcome this difficulty, *electrochemical process* for ozone production which is also called *cold plasma* or *electrolysis* has been developed (Da Silva et al., 2003). In the electrochemical method of ozone production, an electrical current is applied between an anode and cathode in electrolytic solution containing water and a solution of highly electronegative anions. Electric current passed through a liquid causes a chemical reaction, resulting in the evolution of gases. A mixture of oxygen and ozone is produced at the anode. In relation to ozone production, water can be used as the electrolyte leading to direct diffusion, or special electrolytes such as sulfuric acid or aqueous phosphate solutions can be used and ozone gas can be drawn off and diffused and contacted by the usual methods. The advantages associated with this method are use of low-voltage DC current, no feed gas preparation, reduced equipment size, possible generation of ozone at high concentration, and generation in the water (Mahapatra et al., 2005). The use of electrolysis for ozone production is presently limited to small units used for applications that require high concentrations of ozone and in situ diffusion of ozone into pure water.

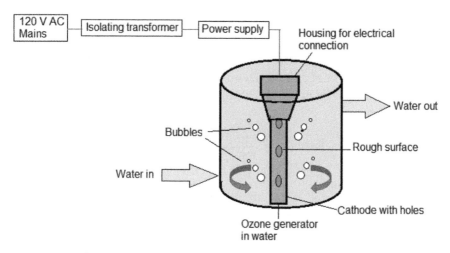

FIGURE 8.2   (See color insert.) Schematic housing of ozone generation probe in water (Salama and Salama, 2016; reprinted with permission).

TABLE 8.1   Ozone Production from Water.

Water conductivity TDS (ppm)	Ozone concentration in 2 L after 5 min ozonation O_3 (ppm)
200	0.2
400	0.4
900	0.8

(Salama and Salama, 2016; reprinted with permission.)

In *radiochemical ozone production*, the dissociation step is initiated by high-energy products of radiochemical decay reactions. The reaction is independent of the surrounding pressure and temperature. Usually, isotopes such as ^{137}Cs, ^{60}Co, or ^{90}Sr are used for excitation of circulating air in a water-cooled closed system. The thermodynamic characteristics of this kind of ozone production are very favorable, since about 35% of the available energy is used for decomposition of oxygen molecules (Heim and Glas, 2011).

8.2.4 OZONE CONTACTING

After generation, ozone is fed into a down-flow contact chamber containing the wastewater to be disinfected. The main purpose of the contactor is to transfer ozone from the gas bubble into the bulk liquid while providing sufficient contact time for disinfection. Ozone must be injected into the water in fine gas bubbles. During this procedure, turbulent mass transfer by diffusion occurs. Since the bubbles contain a few percentage of ozone, and the solubility is low, the mass transfer and, in consequence, the reaction rates are moderate (Tizaoui et al., 2008). Dissolution of ozone in water is expressed by solubility ratio (Khadre et al., 2001):

$$\text{Solubility} = \frac{\frac{mg}{L} \, O_3 \text{ in water}}{\frac{mg}{L} \, O_3 \text{ in the gas phase}}$$

Factors affecting the ozone solubility in water are partial pressure and flow rate of ozone, water temperature, bubble size, contact time and design of ozone contactor, purity and pH of water, and presence of organic matter and minerals (Hoigné and Bader, 1985; Bablon et al., 1991; Katzenelson et al., 1976).

There are two common methods for ozone injection: (1) Ozone pump—a positive displacement injector that pumps the ozone gas into the water similar to a chemical feed pump for chlorine and (2) Venturi—device that restricts flow and pressure to produce a vacuum (Anonymous II, n.d.). This vacuum device sucks the ozone gas into the water. Improper sizing will result in insufficient ozone suction, which will cause insufficient oxidation. The ozone injection system works with the Venturi principle. The components of Venturi system are water tank, measurement probe for dissolved ozone, pumps, pressure sensors, injector and bypass, and degas/contact chamber (Anonymous III, n.d.). Water can flow into and out of the water tank during the treatment process but the minimum contact time for the water to ensure treatment will depend on the size of the tank and the volume of water moving through it. The ozone concentration within the tank will depend on the volume of water moving through the tank, the organic or chemical load in the water, and the capacity of the ozone system. The pump is used to circulate water through the injection system and back to the tank. To operate correctly, the injector must have enough water flowing through it at sufficient inlet and outlet pressure. All injectors come with sets of tables to enable correct selection. Ozone is a corrosive product and the pump must be constructed of materials that can withstand that corrosive effect. The simplest way to determine that the pump is operating is to test the pressure of water leaving it. With ozone systems using oxygen or forced air feed, it is important that the ozone generator is stopped if the pump stops. This is not so important with air feed systems using only the suction from the injector. In that instance if the pump stops, then suction also stops and no ozone gas flows even if the ozone generator remains on. Switching the ozone generator off in the case of a pump fail is still desirable to ensure that other by-products are not formed that can block the fine airways of the ozone cells. All injectors have tables that detail their air suction performance under different pressures and flows.

The off gases from the contact chamber must be treated to destroy any remaining ozone before release into the atmosphere. Therefore, it is essential to maintain an optimal ozone dosage for better efficiency. When pure oxygen is used as the feed gas, the off gases from the contact chamber can be recycled to generate ozone or for reuse in the aeration tank. The ozone off gases that are not used are sent to the ozone destruction unit or are recycled.

8.2.5 MONITORING AND CONTROLLING

Accurate and reliable monitoring of dissolved and ambient ozone is critical to ensure that the production of ozone matches the requirements of the application (Rombach, 2009). Inadequate ozone production may yield inadequate oxidation and disinfection. Too much ozone may damage the product or process equipment. Equally as important as ozone residual to achieve treatment goals are ambient (airborne) ozone levels in the workplace. Ambient ozone should also be monitored to limit worker exposure and automatically shut down the system if necessary. The first step in harnessing the power of ozone is accurate and reliable monitoring of ozone potential or concentration as well as airborne ozone. Ozone dissolved in water could be measured in two ways as *colorimetric* and *instrumental*.

Three main *calorimetric methods* of measuring ozone in water are *iodometric titration*, *N,N-diethyl-p-phenylenediamine*, and *indigo trisulfonate*

(Anonymous IV, 2015). Standard *calorimetric method* provides the procedure for the *indigo trisulphonate* method. This method is based on the decolorization of blue dye by ozone, where the loss of color is directly proportional to the ozone concentration. The greater the reduction in color at a specific wavelength, more the ozone will be present. This method is considered the most accurate wet chemistry method and is widely employed in the drinking water industry and approved by the EPA. Various forms of kits are available in the market. However, these methods do not provide real-time information that can be automatically logged and applied to a control system.

Instrumental measurement of dissolved ozone includes the *oxidation–reduction potential (ORP)*, *UV absorbance*, and *membrane probe*.

ORP is the activity or strength of oxidizers and reducers in relation to their concentration in a solution. It is measured as the voltage potential at which oxidation/reduction occurs at the electrodes of an electrochemical cell. This voltage potential, typically expressed in millivolts (mV), is measured by a sensor with a platinum or gold surface that accumulates charge without reacting chemically (Rombach, 2009). A voltage is generated, which is compared to a reference electrode similar to a pH probe. The more the oxidizer available, the greater is the voltage difference between the solutions. The measurement of ORP is similar to pH measurement in that both methods determine relative activity and are not direct measures of concentration. ORP provides a relatively rapid and single-value assessment of water disinfection potential. It also provides robust performance in turbid or salt water; however, in complex and turbid solutions, ORP values can read far below expected values. One disadvantage of ORP is that it cannot be used as a direct indicator of dissolved ozone due to the effect of pH and temperature on the reading. In addition, the response time of ORP probes can be reduced in weak solutions or if the probe is saturated by overinjection of ozone. ORP is a good indicator of antimicrobial conditions. Greater than positive 650 mV is a good indication of disinfected water. Lower values have been used to gauge the quality of cooling water and other water treatment applications.

UV absorption method was developed mainly for measuring the ozone concentration in air, but it is also applicable to dissolved ozone in water—because the UV radiation at 254 nm wavelength is absorbed only by ozone; air, water, or oxygen is "transparent" to this radiation of low-pressure mercury lamp (Majewski, 2012).

Membrane probe has a gas permeable membrane over the platinum electrode. Ozone diffuses from the sample through the membrane and reacts with the electrolyte solution to form an intermediate compound. The intermediate is reduced by a polarizing voltage, which produces a current between the cathode and the anode. The current is proportional to the concentration of ozone in the sample, and the result is detected and reported by the electronic analyzer (Rombach, 2009; Schmidt, n.d.). Both ORP and amperometric analyzers require periodic calibration as per the manufacturer's instructions.

Control parameters considered for monitoring are water flow rate, water temperature, pH, oxygen concentrator performance, feed gas flow rate and back pressure, Venturi vacuum, ozone generator temperature, and ambient and cooling temperature (Rombach, 2009).

In addition to monitoring the dissolved ozone concentration in the process, airborne ambient ozone in the workplace should be monitored as well to ensure worker safety. Handheld and personal wearable ozone monitors and accumulated exposure badges are convenient and useful for spot-checking, monitoring, and documenting employee exposure to ozone gas.

When the ozone is dissolved in water, there is a need of measuring the ozone concentration at the inlet. The concentration level of the applied ozone depends on the kind of microorganisms or inorganic pollutants to be destroyed and on the required reduction ratio for microorganisms (Majewski, 2012).

8.2.6   SAFETY REQUIREMENTS

Ozone is a toxic gas and can cause severe illness, and even death, if inhaled in high quantity. The primary target for ozone toxicity is the pulmonary system (Mehlman and Borek, 1987). Primary exposure occurs when people breathe ambient air containing ozone. The rate of exposure for a given individual is related to the concentration of ozone in the surrounding air and the amount of air the individual is breathing per minute (Anonymous V, 2001). The concentration of ozone at which effects are first observed depends upon the level of sensitivity of the individual as well as the dose delivered to the respiratory tract. The dose, in turn, is a function of the ambient concentration, the minute ventilation, and the duration of exposure. According to the Occupational Safety and Health Administration of United States, ozone levels for 8 h/day exposure should never exceed 0.1 ppm for doing light work, 0.08 ppm for doing moderate work, and 0.05 ppm for doing heavy work (Anonymous VI, n.d.). Ozone in water above 1 $\mu g/mL$ can liberate ozone into the air that exceeds safe levels of 0.1 ppm (Smilanick et al., 1999). Toxicity symptoms, such as sharp irritation to the nose and throat could, result instantly at 0.1

ppm dose. Maximum time of exposure is no more than 2 h at 0.2 ppm. Ozone level of 50 ppm or more is potentially fatal (Muthukumarappan et al., 2000).

8.3 POSTHARVEST APPLICATION OF OZONE AND ITS IMPACT ON QUALITY AND SAFETY PROPERTIES OF FRESH FRUITS AND VEGETABLES

Primary objectives of fresh fruits and vegetables processing, whether for additional treatment or direct distribution to consumers, are sanitization, preventing spoilage and over-ripening, and removing contaminants. Usage of ozonation in perishable fruits and vegetables has been especially increased after 1982, when the US Food and Drug Administration recognized safe status for ozone usage as a disinfectant or sanitizer in the gas or liquid phase on food and for direct contact use as an antimicrobial for treatment, storage, and processing on diverse foods including raw and minimally processed fruits and vegetables. The increase in food-borne outbreaks linked with consumption of raw fruits and vegetables has motivated new research focusing on prevention of preharvest produce contamination by ozonation. Mechanism of microbicidal action of ozone has been reviewed in detail by Khadre et al. (2001). Below studies about ozone treatment of fresh fruits and vegetables are summarized in chronological order. It should be also noted that most of cited studies have been carried out in laboratory scale. Scaling up such results to industrial scale is not an easy task and needs another effort.

8.3.1 EFFECT OF AQUEOUS OZONE TREATMENT ON QUALITY AND SAFETY PROPERTIES OF FRUITS AND VEGETABLES

Ozone in water is often described as an alternative to chlorine-based disinfectants. The solubility of ozone in water is only partial and is governed by Henry's Law (the solubility of the gas in water is directly proportional to its partial pressure in the gas phase). Consideration of Henry's Law leads to the obvious conclusion that the higher the ozone concentration, the greater will be the solubility of ozone in water. The more ozone is dissolved in water, the more effective it is as a disinfectant. Maximum solubility of ozone in water is 29.9 μg/mL at 20°C (Smilanick et al., 1999). In practice, much lower solubility of ozone in water could be expected depending on the environmental conditions. Primary objectives of ozonation of fruits and vegetables for any application, whether for additional processing or direct distribution to consumers, are sanitization, preventing spoilage or over-ripening, and removing contaminants.

8.3.1.1 EFFECT OF AQUEOUS OZONE TREATMENT ON MICROBIAL POPULATION

Apples were inoculated with *Escherichia coli* O157:H7 and treated with ozonated water by Achen and Yousef (2001). Two ozone-delivery methods that are potentially applicable to commercial sanitization of apples were compared in this study: (1) dipping in ozonated water (22–24 mg O_3/L) and (2) washing in bubbling ozone water (~21, 25, and 28 mg O_3/L residual ozone). In both methods, apples were treated for 1, 3, and 5 min. Maximum decrease in surface counts of *E. coli* O157:H7 were 3.7 and 2.6 $\log_{10}$ CFU/g when apples were treated for 3 min by washing in water with bubbling ozone or dipping in ozonated water, respectively, compared to unwashed controls. In both delivery methods, counts of *E. coli* O157:H7 in the stem–calyx region did not decrease appreciably; the bubbling ozone wash and dip method decreased these populations 0.6 and 0.5 $\log_{10}$ CFU/g, respectively. Differences in counts among the three exposure times, in both the delivery methods, were not significant. An earlier study (Farooq et al., 1990) suggested that ozone concentration in the liquid film at the gas–liquid interface is higher than in the surrounding solution, which may account for the greater efficacy of bubbling, compared to dipping treatments. To observe the influence of temperature on the ozonation process, water at 4, 22, or 45°C was used. When apples were treated with bubbling ozone for 3 min at 4, 22, and 45°C, counts of *E. coli* O157:H7 decreased 3.3, 3.7, and 3.4 at the surface and 0.4, 0.4, and 0.2 $\log_{10}$ CFU/g on the stem–calyx regions, respectively. Statistical analysis, however, showed no significant difference between the three treatments. The residual ozone concentration was greatest at the lowest temperature (4°C) and decreased with increasing temperature. At colder temperatures, ozone is relatively stable, but as the temperature increases, the decomposition rate increases. Ozone inactivated *E. coli* O157:H7 effectively on the surface, but its efficacy was limited in the inaccessible areas (the stem and the calyx regions) of the apple. Therefore, inoculated apples were subjected in this study to pretreatments to help expose cells in these areas to the ozone wash. Spraying the apple's stem–calyx region with water prior to the ozone wash decreased the count of *E. coli* O157:H7 in core samples by 1.5 $\log_{10}$ CFU/g.

The antimicrobial effects of ozonated deionized water on *E. coli* O157:H7 inoculated onto lettuce and baby carrots were studied by Singh et al. (2002). The results obtained in this study have shown that treatment with ozonated water (5.2 mg/L) of lettuce did not result in any significant reduction in *E. coli* O157:H7 populations. However, a significant reduction in microbial populations on baby carrots was observed after 10 min of exposure to 5.2 mg/L ozonated water compared to 1 or 5 min. The reduced efficacy of ozonated water during lettuce washing might be due to more ozone demand of organic material in the medium. The study of Restaino et al. (1995) also indicated that the type of organic material present during ozonation is more important than the amount present.

The treatments for reduction of *E. coli* O157:H7 and *Listeria monocytogenes* in an aqueous model system and on inoculated apples have been studied by Rodgers et al. (2004). Samples were immersed in water containing ozone (3 ppm) for up to 5 min. In the model system study, both pathogens decreased >5 $\log_{10}$ following 2–5 min of exposure, with ozone being the most effective (15 s), followed by chlorine dioxide (19–21 s), chlorinated trisodium phosphate (25–27 s), and peroxyacetic acid (70–75 s). On apples, ozone and chlorine dioxide (5 ppm) were most effective, reducing populations approximately 5.6 log. After treatment, produce samples were stored at 4°C for 9 days and quantitatively examined for *E. coli* O157:H7, *L. monocytogenes*, mesophilic aerobic bacteria, yeasts, and molds. Populations of both pathogens remained relatively unchanged, whereas numbers of mesophilic bacteria increased 2–3 $\log_{10}$ during storage.

Blueberries inoculated with *E. coli* O157:H7 and *Salmonella* were treated with ozonated water at 20°C and 4°C by Bialka and Demirci (2007). Final ozone concentrations were measured in the treatment flask with no fruits and were 1.7, 1.8, 3.7, 7.6, 7.9, and 8.9 mg/L for 2, 4, 8, 16, 32, and 64 min, respectively, at 20°C, whereas the final ozone concentration at 4°C was 21 mg/L for a 64-min treatment time. At 20°C, reductions of *E. coli* O157:H7 ranged from 1.3 to 4.9 $\log_{10}$ CFU/g for 2- and 64-min treatments, respectively. The reduction after 64 min of treatment was significantly higher than the other treatments. When the temperature was decreased to 4°C, the $\log_{10}$ reductions increased to 5.2 $\log_{10}$ CFU/g; furthermore, this treatment resulted in plate counts less than the minimum detection limit (1.5×10^1 CFU/g) of *E. coli* O157:H7 with two out of three enrichments negative for *E. coli* O157:H7. At both 20°C and 4°C, the $\log_{10}$ reductions achieved after treatment with ozone were significantly higher than the treatment with air; 4.9 compared with 2.1 $\log_{10}$ CFU/g at 20°C and 5.2 compared with 2.3 $\log_{10}$ CFU/g at 4°C. Reductions of *Salmonella* were between 0.7 and 4.9 $\log_{10}$ CFU/g for times of 2 and 32 min, respectively. The treatment times of 32 and 64 min had significantly higher $\log_{10}$ reductions, 4.9 and 4.7 CFU/g, respectively, than treatments less than 8 min. When the temperature was decreased to 4°C, the $\log_{10}$ reduction increased to 6.2 CFU/g. Both the 20°C and 4°C ozone treatments resulted in significantly higher reductions than the air treatments that resulted in reductions between 1.1 and 2.1 $\log_{10}$ CFU/g.

Smilanick et al. (2002) determined the impact of sporocidal or higher O_3 doses on fruit shelf-life and quality. Green mold and sour rot on citrus fruit, caused by *Penicillium digitatum* and *Geotrichum citri-aurantii*, respectively, were not reduced by 20 min immersion in 10 ppm O_3. These fungi infect through wounds; their spores were placed in shallow wounds (1 mm wide by 2 mm deep) 24 h before treatment. On five peach varieties, the average natural incidence of brown rot, caused by *Monilinia fructicola*, was reduced from 10.9% to 5.4% by 1 min immersion in 1.5 ppm O_3. A treatment of 15 min with 5 ppm O_3 further reduced decay to 1.7%, but consistent control of brown rot was associated only with this severe treatment and it caused shallow pits on the fruit. Brown rot caused by spores placed in wounds before treatment was not controlled. Immersion for 1 or 5 min in 5 ppm O_3 reduced natural aerobic bacteria populations by 1.1 and 1.6 $\log_{10}$ units, respectively, and yeast and filamentous fungal populations by 0.7 and 1.3 $\log_{10}$ units, respectively. Spores of *Botrytis cinerea*, cause of gray mold, were sprayed on table grape clusters, the clusters were dried and then immersed for 1–6 min in 10 ppm O_3. In two tests, immersion for 1 min in O_3 reduced gray mold from 35% among untreated grapes to about 10%, while in two other tests, the incidence was only reduced from 35% to 26%. Minor injury to the rachis of grape clusters occurred at high O_3 rates. Immersion in ozonated water did not control postharvest decay of citrus fruit, injured peaches, and nectarines at doses that reliably controlled decay, and on table grapes, control was irregular and caused minor rachis injury at high rates.

Carrots were placed in baskets and immersed in water containing ozone at 1 ppm for 5 min at 5°C (Alegria et al., 2009). Ozonated water promoted a microbial reduction up to 0.4 log for aerobic mesophilic bacteria and slightly higher values for yeasts/molds (0.6–0.7 log reductions).

In the study of Karaca (2010), the effects of distilled water, ozonated water (12 ± 0.5 ppm), and chlorinated water (100 ppm) on reductions of *E. coli* and *Listeria innocua* inoculated on lettuce, spinach, and parsley and on some chemical parameters (chlorophyll a, chlorophyll b, ascorbic acid, and total phenolic contents and antioxidant activity) of these vegetables were investigated. Ozonated water washes for 5 and 15 min were both effective in microbial inactivation in all vegetables tested. Washing with distilled water for 15 min had a limited effect ($\sim$1 $\log_{10}$ reduction) in $E.$ $coli$ and $L.$ $innocua$ inactivation whereas chlorinated water wash was the most effective treatment resulted in nearly 3 and 2.3 $\log_{10}$ reductions, respectively. The maximum reductions achieved after ozonated water treatment for 15 min were 2.22 and 2.24 $\log_{10}$ units for $E.$ $coli$ and $L.$ $innocua$, respectively, and observed in lettuce. For all three vegetables tested, the efficiency of ozonated water in inactivation of $L.$ $innocua$ was so close to that of chlorinated water.

The inactivation of tomatoes employing ozonated water presented higher values and was significantly different with respect to the unwashed tomatoes and those washed with tap water (Venta et al., 2010). Additionally, the inactivation values indicated statistically significant differences between the washing conditions with ozonated water. For example, for treatment A (the reactor with mechanical stirring) using a dissolved ozone concentration of 1 mg/L during 15 and 30 min, 4 $\log_{10}$ reduction was achieved. Under the same conditions, for treatment B (the reactor placed on a shaker), up to 6 $\log_{10}$ reduction was reached. The infected tomatoes presented a high initial load of $E.$ $coli$, about 10^6 and 10^8 CFU/mL before treatment A and B, respectively. Therefore, the ozone disinfection effectiveness for tomatoes in the aqueous phase was observed. Taking into account the obtained results and that the harvested tomatoes presented an aerobic mesophilic concentration about 102 CFU/mL, the application of a dissolved ozone concentration of 1 mg/L for 15 min is recommended. This security factor would guarantee the disinfection of tomatoes with a higher pathogen load, as well as when tomatoes are infected with more resistant microorganisms during harvest, postharvest, or transportation to the market or factories. The effect of the dissolved ozone concentration on bacterial inactivation indicated to be more important than the effect of contact washing time, mainly when 1 mg/L was applied.

The effectiveness of ozone in aqueous solution treatment on microbial inactivation was studied for three combinations microorganism/food: $L.$ $innocua$/red bell peppers (artificially inoculated), total mesophiles/strawberries, and total coliforms/watercress, with two concentrations (0.3 and 2.0 ppm) and three treatment time (Alexandre et al., 2011). The highest microbial reductions were obtained for the highest concentration with the highest treatment time (3 min). Under those conditions, $L.$ $innocua$/peppers, total mesophiles/strawberries, and total coliforms/watercress were reduced, respectively, 2.8 ± 0.5, 2.3 ± 0.4, and 1.7 ± 0.4 log cycles. When ozone is applied in aqueous solution for reducing microbial loads in fruits and vegetables, its effectiveness is dependent on the combination of microorganism/ food. However, for all cases studied, ozonated water washings at the highest concentration were more effective than simple water washings. The presence of ozone generally added an additional reduction of 0.5–1.0 log-cycles and can be considered an adequate sanitizer method for low-contaminated fruits and vegetables.

Two brands of ozone produce washers Washer-A and Washer-B were used by Long et al. (2011). Washer-A washes produce with ozone and mid-range ultrasonic waves to facilitate cleaning, whereas Washer-B utilizes a rotational movement for washing produce in conjunction with ozone decontamination. Both washers allow consumers to adjust wash setting to either light or heavy. Ozone levels in wash water from Washers-A and -B were in the ranges of 0.05–0.1 and 0.1–0.2 ppm, respectively. Ozone washings reduced total aerobic mesophile counts on whole tomatoes but not green onions when compared to unwashed samples. Washer-B at heavy setting reduced coliform numbers on whole tomatoes by 1.9 CFU/g. Furthermore, Washer-B yielded more than $2.0 \log_{10}$ reductions of *E. coli* (inoculation level at ~6.3 log CFU/g) on the edible portions of tomatoes than Washer-A at comparable settings. A follow-up study using Washer-B showed that ozone application can significantly reduce *E. coli* and *Salmonella* in produce wash water to prevent cross-contamination. It is recommended that the removing of nonedible portions (stem scars or root bulbs) of washed tomatoes and green onions is beneficial for microbial decontamination.

P. digitatum, *Penicillium italicum*, and *B. cinerea* attack fresh fruit and cause significant postharvest decay losses. The toxicity of ozone gas at different relative humidity (RH) to control their conidia was determined by Ozkan et al. (2011). Conidia distributed on cover glasses were exposed to an atmosphere containing 200–350 µL/L of O_3 gas at 35%, 75%, and 95% RH at 25°C. After exposure to O_3 for varying periods, the *conidia* were removed from the chamber, placed on potato dextrose agar, and their germination was observed. Conidia died more rapidly at higher humidity than at lower humidity, and *P. digitatum* and *P. italicum* were more resistant to O_3 than *B. cinerea*. At 95% RH, 99% of the conidia of *P. digitatum*, *P. italicum*, and *B. cinerea* were incapable of germination after O_3 exposures of 817, 732, and 702 µL/L h, respectively. At 75% RH, similar inhibition required exposures of 1781, 1274, and 1262 µL/L h, respectively. At 35% RH, O_3 toxicity declined markedly, and 99% mortality required 11, 410, 10, 775, and 7713 μL/L h, respectively. These values can be used to select O_3 gas exposures needed to control these conidia. Conidia of *B. cinerea* were sprayed on to the surface of table grapes and 2 h later, the grapes were exposed to 800–2000 μL/L h of O_3. O_3 at 800 μL/L h or more reduced the incidence of infected berries by 85% and 45% on "Autumn Seedless" and "Scarlet Royal" grapes, respectively.

Kechinski et al. (2012) examined the effects of O_3 and hot water treatments on the epidermis of Golden papaya fruit. Heat treatments were applied in a hot-water brushing (HWB) system. Papayas were brushed under a pressurized hot water rinse stage at 45, 55, and 65°C for 60 s. In the HWB treatment, 4 ppm ozone was applied to the papayas for 1 or 2 min. No mold was observed under the wax film of fruits treated with hot water, ozonated water, and wax, indicating that the combined treatment effectively disinfected the papaya fruits.

The effect of nonthermal technologies (ozone in aqueous solution, ultrasound, and UV-C radiation) and washings with chemical solutions (sodium hypochlorite and hydrogen peroxide) on safety and quality features of strawberries was studied by Alexandre et al. (2012). Among the technologies, ozonated water washing was the most effective. On average, a 1.21 ± 0.33-log_{10} unit reductions occurred when samples were washed with water containing ozone at 0.3 ppm for 2 min.

The effects of nonthermal technologies and sanitizer solutions on *L. innocua* inoculated in red bell peppers (before treatments, after rapid freezing, and during storage under frozen conditions) were studied by Alexandre et al. (2013). *L. innocua* was selected as an indicator microorganism. This bacterium is often used as a surrogate of the pathogenic *L. monocytogenes*. Results showed that untreated samples presented the highest incidence of microorganisms before and during storage at −7 and −30°C. Simple water washings allowed microbial log reductions of 1.43 ± 0.04. Among the technologies applied, ozone and ultrasounds (Fig. 8.1) were the most effective processes. On average, 2.27 ± 0.25 and 1.98 ± 0.21 log cycles reductions occurred when samples had been washed with aqueous ozone or ultrasounds, respectively. During storage, both treatments allowed similar microbial reductions. At the end of storage at −7°C, all treated samples had undetectable *L. innocua* loads. At −30°C, the observed log-reductions were 2.11 ± 0.32 (in ozonated samples) and 1.93 ± 0.51 (in ultrasonicated samples). However, only ozonated samples presented microbial reduction significantly higher than the ones obtained with water washings. Overall, it was concluded that throughout frozen storage, samples pre-washed with all sanitizer solutions and with aqueous ozone presented lower microbial loads than untreated, UV-C irradiated and water-washed samples.

The work of Jemni et al. (2014) aims to contribute with a sustainable alternative to chemicals for avoiding deterioration of harvested dates by evaluating the single or combined use of UV-C radiation and ozonated or electrolyzed water (EW). The obtained ozonated water with 390 mV of ORP and at a concentration of 0.55 ± 0.5 mg/L O_3 was used for washing the dates for 2 min at about 15°C followed with 1 min rinsing in tap water at about 15°C. In the study, the effects different combinations of treatments on overall quality of *Deglet Nour* dates stored for 30 days at 20°C were studied. As expected, microbial counts increased with storage time. All sanitizers reduced microbial growth which in turns decreased with the increased UV-C dose applied before storage. Since 6.22 kJ/m² gave the best results and appeared as the optimum dose for disinfection of dates, combinations of this dose with neutral electrolyzed water (NEW), alkaline electrolyzed water (AEW), and O_3 were tested. All these combinations reduced the microorganisms load compared to 6.22 kJ/m² UV-C alone, showing between them a synergistic effect. The combined O_3 + 6.22 kJ/m² UV-C treatment was the most effective against molds and yeasts, with a reduction of 1.63 log CFU/g after storage. The mesophilic and coliforms total counts reached the lowest values with the O_3 + 6.22 kJ/m² UV-C, NEW + 6.22 kJ/m² UV-C, and AEW + 6.22 kJ/m² UV-C showing a reduction of about 1.05 and 0.82 log CFU/g, respectively. The low microbial counts found after UV-C treatments could be attributed to direct microorganism elimination by causing DNA denaturation.

Aqueous ozone has been widely used as an alternative sanitizer in several fresh-cut products which achieves microbial reductions and extends the produce's shelf-life (Miller et al., 2013). Liu et al. (2016) have evaluated aqueous ozone (1.4 mg/L) treatments for their effectiveness on microbial growth, quality attributes, and shelf-life of fresh-cut apples. Total bacteria counts of the control samples significantly increased 3.26 $\log_{10}$ CFU/g from 0 to 12 days of storage and reached 5.33 $\log_{10}$ CFU/g on the 12th day. By making comparison with the control samples, the initial total bacteria counts were, respectively, reduced by 0.09, 0.87, and 0.87 $\log_{10}$ CFU/g, and the final total bacteria counts were reduced by 0.76, 1.83, and 2.13 $\log_{10}$ CFU/g in those samples treated with O_3 2, 5, and 10 min. Aqueous ozone treatments not only could decrease bacteria load but also could reduce molds and yeasts load in fresh-cut apples. According to these results, all treatments have shown a similar mold level of <0.05 (in reducing the final mold counts <0.05) in decreasing the yeasts load. After 12 days of storage, O_3 5- and 10-min treatments reached values of 1 $\log_{10}$ CFU/g lower than that of control. A conclusion could be drawn that 5- and 10-min O_3 treatments had made remarkable disinfecting and reducing effects on bacteria, molds, and yeasts in fresh-cut apples. Besides, the bacteria counts of samples treated with O_3 for 5 and 10 min were kept below 5 $\log_{10}$ CFU/g and did not exceed the China Shanghai local standard (DB 31/2012-2013) during the entire storage period. The results indicated that treating fresh-cut apples with 1.4 mg/L aqueous ozone for 5 and 10 min in combination with modified atmosphere packaging (MAP) could effectively extend the shelf-life up to 10 days. Based on the study, it was underlined that the bacterial reduction in all samples was greater than the fungal reduction. It is also of much importance to note that the disinfection effects of aqueous ozone had positive correlation with the treatment time. Aguayo et al. (2014) reported the similar results that the longer the aqueous ozone treatment time was, the bigger number of microbial flora decreased in products. Nevertheless, industrial production needs short treatment time to reduce the cost, and thus, O_3 5 min was more suitable for fresh-cut apple sanitization.

8.3.1.2   EFFECT OF AQUEOUS OZONE TREATMENT ON VISUAL APPEARANCE

Kechinski et al. (2012) examined the effects of O_3 and hot water treatments on the epidermis of Golden papaya fruit. Heat treatments were applied in a HWB system. Papayas were brushed under a pressurized hot water rinse stage at 45, 55, and 65°C for 60 s. In the HWB treatment, 4 ppm ozone was applied to the papayas for 1 or 2 min. The results showed that ozone applications did not affect the fruit's cuticular surface, while heat treatments allowed natural fissures on the fruit epidermis to recover. Several crystalloid forms were identified on the epidermis of the papayas after the heat treatments. The predominant crystalloid forms on papayas are tubular and there is a positive response to temperature; the higher the temperature, the larger and more frequent the tubular crystalloid.

8.3.1.3   EFFECT OF AQUEOUS OZONE ON ENZYME ACTIVITIES

Higher activities of scavenger antioxidant enzymes may help in protecting plants from oxidative stress and can slow the senescence process of vegetables. The changes in lignification, antioxidant enzyme activities, and cell-wall compositions of fresh-cut green asparagus (*Asparagus officinalis* L.) in 1 mg/L aqueous ozone pretreated, and subsequent MAP during storage at 3°C for 25 days was investigated by An et al. (2007). The enzyme activities in fresh-cut asparagus including phenylalanine ammonia lyase, superoxide dismutase, ascorbate peroxidase, and glutathione reductase were inhibited by aqueous ozone treatment and subsequent MAP. Changes in lignin, cellulose, and hemicellulose contents were also monitored during storage. Similarly, the increase of the cell-wall compositions under the aqueous ozone treatment or/and MAP were significantly reduced.

The purpose of the study of Demirkol et al. (2008) was to examine the impact of ozone (aqueous-phase) with treatment times (1, 5, 30, 60, or 120 min) on the bulk concentrations of γ-glutamylcysteinylglycine (GSH) or cysteine (CYS) in strawberry samples. The treatment of liquid ozone could not significantly reduce the concentrations of γ-GSH and CYS in strawberry samples at any time intervals. However, 15 and 60 min exposure to liquid ozone insignificantly increased the GSH level in strawberries from 27.7 (the control) to 34.4 and 32.3 nmol/g, respectively. Similarly, the concentration of CYS in strawberry samples was insignificantly raised with all tested treatment times. A maximum increase was observed from 37.5 (the control) to 45.84 nmol/g at 30 min. The results also showed that 15-min liquid ozone treatment increased the ratio of GSH to oxidized glutathione (GSSG) by approximately 7.6%.

8.3.1.4   EFFECT OF AQUEOUS OZONE TREATMENT ON pH

Treatment of red bell pepper with ozonated water did not significantly alter the pH (4.96 ± 0.08) of the fresh samples as reported by Kechinski et al. (2012). At selected temperatures, the storage effect was not evident.

8.3.1.5   EFFECT OF AQUEOUS OZONE TREATMENT ON COLOR

Efficacy of aqueous ozone treatment on blueberries was evaluated for the purpose of decontaminating blueberries artificially contaminated with either *E. coli* O157:H7 or *Salmonella* (Bialka and Demirci, 2007). Fruits from the most effective treatment times (64 min) were analyzed for color to determine whether ozone treatment had any negative effects. Blueberries treated with ozone had mean L^*, a^*, and b^* values of 35.25, −1.71, and −2.41 compared to the untreated values of 35.95, −1.32, and −2.56, which were not significantly different.

The a^* value of lettuce in the L^*, a^*, and b^* color scale, which reflects the extent of browning, increased dramatically in lettuce treated with 10 ppm ozonated water (Koseki and Isobe, 2006). Treatment with 3 or 5 ppm ozonated water resulted in more rapid changes in a^* value than after the water treatment. The combined treatment of hot water (50°C, 2.5 min) followed by ozonated water (5 ppm, 2.5 min) greatly suppressed increases in the a^* value, thus retarding the progress of browning compared with other treatments throughout the 6-day storage.

The control of lignification by ozone in synergy with CA storage was characterized by decrease in L^* values (Chauhan et al., 2011). The results highlighted the positive role of ozonation (10 ppm for 10 min) in combination with CA storage in controlling lignification and microbial spoilage of carrot sticks.

Kechinski et al. (2012) have found that L^* color parameter for papayas, the temperature of the hydrothermal treatment, and ozonated water concentration exerted a significant effect; a^* color parameter, the temperature of the hydrothermal treatment, ozonated water concentration, and interaction between these two variables exerted a significant effect; and b^* color parameter showed that the temperature of the hydrothermal treatment and the interaction between the temperature and ozonated water concentration exerted a significant effect.

Throughout storage, strawberries treated with nonthermal technologies (ozone, UV-C radiation, and ultrasound) presented higher anthocyanins content than the samples washed with chemical solutions (hydrogen peroxide or sodium hypochlorite) (Alexandre et al., 2012). Results showed that under storage for 13 days at room temperature (15°C) or under refrigerated conditions (4°C), the anthocyanins content of strawberries was better retained if strawberries were previously treated with ozone at 0.3 ppm for 2 min. After 13 days of storage at 4 ± 1°C, ozonated strawberries preserved on average 82% of anthocyanins (when compared to fresh samples), while untreated and water-washed samples only retained 55%. Temperature has a considerable influence on anthocyanins degradation. If strawberries were kept at room temperature, anthocyanins content decreased rapidly. For all sanitizer solutions, and after 6 days of storage at room temperature, strawberries lost more than 90% of their initial anthocyanins content. However, if samples had been ultrasonicated, ozonated, UV-C radiated, or water washed, losses were 44%, 69%, 84%, and 95%, respectively.

The effect of nonthermal technologies (ozone in aqueous solution, ultrasounds, and UV-C radiation) and washings with chemical solutions (sodium hypochlorite and hydrogen peroxide) on ascorbic acid content was studied in red bell peppers (Alexandre et al., 2013). All pretreatments had a negative impact on red bell pepper color changes. However, nonthermal technologies were less severe than chemical solutions. Throughout the storage at −7°C, the samples pretreated with all technologies presented better color preservation than the untreated samples or water-washed samples. UV-C-radiated samples and ozonated samples did not differ throughout frozen storage time and were the ones with better color retention. During storage at −30°C, all technologies induced similar color alterations and results were significantly better than the ones observed in untreated samples. The storage temperature effect on color is significant: at −7°C, samples suffered considerable changes; at −30°C, color was better preserved.

$L*$ values of dates were affected by the kind of treatment and the storage period while chrome and hue were affected only by the storage period (Jemni et al., 2014). Immediately after treatment, $L*$ and chrome values decreased in UV-C alone treated samples and increased in samples treated by combined UV-C and O_3 or EW. Hue values decreased with 2.37 and 6.22 kJ/m^2 UV-C remaining stable for the other treatments.

Browning, as a particular problem, is not only for white flesh fruits such as apples and pears but also for many other fresh-cut products (Toivonen and Brummell, 2008). Browning is generally considered to be caused by a range of endogenous phenolic compounds, with subsequent reactions leading to the formation of brown, black, or red pigments. Browning index (BI), defined as brown color purity, is one of the most common indicators of browning in sugar-containing food products (Perez-Gago et al., 2006). The visual qualities of O_3-treated (dipping into water containing ozone at 1.4 mg/L for 2, 5, and 10 min and storage for 12 days at 4°C) fresh-cut apples were higher than the control samples in the early storage period, which was probably caused by ozone's rapidly oxidizing browning enzymes (Liu et al., 2016). BI of all samples observably increased along with the storage time extension that is indicated darkening of the product surfaces, which could be attributed to enzymatic browning, surface desiccation, and microbial growth. The undesirable surface discoloration was observed in the control samples at the end of storage, while aqueous ozone treatments markedly delayed the visual quality deterioration, which might be due to aqueous ozone treatments that decreased microbial growth. In the study, the BI of samples treated with O_3 5 and 10 min was still <40 on the 10th day of storage, which were acceptable in visual quality. Some researchers have been in attempts to minimize the browning in fresh-cut products by aqueous ozone treatment. Beltran et al. (2005) detected that shredded iceberg lettuce, which was washed with aqueous ozone and stored in air, maintained an excellent visual quality during 13-days storage. Chauhan et al. (2011) concluded that the combination of controlled atmosphere (CA) storage and aqueous ozone treatment for 10 min had better stabilized color of carrot sticks than solely CA at 6°C for up to 30 days.

8.3.1.6 EFFECT OF AQUEOUS OZONE TREATMENT ON WEIGHT LOSS AND TEXTURE

In the study of Alexandre et al. (2013), the applied treatments (nonthermal technologies and sanitizer solutions) had no considerable impact on firmness of red bell peppers. However, during storage at −7°C and −30°C, samples that had suffered ozonation, ultrasonication, and UV-C radiation were equivalent to the controls. After 80 days of frozen storage, ozonated ultrasonicated and UV-C radiated samples preserved 39%, 36%, and 45% of initial firmness, respectively. Untreated and water-washed samples retained 42% of firmness.

In the study of Jemni et al. (2014), it was found that the weight loss on dates after the storage period depended on sanitizer's treatment. The losses ranged from 0.75 ± 1.35 to 6.2 ± 1.35 g/kg of fresh weight (FW). The highest weight loss was found in the O_3 + UV-C (6.22 kJ/m^2)-treated dates. They explained it by the fact that during light exposure, water migrates from the inner parts of the sample to the dried surface, leading to moisture loss which caused a higher overall weight loss than in control (untreated sample). In addition, within the storage time, the high temperature and the moderate RH of conservation affect considerably the weight loss. Firmness of date fruits was affected by both, kind of sanitizing treatment and storage time. In fact, dates treated with UV-C light combined with electrolyzed or ozonated water showed the highest firmness. Values decreased at the end of storage from 1.61 ± 0.39 to 2.91 ± 0.39 N to between 1.52 ± 0.39 and 2.39 ± 0.39 N.

The results of Liu et al. (2016) showed that 5- and 10-min ozonated water treatments (dipping into water containing ozone at 1.4 mg/L) significantly reduced the weight loss of fresh-cut apples after 2-day storage. On the 12th day of storage, the weight loss of control sample was 2.99%, while the weight loss of samples treated with O_3 2, 5, and 10 min was, respectively, 2.71%, 2.21%, and 2.20%, which were lower than those in the control samples. The increase in weight loss of fresh-cut apples might be caused by water loss, respiration, and microbial growth. On one hand, ozone treatment could slow the weight loss reduction and ozone could decrease microorganisms, reduce tissue destruction, and keep the cell integrity and thus reduce moisture transpiration. On the other hand, negative oxygen ion was produced by O_3 permeation into the cells and hindered the normal conduct of sugar metabolism and apple tissue metabolism. The firmness of samples was not changed by O_3 on the first day of storage, and the firmness of aqueous ozone-treated samples was higher than that of the control samples after 2-day storage (Liu et al., 2016). Texture of fruit is determined by cell-wall composition, cell turgor, cellular anatomy, and water content. The firmness of fresh-cut apples gradually decreased along with the extension of storage period, which might be attributed to the tissue injury caused by cut or reduction in cell turgor. In the majority of published works, ozone was referred by its efficiency in maintaining firmness of products. However, the firmness of products inevitably decreased over storage time. In general, ozone has made little impact on such reduction. Ozone did not affect the texture of fresh-cut melons and fresh-cut lettuces during storage time (Bolin and Huxsoll, 1991; Selma et al., 2008).

8.3.1.7 EFFECT OF AQUEOUS OZONE TREATMENT ON RESPIRATION AND ETHYLENE PRODUCTION

Fresh-cut carrots were ozonized in water (10 ppm) for 10 min and stored under CA conditions (2% O_2, 5% CO_2, and 93% N_2) at 6 ± 1°C and 85% RH for up to 30 days (Chauhan et al., 2011). Ozonation was found to reduce lignification and maintained the keeping quality of fresh-cut carrots during CA storage. The maximum decrease in respiration and ethylene emission rates were obtained by the combination of CA with ozone followed by CA alone and ozonation compared with the control samples kept under low temperature (6 ± 1°C). Significant reduction in ascorbic acid, carotenoids, and oxidative enzymes such as polyphenol oxidase (PPO) and POD was observed due to ozonation and CA storage.

8.3.1.8 EFFECT OF AQUEOUS OZONE TREATMENT ON REMOVAL OF PESTICIDE RESIDUES

The key elements in the effectiveness of washing in removing residues are location and age of residue, water solubility of the pesticide and temperature, and type of sanitizer used in washing (Holland et al., 1994). Studies on a variety of pesticides on whole food stuffs under cold storage often have shown that residues are stable or decay only slowly. Surface residues are amenable to simple washing operations whereas systemic residues present in tissues was little affected. For example, the highly polar and systemic meth-amidophos were the only pesticide of a number tested whose residues could not be removed from field tomatoes by washing. There is evidence for a variety of crops and pesticides that the proportion of residue can be removed by washing declines with time. This has been interpreted as being due to residues tending to move into cuticular waxes or deeper layers. Polar, water-soluble pesticides are more readily removed than low polarity materials.

Although numerous research concerning sanitizing of food products with ozone were conducted, there are only few studies concerning the use of ozonated water for reduction of pesticide residues in food of plant origin (Balawejder et al., 2013).

Ong et al. (1996) have demonstrated that even as low concentration as 0.25 ppm, ozone in water enables reduction of over 50% of residue levels of substances such as azinphos-methyl, phthalimide fungicide, and formeta-nate hydrochloride in fresh apples. The ozonation process for removal of residues from treated material was more efficient at pH 4.5 compared to higher pH values.

Hwang et al. (2002) reported that a 3-ppm solution of ozone was as effec-tive as 500 ppm of calcium hypochlorite $(Ca(OCl)_2)$, and more effective than 10 ppm of ClO_2, that is, for total degradation of mancozeb (dithiocarbamate fungicide) and ethylenethiourea in fresh apples and their products.

Degradation of the four pesticides by dissolved ozone was investigated to establish the effect of operational parameters: methylparathion, parathion, diazinon, and cypermethrin (Wu et al., 2007). They were commonly used as broad-spectrum insecticides in pest control, and high residual levels had been detected in vegetables. Dissolved ozone (1.4 mg/L) was effective to oxidize 60–99% of methyl-parathion, cypermethrin, parathion, and diazinon in aqueous solution in 30 min and the degradation was mostly completed in the first 5 min. Trace amounts and unstable paraoxon and diazoxon were tentatively identified as primary ozonation by-products of parathion and diazinon. The feasibilities of using low level of dissolved ozone (1.4–2.0 mg/L) for removal of the four pesticides residue on vegetable surface (*Bras-sica rapa*) were also tested. Ozone was most effective in cypermethrin removal (>60%). The removal efficiency of pesticides was highly depended on the dissolved ozone levels and temperature. This chapter validated that ozonation is a safe and promising process for the removal of the tested pesticides from aqueous solution and vegetable surface under domestic conditions.

Ikeuraa et al. (2011) have used ozone microbubbles that were less than 50 μm in diameter and have special properties such as generation of free radicals, self-pressurization, and negative charge, and their use in the field of food science and agriculture is attracting attention. Ozone introduced in this form can reach higher concentration than in conventional systems where ozone millibubbles diameter is 2–3 mm. In that way, efficiency of treatment with aqueous ozone solution can be enhanced.

8.3.1.9 EFFECT OF AQUEOUS OZONE TREATMENT ON NUTRITIONAL VALUE

Kechinski et al. (2012) examined the effects of O_3 and hot water treatments on the epidermis of Golden papaya fruit. In the HWB treatment, 4 ppm ozone was applied to the papayas for 1 or 2 min. Fruits subjected to the hydrothermal treatments, 4 mg/L ozonated water, and wax treatment had approximately 40% higher vitamin C contents when compared to the control samples and fruits subjected to a similar treatment using 2 mg/L of ozonated water (Kechinski et al., 2012). The increase of ascorbic acid levels in leaves in response to ozone exposure has been previously reported by several authors (Chen and Gallie, 2005; Luwe et al., 1993; Perez et al., 1999; Ranieri et al., 1996). The increase in ascorbic acid occurs in response to oxidative stress caused by ozone. Due to the ability of plants to generate toxic molecular species, ozone acts as a potent phytotoxic agent that elicits plant defense reactions (Sandermann et al., 1998). Therefore, the ozone and ozone-derived oxyradicals may be scavenged by low-molecular weight antioxidants in the plant cell, such as ascorbic acid or polyamines (Schraudner et al., 1992). Elevated levels of vitamin C in regions of high metabolic activity other than the chloroplast may perform a similar function. Thus, changes in sugars and vitamin C contents in ozone-treated strawberries may be the result of an antioxidative system that promotes the biosynthesis of vitamin C from carbohydrate reserves of the fruit (Perez et al., 1999). The study of Kechinski et al. (2012) in which papayas were treated with the hydrothermal, ozone, and wax treatment showed that for vitamin C content, the temperature of the hydrothermal treatment, the concentration of ozonated water, and the interaction between these variables exerted a significant effect; for ratio, the temperature of hydrothermal treatment, ozonated water concentration, and interaction between these two variables exerted a significant effect. The decrease in vitamin C content, ratio, and a^* value indicated increased postharvest longevity of the papaya fruits. Fruits subjected to hydrothermal treatment, 4 mg/L ozonated water, and wax application had higher vitamin C contents compared to control fruits and fruits subjected to a similar treatment using 2 mg/L ozonated water. Based on these results, the use of ozone combined with hydrothermal treatment was recommended followed by a wax treatment to disinfect papaya fruits postharvest.

The effect of nonthermal technologies (ozone in aqueous solution, ultrasounds, and UV-C radiation) and washings with chemical solutions (sodium hypochlorite and hydrogen peroxide) on ascorbic acid content was studied in red bell peppers (Alexandre et al., 2013). The pretreatments and the freezing process had no significant impact on ascorbic acid content of fruits. Even though samples treated with aqueous ozone, ultrasounds, or UV-C radiation lost 6% (in average) of their initial ascorbic acid content, samples washed with sanitizer solutions lost 12% (in average). After the freezing process, and in samples treated with nonthermal technologies, an additional decrease of 2% was observed. In samples washed with sanitizers, an additional loss of 7% (in relation to initial levels) was observed. At $-7°C$, the degradation of ascorbic acid was rapid. Significant decreases in ascorbic acid contents for all treated samples (including controls) occurred during storage. However, no significant differences were found between treated samples and controls, except for ozonated and ultrasonicated samples that retained more ascorbic acid than water-washed ones.

Fresh-cut celery was dipped with ozonated water and evaluated for changes of microbiological population and physiological quality during storage at 4°C (Zhang et al., 2005). The PPO activity and respiration rate of fresh-cut celery was much inhibited by treatment of ozonated water and sensory quality of fresh-cut celery treated with ozonated water was better than that of non-treated. However, there is no significant difference between vitamin C and total sugar of fresh-cut celery treated with ozonated water and non-treated. The best effect of preservation of fresh-cut celery appeared to be the treatment of water ozonated to 0.18 ppm of concentration, with which the microbial population was lowered to 1.69 log CFU/g.

The effect of food disinfection on the beneficial biothiol contents in a suite of vegetables consumed daily, including spinach, green bean, asparagus, cucumber, and red pepper, was investigated by Qiang et al. (2005). Because majority of the biothiols exist in the interior body of the vegetables, a possible hypothesis is that all of the oxidants should be consumed near the vegetable surface, thereby protecting most of the biothiols against oxidation. Results indicated that the common disinfection technologies may result in significant loss of beneficial biothiols in vegetables which are essentially important to human health. For example, as much as 70% of biothiols were lost when spinach was treated with hydrogen peroxide for 30 min. Approximately 48–54% of biothiols were destroyed by free chlorine and aqueous-phase ozone (dipping in water containing ozone 8.0 ppm for 30 min). In red pepper, about 60–71% of reduced glutathione was oxidized by the disinfectants. The potential decrease in biothiols during disinfection was dependent upon the biothiol type, the disinfectant, and the vegetable. The mean biothiol (GSH) concentration in green bean decreased by 11% with hydrogen peroxide and 6% with free chlorine, while neither gaseous-phase nor aqueous-phase ozone decreased GSH.

8.3.2 EFFECT OF GASEOUS OZONE TREATMENT ON QUALITY AND SAFETY PROPERTIES OF FRUITS AND VEGETABLES

Gaseous ozone would be good alternative to chemical fumigants such as SO_2 or fungicides used to reduce decay in fresh and cold stored fruits and vegetables. Main demerit of such chemicals is that they are allergenic for part of the population. Besides, the widespread use of these fumigants in commercial packing houses has led to the proliferation of resistant strains of the pathogens. There are numerous reports on both the benefits and the lack of benefits of ozone in air of cold storage rooms. It is recommended that the efficacy of ozone must be individually assessed for each commodity because contradictory results reported on effects of ozone on quality of fresh fruits and vegetables.

8.3.2.1 EFFECT OF GASEOUS OZONE TREATMENT ON MICROBIAL POPULATION

Ozone is an antimicrobial gas that can extend shelf-life and protect fruit from microbial contamination. In the study of Barth et al. (1995), ozone exposure was assessed for storage of thornless blackberries which are prone to fungal decay. Blackberries were harvested and stored for 12 days at 2°C in 0.0, 0.1, and 0.3 ppm ozone. Ozone storage suppressed fungal development for 12 days, while 20% of control fruits showed decay. The main mold was *B. cinerea* in this study. Ozone storage did not cause observable injury or defects.

Ozone exposure could be considered as a possible substitute for SO_2 fumes in table grapes. In the study of Sarig et al. (1996), a significant decrease in decay was observed in berries that were treated with ozone either before or after being inoculated with *Rhizopus stolonifer*. Exposing berries to ozone was almost as effective as SO_2 fumigation for the control of storage decay caused by *R. stolonifer* and no deleterious effects were observed on the appearance of the grape cluster. This finding indicates that in addition to its sterilizing effect, ozone also induced resveratrol and pterostilbene–phyto-alexins in table grapes. These made the berries more resistant to subsequent infection.

Change in fungal decay was assessed during the postharvest life of strawberries stored in an ozonated atmosphere by Perez et al. (1999). Fruits are placed in two cold rooms controlled at 2°C, 90% RH, and ozone concen-trations of 0 (control) and 0.35 ppm, respectively. After 3 days in the cold room, simulating transport at 2°C, fruits were moved to a 20°C room for 4 days to mimic retail conditions (shelf-life). Ozone treatment (0.35 ppm) for 3 days at 2°C was partially effective in preventing fungal growth after 2 days at 20°C (day 5), with 15% less fungal decay in treated fruits. Probably, a higher ozone concentration could have caused a greater reduction of fungal growth. Nevertheless, after 4 days at 20°C (day 7), a higher incidence of *B. cinerea* rot was observed in ozone-treated strawberries, with similar rates of gray mold proliferation in ozonated and nonozonated fruits.

Postharvest green mold, caused by *P. digitatum*, and postharvest blue mold, caused by *P. italicum Wehmer*, are among the most economically important postharvest diseases of citrus worldwide. Blue mold is especially important on citrus fruit kept under cold storage for long time periods. Both diseases are controlled mainly by application of the fungicides. The effects of gaseous ozone exposure on in vitro growth of *P. digitatum* and *P. italicum* and development of postharvest green and blue molds on artifi-cially inoculated citrus fruit were evaluated by Palou et al. (2001). Valencia oranges were continuously exposed to 0.3 ± 0.05 ppm (v/v) ozone at 5°C for 4 weeks. Eureka lemons were exposed to an intermittent day–night ozone cycle (0.3 ± 0.01 ppm ozone only at night) in a commercial cold storage room at 4.5°C for 9 weeks. Both oranges and lemons were continuously exposed to 1.0 ± 0.05 ppm ozone at 10°C in an export container for 2 weeks. Exposure to ozone did not reduce final incidence of green or blue mold, although incidence of both diseases was delayed about 1 week and infections developed more slowly under ozone. Sporulation was prevented or reduced by gaseous ozone without noticeable ozone phytotoxicity to the fruit. A synergistic effect between ozone exposure and low temperature was observed for prevention of sporulation.

The gaseous ozone exerts a lethal effect toward *E. coli* O157:H7 inoculated onto lettuce and baby carrots (Singh et al., 2002). Ozone treatments (2.1–7.6 mg/L) inactivated *E. coli* O157:H7 by 0.79–1.79 and 1.11–2.64 $\log_{10}$ CFU/g on lettuce and baby carrots, respectively. The bactericidal effect increases with concentration and length of exposure to gaseous ozone on lettuce and baby carrots. Ozone treatment of lettuce leaves at 2.1, 5.2, or 7.6 mg/L did not decrease the population of *E. coli* O157:H7 significantly during 5 or 10 min of exposure; however, a significant reduction (1.42–1.79 $\log_{10}$ CFU/g) in populations of *E. coli* O157:H7 was observed after 15 min of exposure. Decolorization of lettuce leaves was also observed during 10 or 15 min exposure at 5.2 and 7.6 mg/L of ozone concentration. Similarly, ozone treatment of baby carrots did not result in any significant change in population of *E. coli* O157:H7 during 5 or 10 min exposure time at 2.1, 5.2, or 7.6 mg/L concentration level. However, the increase in exposure time (15 min) resulted in a significant decrease (1.84–2.64 $\log_{10}$ CFU/g) in population.

Continuous ozone exposure at 0.3 ppm (v/v) inhibited aerial mycelial growth and sporulation on "Elegant Lady" peaches wound inoculated with *M. fructicola*, *B. cinerea*, *Mucor piriformis*, or *Penicillium expansum* and stored for 4 weeks at 5°C and 90% RH (Palou et al., 2002). Aerial growth and sporulation, however, resumed afterward, in ambient atmospheres. Ozone exposure did not significantly reduce the incidence and severity of decay caused by these fungi with the exception of brown rot. Gray mold nesting among "Thompson Seedless" table grapes was completely inhibited under 0.3 ppm ozone when fruits were stored for 7 weeks at 5°C. Gray mold incidence, however, was not significantly reduced in spray-inoculated fruit.

Aguayo et al. (2006) observed that cyclic treatment of tomatoes with 4 ppm of gaseous ozone inhibited microbial infection of tomato fruits for 15 days. Such procedure resulted in overall better sensory quality of treated tomato fruits in comparison to the control fruits.

The effect of ozone treatment (5–30 mg/L ozone gas for 0–20 min) was considered for surface sanitation before storage (Das et al., 2006). Gaseous ozone treatment has bactericidal effect on *S. enteritidis*, inoculated on the surface of the tomatoes, and can be used for surface sanitation of *S. enteritidis* on tomatoes before storage at different conditions. Around 10 mg/L ozone gas treatment with different time intervals of 5 and 15 min was found to be effective, respectively, on low- and high-dose inoculum levels of *S. enteritidis* attached for 1 h. Another variable considered during ozone treatment was the 4-h attachment time.

The efficacy of gaseous ozone on *E. coli* O157:H7 and *Salmonella* inoculated on the surfaces of blueberries was evaluated at treatment times of 2, 4, 8, 16, 32, and 64 min (Bialka and Demirci, 2007). Reductions of *Salmonella* ranged from 0.3 to 1.0 $\log_{10}$ CFU/g for treatment times of 4 and 64 min, respectively. A treatment time of 64 min resulted in a significantly higher reduction of *Salmonella* than the lower treatments except for 32 min. Reductions of *E. coli* O157:H7 were between 0.4 and 2.2 $\log_{10}$ CFU/g for times of 4 and 64 min, respectively. The 64-min treatment resulted in significantly higher $\log_{10}$ reductions than the lower treatment times. Greater $\log_{10}$ reductions for *E. coli* O157:H7 were observed after continuous treatment compared to reductions of *Salmonella*. The pressurization of ozone gas increased reductions of *Salmonella*. Ozone gas was pressurized to 83 kPa and held for 2, 4, 8, 16, 32, or 64 min. Reductions of *Salmonella* were between 0.3 and 3.0 $\log_{10}$ CFU/g for times of 2 and 64 min, respectively. Treatment times of 32 and 64 min resulted in significantly higher reductions than the lower treatment times with reductions of 2.2 and 3.0 $\log_{10}$ CFU/g, respectively. Reductions of *E. coli* O157:H7 ranged from 0.4 to 1.4 $\log_{10}$ CFU/g for treatment times of 4 and 64 min, respectively. The 64-min treatment resulted in significantly higher reductions than other treatment times except for the 32-min treatment. Continuous ozone exposure and pressure as a combined treatment was evaluated to determine whether exposing the surface microbial populations to ozone and then following that with a pressurized treatment intended to force the ozone into any crevices was efficacious. The final treatment was to evaluate whether replacing air with ozone through a vacuum was effective.

Selma et al. (2008) have investigated the efficacy of gaseous ozone, applied under partial vacuum in a controlled reaction chamber, for the elimination of *Salmonella* inoculated on melon rind. The performance of high-dose, short-duration treatment with gaseous ozone, in this pilot system, on the microbial and sensory quality of fresh-cut cantaloupes was also evaluated. Gaseous ozone (10,000 ppm for 30 min under vacuum) reduced viable, recoverable *Salmonella* from inoculated physiologically mature non-ripe and ripe melons with a maximum reduction of 4.2 and 2.8 log CFU/rind-disk (12.6 cm^2), respectively. The efficacy of ozone exposure was influenced by carrier matrix. *Salmonella* adhering to cantaloupe was more resistant to ozone treatment when suspended in skim-milk powder before aqueous inoculation to the rind. This indicated that organic matter interferes with the contact efficiency and resultant antimicrobial activity of gaseous ozone applied as a surface disinfectant. Conversely, in the absence of an organic carrier, *Salmonella* viability loss was greater on dry exocarp surfaces than in the wetted surfaces, during ozone treatment, achieving reductions of 2.8 and 1.4 initial log CFU/rind-disk, respectively. Gaseous ozone treatment of 5000 and 20,000 ppm for 30 min reduced total coliforms, *Pseudomonas fluorescens*, yeast, and lactic acid bacteria recovery from fresh-cut cantaloupe. A dose CT value (concentration exposure time) of 600,000 ppm/min achieved maximal log CFU/melon-cube reduction, under the test conditions. Finally, fresh-cut cantaloupe treated with gaseous ozone maintained an acceptable visual quality, aroma, and firmness during 7-day storage at 5°C. Conclusions derived from this study illustrate that gaseous ozone is an effective option to risk reduction and spoilage control of fresh and fresh-cut melon. Moreover, depending on the timing of contamination and postcontamination conditions, rapid drying combined with gaseous ozone exposure may be successful as combined or sequential disinfection steps to minimize persistence of *Salmonella* on the surface of cantaloupe melons and transference during fresh-cut processing of home preparation. Based on these results, greater efficacy would be anticipated with mature but nonripe melons while ripe tissues reduce the efficacy of these gaseous ozone treatments, potentially by oxidative reaction with soluble refractive solids.

Tzortzakis et al. (2007a,b; 2008) investigated development of microbial infection in clementines, grapes, tomatoes, and plums in ozone-enriched atmosphere. In that experiment, ozone concentration of 0.1 ppm suppressed development of fungal infection for 13 days in comparison to the samples stored in atmosphere without ozone. Low-level atmospheric ozone enrichment resulted in a modest, but statistically significant, reduction in fungal lesion development, higher concentrations of the gas resulting in greater effects. This finding implies concentration-specific impacts on fungal lesion development. A fluorescent lection assay revealed that the ozone-induced inhibition of visible lesion development was reflected in a similar reduction in fungal biomass below the fruit surface. Fungal spore production in vivo was markedly reduced when fruits were stored in an ozone-enriched atmosphere. Higher concentrations/duration of exposure resulted in greater reduction in spore production, with considerable benefits resulting from exposure to low levels of ozone. Quality of tomatoes could also be maintained for a longer period of time in comparison to the control product.

In the study of Najafi and Khodaparast (2009), ozone was applied in gas form at three concentrations (1, 3, and 5 ppm) for four different periods (15, 30, 45, and 60 min) on date fruits, and the reduction in the total bacterial count, coliform, *Staphylococcus aureus* as well as yeast/mold counts was examined. The promising results indicated the efficacy of ozone to reduce the microbial populations in date fruits. *E. coli* and *S. aureus* were not found on cultured plates inoculated with the treated samples after treatment with 5 ppm in 60 min. One-hour ozone treatment at the level of 1, 3, and 5 ppm reduces the total mesophilic microorganism counts of date fruit with initial values of 4.06–3.8; 3.6, and 3.5 $\log_{10}$ CFU/g, respectively. The initial mean value of coliform bacteria was 3.54 $\log_{10}$ CFU/g. After 30 and 45 min ozone treatment at the 5-ppm concentration level, coliform counts were reduced to 0.89 and 0.44 $\log_{10}$ CFU/g, respectively. At the 5-ppm level, no coliform bacteria were found. The initial yeast/mold count was 3.93 $\log_{10}$ CFU/g. It decreased significantly as ozone concentration level increased. One hour of ozone treatment at concentration levels of 1, 3, and 5 ppm reduced the yeast/ mold count to 3.80; 3.63 and 3.50 $\log_{10}$ CFU/g, respectively. The initial mean value of *S. aureus* was 3.52 $\log_{10}$ CFU/g. After 30 and 45 min ozone treatment at the 5-ppm concentration level, *S. aureus* counts were reduced to 0.85 and 0.41 $\log_{10}$ CFU/g, respectively. At the 5-ppm level, no *S. aureus* were found.

The effect of gaseous ozone on spore viability of *B. cinerea* and mycelial growth of *B. cinerea* and *Sclerotinia sclerotiorum* were investigated by Sharpe et al. (2009). Spore viability of *B. cinerea* was reduced by over 99.5% and height of the aerial mycelium was reduced from 4.7 mm in the control to less than 1 mm after exposure to 450 or 600 ppb ozone for 48 h at 20°C. Sporulation of *B. cinerea* was also substantially inhibited by ozone treatments. However, ozone had no significant effect on mycelial growth of *S. sclerotiorum* in vitro. Decay and quality parameters including color, chlorophyll fluorescence, and ozone injury were further assessed for various horticultural commodities (apple, grape, high-bush blueberry, and carrot) treated with 450 ppb of ozone for 48 h at 20°C over a period of 12 days. Lesion size and height of the aerial mycelium were significantly reduced by the ozone treatment on carrots inoculated with mycelial agar plugs of *B. cinerea* or *S. sclerotiorum*. Lesion size was also reduced on treated apples inoculated with *B. cinerea*, and decay incidence of treated grapes was reduced.

The aim of the study carried out by Vurma et al. (2009) was to integrate an ozone-based sanitization step into existing processing practices for fresh produce and to evaluate the efficacy of this step against *E. coli*

O157:H7. Baby spinach inoculated with *E. coli* O157:H7 (107 CFU/g) was treated in a pilot-scale system with combinations of vacuum cooling and sanitizing levels of ozone gas. The contribution of process variables (ozone concentration, pressure, and treatment time) to lethality was investigated using response-surface methodology. SanVac processes decreased *E. coli* O157:H7 populations by up to 2.4 log CFU/g. An optimized SanVac process that inactivated 1.8 $\log_{10}$ CFU/g with no apparent damage to the quality of the spinach had the following parameters: O_3 at 1.5 g/kg gas mix (935 ppm, vol/vol), 10 psig of holding pressure, and 30 min of holding time. In a separate set of experiments, refrigerated spinach was treated with low ozone levels (8–16 mg/kg; 5–10 ppm, vol/vol) for up to 3 days in a system that simulated sanitization during transportation (SanTrans). The treatment decreased *E. coli* populations by up to 1.4 $\log_{10}$ CFU/g, and the optimum process resulted in a 1.0-log inactivation with minimal effect on product quality. In a third group of experiments, freshly harvested unprocessed spinach was inoculated with *E. coli* O157:H7 and sequentially subjected to optimized SanVac and SanTrans processes. This double treatment inactivated 4.1–5.0 $\log_{10}$ CFU/g, depending on the treatment time. These novel sanitization approaches were effective in considerably reducing the *E. coli* O157:H7 populations on spinach and should be relatively easy to integrate into existing fresh produce processes and practices.

The effect of electrolyzed oxidizing (EO) water in combination with ozone to control postharvest decay of tangerine was investigated by Whang-chai et al. (2010). When the fruits inoculated with *P. digitatum* were washed in EO water for 4, 8, and 16 min and stored at 5°C for 18 days, it was found that immersion of the fruit in EO water for 8 min was the most effective to reduce disease incidences. Moreover, washing fruit in EO water and keeping in a refrigerated chamber at 5°C with continuous ozone exposure at a concentration of 200 mg/L for 2 h/day to extend storage life suppressed the disease incidence until 28 days. However, none of the treatments had any effect on the quality of fruit such as total soluble solids (SS), titratable acidity (TA), percent weight loss, and peel color. Therefore, EO water may be useful for surface sanitation and ozone has potential to control the recontamination of postharvest diseases in tangerine fruit in storage room.

Gaseous ozone applied directly to the parsley samples resulted in 1.31 and 1.14 log reductions in the numbers of *E. coli* and *L. innocua*, respectively (Karaca, 2010). This treatment did not affect chlorophyll a and chlorophyll b levels in parsley, however caused 40%, 12%, and 41% of reductions in ascorbic acid, total phenolic contents, and antioxidant activity, respectively.

Tomatoes harvested at the breaker stage were exposed to ozone concentrations of 25 and 45 mg/m^3 for 2 h/day during 16 days, at non-controlled temperature and RH (Venta et al., 2010). Control group showed damages since 8 days of storage; however, the ozonated groups did not show this behavior. The main physiological damages were wrinkling, pitting injuries, cracking, and irregular maturation. The control group showed a high affectation with bacterial bland spoilage, which is easily recognized by the softness of damaged tissues and disagreeable odor. Additionally, spoilage by *Rhizopus* was observed. This fungus has a gray color appearance and it can be seen below the damaged tissues. Also, the bacterial spots and speck were observed. This appearance causes a serious impact to commercial fresh-marketing of the tomatoes. The obtained results indicated that the gaseous ozone application, under our experimental conditions of temperature and RH, had an important germicidal action on the microorganisms.

To control postharvest decay, table grapes are commercially fumigated with sulfur dioxide. Gabler et al. (2010) evaluated ozone fumigation with up to 10,000 µL/L of ozone for 2 h to control postharvest gray mold of table grapes caused by *B. cinerea*. Fumigation for 1 h with 2500 or 5000 µL/L of ozone was equal in effectiveness. Both treatments reduced postharvest gray mold among inoculated "Thompson Seedless" grapes by approximately 50% when the grapes were examined after storage for 7 days at 15°C following fumigation. In a similar experiment, "Redglobe" grapes were stored for 28 days at 0.5°C following fumigation for 1 h with 2500 or 5000 µL/L of ozone. Both treatments were equal in effectiveness, but inferior to fumigation with 10,000 µL/L. Ozone was effective when grapes were inoculated and incubated at 15°C up to 24 h before fumigation. The cluster rachis sustained minor injuries in some tests, but berries were never harmed. Ozone was applied in three combinations of time and ozone concentration (10,000 µL/L for 30 min, 5000 µL/L for 1 h, and 2500 µL/L for 2 h) where each had a constant concentration × time product ($c \times t$) of 5000 µL/L h. The effectiveness of each combination was similar. The incidence of gray mold was reduced by approximately 50% among naturally inoculated, organically grown "Autumn Seedless" and "Black Seedless" table grapes, and by 65% among "Redglobe" table grapes, when they were fumigated with 5000 µL/L ozone for 60 min in a commercial ozone chamber and stored for 6 weeks at 0.5°C.

The effects of gaseous ozone on the growth of green mold (*P. digitatum*) and the activity of antioxidant enzymes (superoxide dismutase, catalase, and ascorbate POD) in the peel of artificially inoculated tangerine (*Citrus reticulata* Blanco cv. Sai Nam Pung) fruit were examined (Boonkorn et al., 2012). Ozone (200 μL/L) was applied for 0 (control), 2, 4, or 6 h to inoculated tangerine fruit. Exposing fruit to ozone for 4 and 6 h delayed disease incidence and reduced severity. Scanning electron microscopy confirmed that exposing fruit to ozone for 4 and 6 h reduced growth of fungi on the fruit peel. The activities of superoxide dismutase, catalase, and ascorbate POD were increased after ozone fumigation and remained significantly higher than those of the control fruit through 3 days of storage at 25°C. Throughout the experiment, fruit qualities in all treatments were not affected by ozone exposure and no phytotoxicity occurred in fruit exposed to high doses of ozone.

In the study of Tuffi et al. (2012), cold-stored tomatoes (*Solanum lycopersicum* L.) and strawberries (*Fragaria* × *ananassa* Duch.) were kept at 12°C and 2°C, respectively, under 95% RH, and exposed to air (control), ozone (0.5 ppm), negative air ions (NAI), and a NAI + ozone mixture for 15 days. The effects of the gaseous treatments on control of epiphytic microflora and the development (disease incidence and aerial mycelial growth) of *B. cinerea* Pers. and *P. expansum* link on the artificially inoculated fruits were evaluated. The examined treatments reduced superficial microbial contamination after 15 days of exposure. The best performance was achieved by the NAI + ozone treatment combined with electrostatic filters. Visible mycelia and spore production on the artificially inoculated fruits were also considerably suppressed in the presence of ozone and/or NAI with a variable degree of success. The disease incidence was considerably reduced, and the mean sporulation index never surpassed a value of 1, which completely inhibited spore formation. The fungistatic effect vanished when the fruits were removed from the atmosphere enriched with oxidizing agents.

The aim of the study carried out by Ong et al. (2013) was to evaluate the effectiveness of gaseous ozone as a potential antifungal preservation technique to overcome anthracnose disease of papaya during cold storage. Different concentrations of ozone [0 (control), 0.04, 1.6, and 4 ppm] were applied for various exposure durations (48, 96, and 144 h). Radial mycelia growth and conidial germination were evaluated in vitro after fungal exposure to the different levels and durations of ozone. Significant inhibition in radial mycelia growth of *C. gloeosporioides* was observed in all ozone treatments as compared to the control during 8 days of incubation at room temperature (25 ± 3°C). Ozone treatment of papaya fruit with 1.6 ppm ozone for 96 h delayed and simultaneously decreased the disease incidence to 40% whereas disease severity was rated at 1.7, following 28 days of storage at 12

± 1°C and 80% RH. The scanning electron microscopy showed that 4-ppm ozone caused disintegration of spore structure and did not affect the cuticular surface of fruit. Thus, ozone fumigation can reduce postharvest losses of papaya caused by anthracnose.

Fresh blueberries are commonly stored and transported by refrigeration in CAs to protect shelf-life for long periods of storage. Low concentrations of ozone gas together with proper refrigeration temperature can help protect fresh blueberries quality during storage. Shelf-life of fresh high-bush blueberries was determined over 10-day storage in isolated cabinets at 4°C or 12°C under different atmosphere conditions, including air (control); 5% O_2: 15% CO_2: 80% N_2 (controlled atmosphere storage [CAS]); and ozone gas 4 ppm at 4°C or 2.5 ppm at 12°C, at high RH (90–95%) (Concha-Meyer et al., 2015). Samples were evaluated for yeast and molds growth, weight loss, and firmness. CAS and O_3 did not delay or inhibit yeast and molds growth in blueberries after 10 days at both temperatures. Fruit stored at 4°C showed lower weight loss values compared with 12°C. Blueberries stored under ozone atmosphere showed reduced weight loss at 12°C by day 10 and loss of firmness when compared to the other treatments.

Different biotic contaminations can affect apple production. Among these infections by *P. expansum*, the causal agent of blue-green postharvest rot and patulin production is particularly important (Yaseen et al., 2015). Fruits of the apple varieties, "Royal Gala", "Golden Delicious", and "Fuji" were challenged with a patulin-producing *P. expansum* strain and stored at 1 ± 1°C in presence of gaseous ozone at 0.5 µL/L for 2 months. During the storage period, fungal populations, the biosynthesis of patulin and the activity of some pathogenesis-related proteins (glucanase, POD, and phenylalanine ammonia-lyase) were evaluated. Ozone treatment reduced fungal populations and patulin production. The activity of the assayed enzymes was not directly or clearly correlated with the inhibiting effect of ozone. These results indicate that ozone could be used to increase storage duration of apple varieties to maintain their quality.

In the study of Tabakoğlu (2016), changes in some important quality parameters of fresh black mulberries were investigated during storage at 2°C and 95% RH for 6 days in atmospheres of air and ozone (0.3 and 2.4 ppm). On the first sampling day, the numbers of all microorganisms tested were significantly lower in the samples stored in ozone atmosphere compared to that in air from each other. The case was different for *Entero bacteriaceae* where the counts in samples stored in 2.4 ppm ozone atmosphere (3.67 ±

0.11 log CFU/g) were significantly lower than that in 0.3 ppm ozone (5.38 ± 0.07 log CFU/g) and air (5.87 ± 0.03 log CFU/g) at the end of storage.

8.3.2.2 EFFECT OF GASEOUS OZONE TREATMENT ON WEIGHT LOSS

During the tomato storage process, temperature values below 20°C and RH between 85% and 95% are recommended. These storage conditions allow reducing the vapor pressure difference between the fruit and the environment, decreasing the tomato transpiration and weight loss. The weight loss of tomatoes was increased significantly during the storage period (Venta et al., 2010). This behavior is mainly due to the storage conditions in this study, corresponding to mean values of temperature and relative humidity of 27°C and 60%, respectively. However, under these experimental storage conditions, an ozone effect on weight loss was observed. The ozonated tomatoes tended to have a smaller weight loss than the control tomatoes; the ozone treatment (ozone gas concentration: 25 mg/m^3) was the most effective. Ozone-treated fruits showed less weight loss than the nontreated samples after cold storage. However, this effect was not significant when the fruits were returned to ambient conditions.

The thin skin of fruits such as raspberries makes them susceptible to rapid water loss, resulting in shriveling and deterioration. The weight loss (water loss) parameter is particularly interesting to evaluate the success of storage and losses of 8–6% affect the marketability of the fruits. In the study of Giuggioli et al. (2015), red raspberries exposed to the O$_3$ constant concentration of 500 ppb (S1) and to the O$_3$ concentration between 200 and 50 ppb (S2) were compared with fruits maintained in the normal atmosphere (control). Fruits with the S1 and S2 treatment were maintained respectively under exposure to 500 ppb O$_3$ in a continuous cycle and a constant flushing of 200 ppb (12 h) and 50 ppb (12 h); both flushing were maintained up to the end of storage. The fruits were stored at 1 ± 1°C in a cold room held at 90–95% RH for 3, 6, 9, and 13 days. Every time at low temperature, same berries were held in shelf-life for different additional days (respectively, +6, +3, +9, and +2) at 20 ± 1°C to simulate retailer conditions. As expected, the weight loss of the fruits due to the transpiration and the respiration increased over time and statistically significant differences were observed among all the treatments up to the end of storage. Control fruits and the S2 treatment showed similar weight losses content (respectively, 5.30% and 5.55%) for 3 days at $1 \pm 1°C$ followed by 6 days $20 \pm 1°C$. Similar data were observed with the S1 treatment when raspberry fruits were stored for 6 days at $1 \pm 1°C$ followed by 3 days $20 \pm 1°C$. Storing fruits up to 15 days weight losses limit to the marketability were observed for all the treatments (higher than 6%), in fact independently from the exposure time to the different storage temperatures (9 and 13 days at $1 \pm 1°C$ followed respectively by 6 and 2 at $20 \pm 1°C$); all samples have showed similar weight losses within range of 10.92–7.92%.

Probably due to restricting transpiration, storage in the ozone atmosphere (2°C and 95% RH for 6 days) reduced weight loss from the fruit and this impact increased by ozone concentration (Tabakoğlu, 2016).

8.3.2.3   EFFECT OF GASEOUS OZONE TREATMENT ON COLOR

Anthocyanins are water-soluble vacuolar pigments that give many fruits and berries a blue, red, or dark purple hue and are found in a wide assortment of healthy foods. Anthocyanin acts as an effective antioxidant within the human body. In the study of Barth et al. (1995), blackberries were harvested and stored for 12 days at 2°C in 0.0, 0.1, and 0.3 ppm ozone. A sharp decrease in anthocyanin levels of ozonated blackberries was determined after 4 days of storage. By 12 days of storage, anthocyanin content of fruit juice was similar to initial levels for all treatments. Surface color was better retained in 0.1- and 0.3-ppm stored berries by 5 days and in 0.3-ppm berries by 12 days, by hue angle values.

Gaseous application of ozone during prestorage of strawberries has negative impact on anthocyanin content (Perez et al., 1999). The initial value of anthocyanin content determined in Camarosa strawberries was 959.34 $\pm$ 11.12 nmol/g of FW. After 3 days at 2°C, a decrease in the anthocyanin content of treated and nontreated samples was observed, with a significantly lower value in ozonated fruits (639.08 $\pm$ 11.01 nmol/g of FW) than in nontreated fruits (811.34 $\pm$ 6.81 nmol/g of FW). When strawberries were placed at 20°C, a slight increase in anthocyanin accumulation was determined in both treated and nontreated fruits. No significant differences were found after 4 days at 20°C.

Ten different gaseous treatments, including ozone intermittent (8 ppm of O_3 for 30 min every 2.5 h) and continuous (0.1 ppm O_3 at 0°C and 90% RH for 1 h) applications, were evaluated during 38 days of storage at 0°C followed by 6 days of shelf-life at 15°C in air (Artés-Hernández et al., 2003).

The total anthocyanin content at harvest was 170 (19 µg/g of FW of grapes), which declined in most of the treatments applied, and was reflected in the loss of red color. Peonidin-3-glucoside was detected at all sampling times as the major anthocyanin (always >50% from the total content). Treatments applied kept or decreased the total flavonol content from that measured at harvest (17 ± 1.4 µg/g of FW of berries). However, an increase of up to twofold in total stilbenoid content after shelf-life for CA and O_3 treatments was observed. At all sampling times for almost every treatment, piceid concentration remained unaltered or slightly changed, whereas large increases were observed after shelf-life for resveratrol (1.2 ± 0.6 µg/g of FW of grapes sampled at harvest), even up to three- and fourfold for O_3-treated grapes and twofold for CA-treated ones.

In the study of Malone (2003), the effect of gaseous ozone on lycopene, ascorbic acid, and color in sliced tomatoes was evaluated. This system was designed to produce 0.9 g ozone/h at a flow rate of 2.4 L/min. Samples were treated with ozonated air at treatment levels of 0, 90, 105, 120, and 135 min. No significant differences in lycopene content were found between untreated tomato slices or slices treated for 90, 105, 120, or 135 min. The ascorbic acid content of sliced tomatoes treated with gaseous ozone for 135 min was significantly lower than untreated sliced tomatoes. No significant differences were found in Hunter $L*$, $a*$, or $b*$ values between untreated tomato slices or slices treated for 135 min.

In the study of Bialka and Demirci (2007), blueberries from the 64-min continuous ozone treatment were analyzed for color to determine whether ozone treatment had any negative effects. Significant differences were observed in $L*$, $a*$, and $b*$ values of treated compared with untreated blueberries. The mean $L*$, $a*$, and $b*$ values of treated blueberries were 30.55, +2.38, and −2.85, respectively, and untreated blueberries had values of 34.06, +0.91, and −5.94. More specifically, the berries appeared darker in color after treatment. Clearly, a treatment scenario needs to be optimized to minimize color change but maximize pathogen reduction.

Browning is used as one of the major indicators for the loss of quality of cut lettuce. The lower browning for the lettuce treated by chlorine treatment may be partially attributed to higher CO_2 developed during storage, which would prevent the browning of damaged plant tissues by blocking production of phenolic compounds and inhibiting PPO activity. The browning was less affected by ozone treatment of lettuce (Wei et al., 2007). Nevertheless, the browning was less than 2.5% for a storage period of 1 week. When the ozone concentration was increased up to 10 mg/L, the browning was increased up to about 6% and 9% after 2 and 3 weeks of storage, respectively.

Ozone could be seen as an alternative to refrigeration to enhance tomato shelf-life in areas where cold facilities are not available. However, the effect of ozone on fruit ripening and quality is still unclear. From the other side, it is well known that tomato ripening can be correlated to the development of red color. Therefore, experiments were carried out to develop a redness index to characterize the dynamics of ripening which was further used to characterize the effect of ozone (exposure to 20, 35, and 50 ppm for 10 min and storage at 15, 25, and 35°C with 68% RH) on storage and ripening (Zambre et al., 2010). Several gaseous ozone treatments were applied. Color changes from green to red were monitored. Ozone treatment delayed both the development of red color as well as of rotting. Color development and rotting followed a trend like that described by Hill's equation. Shelf-life was enhanced by 12 days when treated tomatoes were stored at 15°C. The longer shelf-life was mainly due to a reduction in surface microbial count.

In the study of Giuggioli et al. (2015), red raspberries exposed to the O_3 constant concentration of 500 ppb (S1) and to the O_3 concentration between 200 and 50 ppb (S2) were compared with fruits maintained in the normal atmosphere (control). In this study, color in the raspberries was significantly affected by the treatments, especially after 15 days of storage with decreasing brightness (L) and increasing redness as indicated by decreasing chroma (C) and hue angle ($h°$) values. At the harvest time, L value was 42.1 (data not showed) and its decrease reflects the darkening of fruits, probably due the accumulation of anthocyanins, and indicates that the ripening process is occurring in the fruits. Treatments S1 and S2 showed the highest $L*$ values as a consequence of the O_3 concentration showing after 15 days of storage similar L values (32.5 and 31.7 after 9 + 6 days of shelf-life) and (37.6 and 36.1 after 13 + 2 days of shelf-life). Raspberries stored with the S2 treatment has been maintained more vivid (higher chroma) during all the storage time. The $h°$ value is directly related to the humidity during storage and no statistically significant differences were observed between treatments. In this study, no losses of bioactive compounds during the storage time were detected and no statistically significant differences were observed among treatments for each quality control. For all the storage time, both control fruits than raspberries with the S1 and S2 treatments showed similar antioxidant capacity values in a range of 21.7 and 22.3 mmol Fe^{2+}/kg. According to Tabakoğlu (2016), storage in ozonated atmosphere (2°C and 95% RH for 6 days) did not affect total monomeric anthocyanin contents and caused slight changes in ascorbic acid levels of the samples.

8.3.2.4   EFFECT OF GASEOUS OZONE TREATMENT ON RESPIRATION

The effects of ozone treatment on fresh strawberry and shredded lettuce food quality were tested by Wei et al. (2007). At each of the storage time, the O_2 concentration only slightly decreased as the applied ozone concentration increased. However, as the storage time increased, more O_2 concentration was reduced. At an applied ozone concentration of 10 mg/L, the O_2 concentration was decreased by 35% after 1 week of storage and 55% after 2 weeks' storage as compared to its initial levels. Afterward, the O_2 concentration in the treated lettuce packages generally remained around 6% (v/v). Meanwhile, the CO_2 concentration increased substantially when the ozone concentrations were increased up to 1 mg/L, above which the ozone concentrations posed little additional effects on CO_2 concentration. The increase in CO_2 concentration occurred largely within the first week, and then the CO_2 concentration became almost constant for each of the ozone concentrations. In addition, the pH showed little effect on O_2 and CO_2 concentration as long as it was kept below 7 regardless of different temperatures and ozonation contact times. When the pH was up to 8, the O_2 concentration was decreased by as much as 50% and 80% after 1 and 2 weeks of storage, respectively. Afterward, little change in the O_2 and CO_2 concentrations was observed during the storage. Almost identical respiration rates were observed for the strawberries after ozone treatment or chlorine treatment. Regardless of their different treatment conditions, the O_2 concentration was decreased by about 60% while the CO_2 concentration was increased by approximately 25–35% after the first 5 days of storage. As compared with lettuce, strawberries had higher respiration rates. Ozone treatment could greatly extend the shelf-life of postharvested produces beyond 21 days, as is evident by the lower respiration rates.

8.3.2.5   IMPACT OF GASEOUS OZONE TREATMENT ON FIRMNESS

It is well recognized that the crispy texture of fruits and vegetables is an important quality attribute to be considered as freshness by consumers. Adverse impact on produce texture would result in the quality deterioration and consequently shorten the shelf-life. The effect of ozone treatment on the firmness of fresh strawberry and shredded lettuce was evaluated by Wei et al. (2007). In general, ozone treatment caused little change in lettuce firmness throughout 21 days of storage, regardless of difference in ozone concentrations. The firmness values varied within a range of 11–13 kgf/10 g. An exception was that after treated by ozone at pH 8, the firmness of treated samples remained almost the same over 2 weeks period in storage but rapidly declined by about 25% of initial level. Nevertheless, the firmness values of treated lettuces after 21 days of storage were still 8.5–10.5 kgf/10 g regardless of different applied ozone concentrations. Similar trends were observed for strawberry samples treated with ozone. Ozone treatment of strawberries resulted in an increase in firmness during the first 5 days of storage and then remained almost the same for the following 12 days, which may be a consequence of water loss rather than retention of flesh firmness and of relatively less growth of molds on the strawberry surfaces. In comparison with chlorine treatment, ozone treatment at appropriately controlled conditions may be more helpful to keep texture crisp.

Rodoni et al. (2010) evaluated the effect of short-term gaseous ozone treatment (10 μL/L; 10 min) on tomato fruit quality and cell-wall degradation. The treatments did not modify fruit color, sugar content, acidity, or antioxidant capacity but reduced fruit damage and weight loss and induced the accumulation of phenolic compounds. In addition, softening was delayed in ozone-treated fruit. Cell-wall analysis showed that exposure to ozone decreased pectin but not hemicellulose solubilization. Polyuronide depolymerization was also reduced in ozone-treated fruit. While the treatments did not alter the activity of the pectin-degrading enzymes polygalacturonase (PG) and beta-galactosidase, a clear decrease in pectin methyl esterase was found. Results show that short-term ozone treatments might be useful to reduce fruit damage and excessive softening, two of the main factors limiting tomato postharvest life, without negatively affecting other quality attributes. The impact of the treatments on fruit softening might be associated with reduced disassembly (solubilization and depolymerization) of pectic polysaccharides.

An effect of gaseous ozone application on the reduction of firmness decay under the storage experimental conditions was observed by Venta et al. (2010). Firmness of tomatoes is one of the most important factors for acceptance of consumers and it is related to the maturity stage evolution. Firmness has an inverse behavior to the penetration distance (displacement in mm), which was measured using a penetrometer. During the maturation of tomatoes, the synthesis of ethylene and PG is increased. PG is an enzyme responsible for the softening. Tomatoes from the control group showed the smallest firmness, because the penetration distance values were significantly higher than the values obtained for ozonated tomatoes. The gaseous ozone can diminish the reaction rate of these physiological processes related to the ethylene synthesis and enzyme activities, which are responsible for the softening.

8.3.2.6 IMPACT OF GASEOUS OZONE TREATMENT ON SOLUBLE SOLIDS CONTENT AND TITRATABLE ACIDITY

Perez et al. (1999) stored strawberries at 2°C and 20°C in an atmosphere containing ozone and reported significant difference in the pattern of conversion of sucrose into glucose and fructose between treated and nontreated fruits.

During the natural maturation of tomatoes, Venta et al. (2010) have reported a slight increase of the SS content and pH (frequently not significantly), as well as a light decrease of the acidity and the SS/TA ratio. Therefore, postharvest alternatives are necessary to maintain the sensorial quality of the fruits. The effect of gaseous ozone on these organoleptic properties under the studied storage conditions was not significant. The storage temperature (27°C) and the ozone exposure had a competitive effect on SS content of tomatoes. The ozone application did not decrease the level of SS, while the TA loss had a trend to decrease during storage up to 13 days, obtaining higher values with respect to the control group. After storage period, the SS/TA ratio was higher than at the beginning.

In the study of Giuggioli et al. (2015), the soluble solids content (SSC) of the raspberries cv. Grandeaur at harvest was 10.9 °Brix; all treatments showed increases in the values during the storage time due to the weight losses which greatly affect the concentration of the total sugars. But no statistically significant differences were observed between treatments during the storage time unlike what was observed in other soft and perishable red fruits such as strawberries when the SS levels steadily increased in ozonated fruits, reaching significantly higher levels than in controls after 7 days of storage (Skog and Chug, 2001). The sweetness and acidity components affect the flavor and the taste of raspberry fruits, but this ratio (SSC/TA) remained nearly unchanged in all treatments and during the storage period.

8.3.2.7 IMPACT OF GASEOUS OZONE TREATMENT ON SENSORY QUALITY

In the study of Giuggioli et al. (2015), the sensory quality of raspberry fruits at the end of 9 and 15 days of storage was evaluated by a sensory panel. The mean subjects of sensory attributes were taste, appearance, marketing, freshness, and off-flavor. As expected, the sensory scores of berries stored by O_3 were higher than control after 9 and 15 days of storage both. Raspberries control probably suffered from anaerobic respiration, so the off-flavor of fruits reduced the sensorial scores. Generally, fruits stored with the S2 treatment that was explained above were judged of better sensory quality than berries stored with the S1 strengthening the quality analysis.

8.3.2.8 IMPACT OF GASEOUS OZONE TREATMENT ON PESTICIDE RESIDUES REMOVAL

Residues of fenhexamid, cyprodinil, pyrimethanil, and pyraclostrobin were reduced by 68.5%, 75.4%, 83.7%, and 100.0%, respectively, after a single fumigation of table grapes with 10,000 µL/L ozone for 1 h (Gabler et al., 2010). Residues of iprodione and boscalid were not significantly reduced. Ozone is unlikely to replace sulfur dioxide treatments in conventional grape production unless its efficacy is improved, but it could be an acceptable technology to use with grapes marketed under "organic" classification, where the use of SO_2 is prohibited, or if SO_2 use was to be discontinued.

Combined use of oxidants (i.e., ozone + hydrogen peroxide) or oxidant with UV radiation (i.e., ozone + UV) is referred to as advanced oxidation processes. According to Karaca et al. (2012), gaseous ozone treatment during storage has a great potential for degrading contemporary fungicides related to table grape production. In this study, the persistence of residues of some fungicides, commonly applied in table grape vineyards to reduce bunch rot, was investigated during the cold storage of "Thompson Seedless" table grape-stemmed berries in atmospheres of air or 0.3 μL/L ozone-enriched air. Grape berries were sprayed with a mixture of boscalid, iprodione, fenhexamid, cyprodinil, and pyrimethanil solutions, dried in air for 24 h, and packed in plastic clamshell containers in expanded polystyrene boxes. The boxes were stored either in ozone or in ambient air atmosphere (2°C, 95% RH) for 36 days. Storage in the ozone atmosphere markedly accelerated the rates of decline of fenhexamid, cyprodinil, and pyrimethanil, but not those of boscalid or iprodione. At the end of storage, degradation of fenhexamid, cyprodinil, or pyrimethanil was 1.6-, 2.8-, or 3.6-fold higher, respectively, in the ozone atmosphere compared with that in air.

One of the most important problems in the production of fruits and vegetables is protection from pathogens and pests. Despite application of modern plant protection products in accordance with good agricultural practice standards, it is nearly impossible to find any samples of some fruits (apple, raspberry, and blackcurrant) without residues of their active ingredients (Balawejder et al., 2013). Active ingredients of many pesticides seem to be very persistent and their residues decline very slowly. Agricultural products with lower pesticide residues are demanded by buyers in international trade due to increasing public concerns about health risks associated to residues on foods. Recently, many processes have been tested for degrading pesticides on various agricultural products. Gaseous ozone treatment as an alternative for food decontamination has been studied in recent years due to its high oxidation potential. The tests carried out on kiwifruit stored at 0–10°C showed that after 3 months, there was less than 20% decline of the initial residue levels of insecticides such as chlorpyrifos (organophosphorous insecticides, OPI), diazinon (OPI), permethrin (pyrethroid insecticide), phosmet (OPI), pirimiphos-methyl (OPI), or fungicides (iprodione or vinclozolin). Residue levels of benomyl, carbendazim, methyl thiophanate, and thiabendazole in apples stored at 0–2°C were also stable after 140–150 days where active ingredients were at 36–60% of the initial level (Holland et al., 1994).

8.3.2.9   IMPACT OF GASEOUS OZONE TREATMENT ON SUGAR AND ORGANIC ACID CONTENT

Fruit flavor is determined by an equilibrated balance among main sugars and organic acids. Ascorbic acid, better known as vitamin C, is also present in many fresh fruits, giving an added value due to its important nutritional implications. In the study of Perez et al. (1999), sucrose content decreased with time in both gaseous ozone-treated and -nontreated strawberries in cold storage. The pattern of conversion of sucrose into glucose and fructose is significantly different in ozonated and nontreated strawberries. According to Perez et al. (1999), lower content of sucrose in ozonated fruits is not correlated with a higher accumulation of glucose and fructose. The low sucrose, glucose, and fructose contents could be due to an activation of other sucrose degradation pathways in response to oxidative stress caused by ozone. In this sense, data on vitamin C content could provide important information. Ozone and ozone-derived hydroxyl radicals may be scavenged by low-molecular weight antioxidants of the plant cell such as ascorbic acid or polyamines (Schraudner et al., 1992). The increase of ascorbic acid levels in strawberry in response to ozone exposure has been reported by Perez et al. (1999). It can be assumed that elevated levels of vitamin C in regions of high metabolic activity other than the chloroplast may perform a similar function. Thus, changes in sugars and vitamin C contents in ozone-treated strawberries could be the result of an anti-oxidative system that promotes the biosynthesis of vitamin C from carbohydrate reserves of the fruit.

The effect of ozone treatment on total phenol, flavonoid, and vitamin C content of fresh-cut honey pineapple, banana "pisang mas," and guava was investigated (Artés-Hernández et al., 2007). Although O_3 did not completely inhibit fungal development, its application increased the total flavan-3-ol content at any sampling time. Continuous 0.1 μL/L O_3 application also preserved the total amount of hydroxycinnamates, while all treatments assayed maintained the flavonol content sampled at harvest. Total phenolics increased after the retail period in O_3-treated berries. Ozone treatment tested for retaining the quality of "Autumn Seedless" table grapes during long-term storage seems to maintain or even enhance the antioxidant compound content.

Ozone is expected to cause the loss of antioxidant constituents, because of oxidizing activity. However, there are some discrepancies related to the ozone effect on the acid ascorbic content loss of the horticultural commodities according to applied conditions. The gaseous ozone does not seem to penetrate the surface of tomatoes (Venta et al., 2010). Besides, due to the great solubility in water, the ascorbic acid content is inside the tomatoes (in the bulk) and not in the pericarp. For this reason, it is considered that the ozone treatment, under these experimental concentrations, does not affect the content considerably, because the pericarp acts as a physical barrier.

The effect of ozone treatment on total phenol, flavonoid, and vitamin C content of fresh-cut honey pineapple, banana "pisang mas," and guava was investigated (Alothman et al., 2010). The fresh-cut fruits were exposed to ozone at a flow rate of 8 ± 0.2 mL/s for 0, 10, 20, and 30 min. The vitamin C content of the three fruits in this study decreased significantly when treated with gaseous ozone. The anti-oxidation system, of which vitamin C is a part, in the plant tissue is expected to prevent injuries of the plant. The toxicity of ozone is related to its potential to form different reactive oxygen species (ROS) inside the cell. As an antioxidant, ascorbic acid plays a significant role in the detoxification process that results from the formation of different ROS, such as hydrogen peroxide (H_2O_2), superoxide radicals (O_2^-), and hydroxyl radicals (OH^-) inside the plant cell. Ascorbic acid is highly sensitive to various processing and storage conditions. Therefore, the vitamin C content decreased in the three fruits due to its scavenging of the free radicals formed during the decomposition of ozone. Another factor that could have contributed to the degradation of ascorbic acid in the three ozone-treated fruit types is the activation of ascorbate oxidase. This enzyme is activated under stress conditions, such as chemical exposure. Ascorbate oxidase has been reported to promote the degradation of ascorbic acid to dehydroascorbic acid. A significant negative correlation was detected between the exposure time and vitamin C content of the fruits: −0.842, −0.665, and −0.959 for pineapple, banana, and guava, respectively, which indicates that the longer the exposure time, the more vitamin C was oxidized. This correlation was significant at the 0.01 level. Ozone oxidizes organic matter via two different pathways: by direct oxidation with ozone molecules or by the generation of free-radical intermediates such as OH^-, which is considered a powerful, effective, and nonselective oxidizing agent. A study by Garcia-Viguera and Bridle (1999) indicated that the degradation of ascorbic acid by ozone is more likely to be due to the free-radical mechanism.

8.3.2.10 IMPACT OF GASEOUS OZONE TREATMENT ON AROMA QUALITY

To evaluate the effect of ozone treatment on strawberry aroma, Perez et al. (1999) have studied off-flavor formation, headspace volatile composition, and some enzymatic activities related to aroma biosynthesis. As an index of off-flavor formation, concentrations of acetaldehyde, ethanol, and ethyl acetate were analyzed during simulated transport and shelf-life of ozone-fumigated and nontreated strawberries. No significant differences were found in the acetaldehyde and ethyl acetate content of treated compared to control fruits. The levels of these two compounds, ethyl acetate and acetaldehyde, determined in all analyzed samples might be regarded as acceptable (weak off-flavor) according to the concentration limits predicted by Ke et al. (1991). The final ethanol concentration values determined for control and ozone fumigated fruits were clearly above the tolerance limits determined by Ke et al. (1991) and could be correlated with strong and medium off-flavors, respectively. This study indicates a possible effect of ozone treatment on preventing off-flavor formation in strawberry.

The most significant components of strawberry aroma are methyl and ethyl esters. The ratio of methyl/ethyl esters is the characteristic of each strawberry cultivar (Perez et al., 1992). A considerable decrease after 3 days of simulated transport at 2°C was observed in both of ozone-treated and -nontreated strawberries. The percentage of total volatile production by ozone-treated fruits was 65% compared to nontreated fruits. Data presented in the study of Perez et al. (1999) suggest that ozone treatment could have a nonreversible effect on the fruit's ability to produce volatile compounds. To investigate this detrimental effect of ozone on strawberry aroma, Perez et al. (1999) studied the effect of ozone treatment on different biochemical activities. Enzymes such as lipoxygenase (LOX), human pancreatic lipase (HPL), and alpha-1-antitrypsin (AAT) contribute to strawberry aroma biogenesis. Many of the alcohols, aldehydes, and acids found in fruit aroma are generated from the oxidative degradation of linoleic and linolenic acids catalyzed by LOX and HPL. The LOX and HPL activities of ozonated and nontreated strawberries were examined as an index of aroma formation in the fruit. No significant differences in LOX activity levels in strawberries could be attributed to ozone treatment. Contradictory data on the effect of ozone on LOX pathway have been reported in the literature, with an increase of LOX activity caused by ozone treatment of lentils but no effect in other commodities (Sandermann et al., 1998). The effect of ozone treatment on this AAT was evaluated as a parameter of strawberry aroma quality. Because no clear differences were found in any of the aroma biosynthesis-related enzymes under the study of Perez et al. (1999), a physical alteration of the fruit surface could be considered as an alternative explanation of reduced volatile emission of ozonated fruits. In this regard, changes in the lipid composition of cranberry cuticle were first described by Norton et al. (1968). Differences in wax deposition and cuticle thickness caused by ozone treatment have been used to explain differences in the ripening pattern of plum fruits (Crisosto et al., 1993).

8.3.2.11   IMPACT OF GASEOUS OZONE TREATMENT ON NUTRITIONAL VALUE

Application of ozone, free chlorine, or hydrogen peroxide to the surface of foods may also diminish the level of some nutrients (e.g., antioxidants). Similarly, ozone treatment could also result in a 40% decreased emission of volatile esters in postharvest strawberries (Perez et al., 1999) and a 16–25% decrease of ascorbic acid content in potatoes, carrots, and cabbage (Enshina and Voitik, 1989). It is reasonably supposed, therefore, that some biologically important antioxidants (e.g., biothiols) contained in vegetables or fruits may decompose upon exposure to disinfectants. Strawberries contain varied concentrations of important biothiols with strong antioxidant properties. Biothiols are a type of mercaptan having a sulfhydryl functional group and are among the most important antioxidants that protect human cells against oxidative damage which leads potentially to cancers, Alzheimer's disease, and other maladies. These important biothiols include GSH and CYS (Sen and Packer, 2000; Wlodek, 2002). The purpose of the study of Demirkol et al. (2008) was to examine the impact of ozone (gaseous and aqueous phase) and other disinfectants (including free chlorine and hydrogen peroxide) with treatment times (1, 5, 30, 60, or 120 min) on the bulk concentrations of γ-GSH or CYS in strawberry samples. A small ozone generator was used to generate a low-concentration ozone gas stream (i.e., 40 ppm) from compressed air. The flow rate of the compressed air was maintained at 5.0 L/min. Continuous reactions were carried out for gaseous-phase ozone. It was reported that the inactivation rates of *Bacillus* spores by gaseous-phase ozone increased with increasing exposure humidity. The results of the study indicated that gaseous-phase ozone treatment did not significantly affect the level of oxidized glutathione (GSSG) was also examined prior to and after disinfection as an indicator of oxidative stress in terms of the ratio of GSH to GSSG. The results of the study indicated that gaseous-phase ozone treatment did not significantly affect the level of CYS in strawberry for 15, 30, 60, and 120 min, except for 5 min. The response attribute of CYS toward gaseous-phase ozone application for different time durations drew an unusual pattern. In fact, although the CYS level dropped significantly by 35% after 5-min treatment, it started to elevate again with increasing exposure time. The amount of CYS in strawberry decreased by approximately 24% by 15 min, 6% by 30 min, 2% by 60 min, and 9% by 120 min of exposure of gaseous-phase ozone compared to the control (no ozone treatment). On the other hand, unlike CYS, the GSH level was not statistically influenced by gaseous-phase ozone treatment for all tested time intervals. Similarly, the ratio of GSH to GSSG (reduced glutathione to oxidized glutathione), an oxidative stress parameter, was not statistically affected in strawberry samples.

8.3.2.12   IMPACT OF GASEOUS OZONE TREATMENT ON ETHYLENE PRODUCTION

Postharvest ozone application has recently been shown to inhibit the onset of senescence symptoms on fleshy fruit and vegetables; however, the exact mechanism of action is yet unknown. To characterize the impact of ozone on the postharvest performance of kiwifruit (*Actinidia deliciosa* cv. "Hayward"), fruits were cold stored (0°C, 95% RH) in a commercial ethylene-free room for 1, 3, or 5 months in the absence (control) or presence of ozone (0.3 µL/L) and subsequently were allowed to ripen at a higher temperature (20°C), herein defined as the shelf-life period, for up to 12 days (Minas et al., 2012). Ozone blocked ethylene production, delayed ripening, and stimulated antioxidant and antiradical activities of fruits. Ripening induced protein carbonylation in kiwifruit but this effect was depressed by ozone. A set of candidate kiwifruit proteins that are sensitive to carbonylation was also discovered. Overall, the present data indicate that ozone improved kiwifruit postharvest behavior, thus providing a first step toward understanding the active role of this molecule in fruit ripening.

8.4   CONCLUSION

Application of gaseous or aqueous ozone in fresh fruits and vegetables has been deeply studied in previous works as a way of enhancing quality and safety of produces and prolonging the shelf-life. The efficiency of ozone treatment usually depends on the susceptibility of the target organisms, the contacting conditions (e.g., exposure time, disinfectant concentration, pH, etc.), and characteristics of vegetable and fruit surfaces. On the other hand, ozone technology could be used either alone or in conjunction with other treatment methods. Slight change in any operational parameters could cause success or failure toward achieving expected benefit from ozonation. A survey on previous works showed that although the results on ozonation of fresh produces are promising on a laboratory scale, data on large-scale ozonation plants in processing of fresh fruits and vegetables stay confidential business information of manufacturer. In order for ozone disinfection to become commercially viable alternative to other methods, it is necessary to build a balance between overall cost of an ozonation system and its benefits. A commercial scale ozonation system could be only invested by processors if it provides more benefit compared to other treatments. The overall cost of an ozonation system is determined by the capital and operation and maintenance expenses. Capital cost is roughly dependent on the following factors:

- Ozone output of generator
- Required ozone concentration in water or in air
- Hardware configuration of ozone generator
- Site-specific factors
- Safety aspects

The annual operating costs for ozone disinfection include power consumption, and supplies, miscellaneous equipment repairs, and staffing requirements. For operating large-scale ozone systems, technical restrictions given by EPA should be considered (Anonymous VII, n.d.).

KEYWORDS

- **microorganisms**
- **shelf-life**
- **spoilage**
- **ozone**
- **postharvest**

REFERENCES

Achen, M.; Yousef, A. E. Efficacy of Ozone against *Escherichia coli* O157:H7 on Apples. *J. Food Sci.* **2001,** *66* (9), 1380–1384.

Aguayo, E.; Escalona, V. H.; Artes, F. Effect of Cyclic Exposure to Ozone Gas on Physicochemical, Sensorial and Microbial Quality of Whole and Sliced Tomatoes. *Postharvest Biol. Technol.* **2006,** *39,* 169–177.

Alegria, C.; Pinheiro, J.; Goncalves, E. M.; Fernandes, I.; Moldao, M.; Abreu, M. Quality Attributes of Shredded Carrot (*Daucus Carota* L. cv. Nantes) as Affected by Alternative Decontamination Processes to Chlorine. *Innov. Food Sci. Emerg. Technol.* **2009,** *10* (1), 61–69.

Alexandre, E. M. C.; Brandao, T. R. S.; Silva, C. L. M. Impact of Non-thermal Technologies and Sanitizer Solutions on Microbial Load Reduction and Quality Factor Retention of Frozen Red Bell Peppers. *Innov. Food Sci. Emerg. Technol.* **2013,** *17,* 99–105.

Alexandre, E. M. C.; Brandao, T. R. S.; Silva, C. L. M. Efficacy of Non-thermal Technologies and Sanitizer Solutions on Microbial Load Reduction and Quality Retention of Strawberries. *J. Food Eng.* **2012,** *108* (3), 417–426.

Alexandre, E. M. C.; Santos-Pedro, D. M.; Brandao, T. R. S.; Silva, C. L. M. Influence of Aqueous Ozone, Blanching and Combined Treatments on Microbial Load of Red Bell Peppers, Strawberries and Watercress. *J. Food Eng.* **2011,** *105* (2), 277–282.

Alothman, M.; Kaur, B.; Fazilah, A.; Bhat, R.; Karim, A. A. Ozone Induced Changes of Antioxidant Capacity of Freshcut Tropical Fruits. *Innov. Food Sci. Emerg. Technol.* **2010,** *11* (4), 666–671.

An, J. S.; Zhang, M.; Lu, Q. R. Changes in Some Quality Indexes in Fresh-Cut Green Asparagus Pretreated with Aqueous Ozone and Subsequent Modified Atmosphere Packaging. *J. Food Eng.* **2007,** *78* (1), 340–344.

Andrews, J. H.; Harris, R. F. The Ecology and Biogeography of Microorganisms on Plant Surfaces. *Ann. Rev. Phytopathol.* **2000,** *38,* 145–180.

Anonymous I. *What is Ozone?* n.d. Retrieved from https://www3.epa.gov/apti/ozonehealth/what.html (accessed May 20, 2016).

Anonymous II. *Ozone Injection Selecting Ozone Injection Devises—Ozone Pump vs. Ventri.* n.d. Retrieved from http://www.wqpmag.com/ozone-injection (accessed June 01, 2016).

Anonymous III. *Oxygen Injection into Water Tank.* n.d. Retrieved from http://www.oxyzone.com.au/_assets/Batch_injection_system.pdf (accessed June 15, 2016).

Anonymous IV. *The Measurement of dissolved Ozone.* 2015. Retrieved from https://www.chemetrics.com/image/data/product/pdf/Measurement_of_Ozone_White_Paper_Final.pdf (accessed June 1, 2016).

Anonymous V. *Secondary Direct Food Additives Permitted in Food for Human Consumption.* 2001. Retrieved from http://www.fda.gov/OHRMS/Dockets/98fr/062601a.htm (accessed June 02, 2016).

Anonymous VI. *Health Effects of Ozone in the General Population.* n.d. Retrieved from https://www3.epa.gov/apti/ozonehealth/population.html (accessed June 21, 2016).

Anonymous VII. *Wastewater Technology Fact Sheet: Ozone Disinfection. EPA 832-F-99-063 September 1999.* n.d. Retrieved from https://www3.epa.gov/npdes/pubs/ozon.pdf (accessed July 7, 2016).

Artés-Hernández, F.; Aguayo, E.; Artes, F.; Tomas-Barberan, F. A. Enriched Ozone Atmosphere Enhances Bioactive Phenolics in Seedless Table Grapes after Prolonged Shelf Life. *J. Sci. Food Agric.* **2007,** *87* (5), 824–831.

Artés-Hernández, F.; Artes, F.; Tomas-Barberan, F. A. Quality and Enhancement of Bioactive Phenolics in cv. Napoleon Table Grapes Exposed to Different Postharvest Gaseous Treatments. *J. Agric. Food Chem.* **2003,** *51* (18), 5290–5295.

Bablon, G.; Bellamy, W. D.; Bourbigot, M. M.; Daniel, F. B.; Dore, M.; Erb, F.; Gordon, G.; Langlais, B.; Laplanche, A.; Legube, B.; Martin, G.; Masschelein, W. J.; Pacey, G.; Reckhow, D. A.; Ventresque, C. Practical Applications of Ozone: Principles and Case Studies. *Ozone in Water Treatment: Application and Engineering*; Lewis Publishers: Chelsea, MI, 1991; p 569.

Balawejder, M.; Antos, P.; Sadlo, S. Potential of Ozone Utilization for Reduction of Pesticide Residue in Food of Plant Origin. *A Rev. Rocz. Panstw. Zakl. Hig.* **2013,** *64* (1), 13–18.

Baratharaj, V. *Use of Ozone in Food Processing and Cold Storage.* n.d. Retrieved from http://www.otsil.net/articles/Food%20Processing%20-%2002.pdf (accessed July 7, 2016).

Barth, M.; Hankinson, T. R.; Zhuang, H.; Breidt, F. Microbiological Spoilage of Fruits and Vegetables. In *Compendium of the Microbiological Spoilage of Foods and Beverages. Food Microbiology and Food Safety*; Sperber, W. H., Doyle, M. P., Eds.; Springer: Berlin-Heidelberg, 2009; pp 135–183.

Barth, M. M.; Zhou, C.; Mercier, J.; Payne, F. A. I. Ozone Storage Effects on Anthocyanin Content and Fungal Growth in Blackberries. *J. Food Sci.* **1995,** *60,* 1286–1288.

Beltran, D.; Selma, M. V.; Marin, A.; Gil, M. I. Ozonated Water Extends the Shelf Life of Fresh Cut Lettuce. *J. Agric. Food Chem.* **2005,** *53,* 5654–5663.

Bialka, K. L.; Demirci, A. Decontamination of *Escherichia coli* O157:H7 and *Salmonella enterica* on Blueberries Using Ozone and Pulsed UV-Light. *J. Food Sci.* **2007,** *72* (9), M391–M396.

Bolin, H. R.; Huxsoll, C. C. Effect of Preparation Procedures and Storage Parameters on Quality Retention of Salad-Cut Lettuce. *J. Food Sci.* **1991,** *56,* 60–62.

Boonkorn, P.; Gemma, H.; Sugaya, S.; Setha, S.; Uthaibutra, J.; Whangchai, K. Impact of High Dose, Short Periods of Ozone Exposure on Green Mold and Antioxidant Enzyme Activity of Tangerine Fruit. *Postharvest Biol. Technol.* **2012,** *67,* 25–28.

Buzby, J. C.; Wells, H. F.; Hyman, J. *The Estimated Amount, Value, and Calories of Postharvest Food Losses at the Retail and Consumer Levels in the United States.* 2014. Retrieved from http://www.ers.usda.gov/media/1282292/eib121_summary.pdf (accessed June 07, 2016).

Chang, J. S.; Lawless, P. A.; Yamamoto, T. Corona Discharge Processes. *IEEE Trans. Plasma Sci.* **1991,** *19* (6), 1152–1166.

Chang, T. L.; Streicher, R.; Zimmer, H. The Interaction of Aqueous Solutions of Chlorine with Malic Acid, Tartaric Acid, and Various Fruit Juices—A Source of Mutagens. *Anal. Lett.* **1988,** *21,* 2049–2067.

Chauhan, O. P.; Raju, P. S.; Ravi, N.; Singh, A.; Bawa, A. S. Effectiveness of Ozone in Combination with Controlled Atmosphere on Quality Characteristics Including Lignification of Carrot Sticks. *J. Food Eng.* **2011,** *102* (1), 43–48.

Chen, Z.; Gallie, D. R. Increasing Tolerance to Ozone by Elevating Foliar Ascorbic Acid Confers Greater Protection against Ozone than Increasing Avoidance. *Plant Physiol.* **2005,** *138* (3), 1673–1689.

Concha-Meyer, A.; Eifert, J. D.; Williams, R. C.; Marcy, J. E.; Welbaum, G. E. Shelf Life Determination of Fresh Blueberries (*Vaccinium corymbosum*) Stored under Controlled Atmosphere and Ozone. *Int. J. Food Sci.* **2015**. DOI:10.1155/2015/164143.

Crisosto, C. H.; Retzlaff, W. A.; Williams, L. E.; DeJong, T. M.; Zoffoli, J. P. Postharvest Performance Evaluation of Plum (*Prunus salicina* Lindel., 'Casselman') Fruit Grown under Three Ozone Concentrations. *J. Am. Soc. Hortic. Sci.* **1993**, *118*, 497–502.

Da Silva, L. M.; Santana, M. H. P.; Boodts, J. F. C. Electrochemistry and Green Chemical Processes: Electrochemical Ozone Production. *Quím. Nova* **2003**, *26* (6). DOI:10.1590/S0100-40422003000600017.

Das, E.; Gurakan, G. C.; Bayindirli, A. Effect of Controlled Atmosphere Storage, Modified Atmosphere Packaging and Gaseous Ozone Treatment on the Survival of *Salmonella enteritidis* on Cherry Tomatoes. *Food Microbiol.* **2006**, *23* (5), 430–438.

Demirkol, O.; Cagri-Mehmetoglu, A.; Qiang, Z. M.; Ercal, N.; Adams, C. Impact of Food Disinfection on Beneficial Biothiol Contents in Strawberry. *J. Agric. Food Chem.* **2008**, *56* (21), 10414–10421.

Enshina, A. N.; Voitik, N. P. The Effect of Regular Treatment with Ozone on the Chemical Composition of Potatoes and Vegetables. *Vopr. Pitan.* **1989**, *6*, 61–64.

Farooq, S.; Churey, J. J.; Splittstoesseer, D. F. Effect of Processing Conditions on the Microflora of Fresh Cut Vegetables. *J. Food Prot.* **1990**, *53*, 701–703.

Gabler, F. M.; Smilanick, J. L.; Mansour, M. F.; Karaca, H. Influence of Fumigation with High Concentrations of Ozone Gas on Postharvest Gray Mold and Fungicide Residues on Table Grapes. *Postharvest Biol. Technol.* **2010**, *55* (2), 85–90.

Garcia-Viguera, C.; Bridle, P. Influence of Structure on Color Stability of Anthocyanins Flavylium Salts with Ascorbic Acid. *Food Chem.* **1999**, *64*, 21–26.

Giuggioli, N. R.; Briano, R.; Girgenti, V.; Peano, C. Quality Effect of Ozone Treatment for the Red Raspberries Storage. *Chem. Eng. Trans.* **2015**, *44*, 25–30.

Haag, W. R.; Yao, C. C. D. Rate Constants for Reaction of Hydroxyl Radicals with Several Drinking Water Contaminants. *Environ. Sci. Technol.* **1992**, *26* (5), 1005–1013.

Heim, C.; Glas, K. Ozone I: Characteristics/Generation/Possible Applications. *Brew. Sci.* **2011**, *64* (8), 8–12.

Hoigné, J.; Bader, H. Rate Constants of Reactions of Ozone with Organic and Inorganic Compounds in Water. III: Inorganic Compounds and Radicals. *Water Res.* **1985**, *19*, 993–1004.

Holland, P. T.; Hamilton, D.; Ohlin, B.; Skidmore, M. W. Effects of Storage and Processing on Pesticide Residues in Plant Products. *Pure Appl. Chem.* **1994**, *66* (2), 335–356.

Hwang, E. S.; Cash, J. N.; Zabik, M. J. Degradation of Mancozeb and Ethylenethiourea in Apples Due to Postharvest Treatments and Processing. *J. Food Sci.* **2002**, *67*, 3295–3300.

Ikeuraa, H.; Kobayashi, F.; Tamaki, M. Removal of Residual Pesticides in Vegetables Using Ozone Microbubbles. *J. Hazard. Mater.* **2011**, *186*, 956–959.

Janisiewicz, W. J.; Korsten, L. Biological Control of Postharvest Diseases of Fruits. *Annu. Rev. Phytopathol.* **2002**, *40*, 411–441.

Jay, J.; Loesner, M.; Golden, D. *Modern Food Microbiology*, 7th ed.; Springer Science + Business Media Inc.: New York, 2005.

Jemni, M.; Gomez, P. A.; Souza, M.; Chaira, N.; Ferchichi, A.; Oton, M.; Artes, F. Combined Effect of UV-C, Ozone and Electrolyzed Water for Keeping Overall Quality of Date Palm. *LWT—Food Sci. Technol.* **2014**, *59*, 649–655.

Karaca, H.; Walse, S. S.; Smilanick, J. L. Effect of Continuous 0.3 µL/L Gaseous Ozone Exposure. *Postharvest Biol. Technol.* **2012**, *64*, 154–159.

Karaca, H. Effect of Ozonation on Microbial Inactivation and Shelf Life of Lettuce, Spinach and Parsley. *Ph.D. Thesis*, Ankara University Graduate School of Natural and Applied Sciences Department of Food Engineering: Turkey, 2010.

Katzenelson, E.; Kletter, B.; Shuval, H. F. Inactivation Kinetics of Viruses and Bacteria in Water by Use of Ozone. *J. Am. Water Works Assoc.* **1976**, *66*, 725–729.

Ke, D.; Goldstein, L.; O'Mahoney, M.; Kader, A. A. Effects of Short-Term Exposure to Low O_2 and High CO_2 Atmospheres on Quality Attributes of Strawberries. *J. Food Sci.* **1991**, *56*, 50–54.

Kechinski, C. P.; Montero, C. R. S.; Guimaraes, P. V. R.; Norena, C. P. Z.; Marczak, L. D. F.; Tessaro, I. C.; Bender, R. J. Effects of Ozonized Water and Heat Treatment on the Papaya Fruit Epidermis. *Food Bioprod. Process* **2012**, *90* (C2), 118–122.

Khadre, M. A.; Yousef, A. E.; Kim, J. G. Microbiological Aspects of Ozone Applications in Food: A Review. *J. Food Sci.* **2001**, *66* (9), 1242–1252.

Kim, J.; Yousef, A.; Dave, S. Application of Ozone for Enhancing the Microbiological Safety and Quality of Foods—A Review. *J. Food Prot.* **1999**, *62* (9), 1071–1087.

Koseki, S.; Isobe, S. Effect of Ozonated Water Treatment on Microbial Control and on Browning of Iceberg Lettuce (*Lactuca sativa* L.). *J. Food Prot.* **2006**, *69* (1), 154–160.

Langlais, B.; Reckhow, D. A.; Brink, D. R. Fundamental Aspects. *Ozone in Water Treatment, Application and Engineering. Cooperative Research Report*; Lewis Publishers: London, 1990; pp MI: 133–316.

Liu, C.; Ma, T.; Hu, W.; Tian, M.; Sun, L. Effects of Aqueous Ozone Treatments on Microbial Load Reduction and Shelf Life Extension of Fresh-Cut Apple. *Int. J. Food Sci. Technol.* **2016**, *51*, 1099–1109.

Long, W.; Pao, S.; Inserra, P.; Westbrook, E.; Ahn, S. Efficacy of Ozone Produce Washers in Reducing Natural and Artificially Inoculated Microorganisms on Roma Tomatoes and Green Onions. *J. Food Safety* **2011**, *31* (2), 268–275.

Luwe, M. W. F.; Takahama, U.; Heber, U. Role of Ascorbate in Detoxifying Ozone in the Apoplast of Spinach (*Spinacia oleracea* L.) Leaves. *Plant Physiol.* **1993**, *101* (3), 969–976.

Mahapatra, A. K.; Muthukumarappan, K.; Julson, J. L. Applications of Ozone, Bacteriocins and Irradiation in Food Processing: A Review. *Crit. Rev. Food Sci. Nutr.* **2005**, *45*, 447–461.

Majewski, J. Methods for Measuring Ozone Concentration in Ozone-Treated Water. *Przegląd Elektrotechniczny (Electr. Rev.)* **2012**, 253–255. Retrieved from http://pe.org.pl/articles/2012/9b/61.pdf (accessed June 01, 2016).

Malone, S. Effect of Gaseous Ozone on Antioxidant Content and Color of Sliced Tomatoes. *Master's Thesis*; University of Tennessee: Knoxville, TN, 2003. Retrieved from http://trace.tennessee.edu/utk_gradthes/2106 (accessed June 21, 2016).

McKenzie; Sarr, K. S.; Mayura, A. B.; Bailey, K. R. H.; Miller, D. R.; Rogers, T. D.; Norred, W. P.; Voss, K. A.; Plattner, R. D.; Kubena, L. F.; Phillips, T. D. Oxidative Degradation and Detoxification of Mycotoxins Using a Novel Source of Ozone. *Food Chem. Toxicol.* **1997**, *35*, 807–820.

Mehlman, M. A.; Borek, C. Toxicity and Biochemical Mechanisms of Ozone. *Environ. Res.* **1987**, *42* (1), 36–53.

Miller, F. A.; Silva, C. L. M.; Brandao, T. R. S. A Review on Ozone-Based Treatments for Fruit and Vegetables Preservation. *Food Eng. Rev.* **2013**, *5*, 77–106.

Minas, I. S.; Tanou, G.; Belghazi, M.; Job, D.; Manganaris, G. A.; Molassiotis, A.; Vasilakakis, M. Physiological and Proteomic Approaches to Address the Active Role of Ozone in Kiwifruit Postharvest Ripening. *J. Exp. Bot.* **2012,** *63* (7), 2449–2464.

Muthukumarappan, K.; Halaweish, F.; Naidu, A. S. Ozone. In *Natural Food Anti-microbial systems*; Naidu, A. S., Ed.; CRC Press: Boca Raton, FL, 2000; pp 783–800.

Najafi, M. B. H.; Khodaparast, M. H. H. Efficacy of Ozone to Reduce Microbial Populations in Date Fruits. *Food Contr.* **2009,** *20* (1), 27–30.

Norton, J. S.; Charig, A. J.; Demoranville, I. E. The Effect of Ozone on Storage of Cranberries. *Proc. Am. Soc. Hortic. Sci.* **1968,** *93*, 792–796.

Ong, K. C.; Cash, J. N.; Zabik, M. J.; Siddiq, M.; Jones, A. L. Chlorine and Ozone Washes for Pesticide Removal from Apples and Processed Apple Source. *Food Chem.* **1996,** *55*, 153–160.

Ong, M. K.; Kazi, F. K.; Forney, C. F.; Ali, A. Effect of Gaseous Ozone on Papaya Anthracnose. *Food Bioprocess Technol.* **2013,** *6* (11), 2996–3005.

Ozkan, R.; Smilanick, J. L.; Karabulut, O. A. Toxicity of Ozone Gas to Conidia of *Penicillium digitatum, Penicillium italicum*, and *Botrytis cinerea* and Control of Gray Mold on Table Grapes. *Postharvest Biol. Technol.* **2011,** *60* (1), 47–51.

Palou, L.; Crisosto, C. H.; Smilanick, J. L.; Adaskaveg, J. E.; Zoffoli, J. P. Effects of Continuous 0.3 ppm Ozone Exposure on Decay Development and Physiological Responses of Peaches and Table Grapes in Cold Storage. *Postharvest Biol. Technol.* **2002,** *24*, 39–48.

Palou, L.; Smilanick, J. L.; Crisosto, C. H.; Mansour, M. Effect of Gaseous Ozone Exposure on the Development of Green and Blue Molds on Cold Stored Citrus Fruit. *Plant Dis.* **2001,** *85* (6), 632–638.

Perez, A. G.; Rios, J. J.; Sanz, C.; Olias, J. M. Aroma Components and Free Amino Acids in Strawberry Variety Chandler during Ripening. *J. Agric. Food Chem.* **1992,** *42*, 2232–2235.

Perez, A. G.; Sanz, C.; Rios, J. J.; Olias, R.; Olias, J. M. Effects of Ozone Treatment on Postharvest Strawberry Quality. *J. Agric. Food Chem.* **1999,** *47*, 1652–1656.

Perez-Gago, M. B.; Serra, M.; Rio, M. A. D. Color Change of Fresh Cut Apples Coated with Whey Protein Concentrate Based Edible Coatings. *Postharvest Biol. Technol.* **2006,** *39*, 84–92.

Pi, Y.; Schumacher, J.; Jekel, M. Decomposition of Aqueous Ozone in the Presence of Aromatic Organic Solutes. *Water Res.* **2005,** *39*, 83–88.

Qiang, Z. M.; Demirkol, O.; Ercal, N.; Adams, C. Impact of Food Disinfection on Beneficial Biothiol Contents in Vegetables. *J. Agric. Food Chem.* **2005,** *53* (25), 9830–9840.

Ranieri, A.; Urso, G. D.; Nali, C.; Lorenzini, G.; Soldatini, G. F. Ozone Stimulates Apoplastic Antioxidant Systems in Pumpkin Leaves. *Physiol. Planta* **1996,** *97* (2), 381–387.

Razumovski, S. D.; Zaikov, G. E. *Ozone and Its Reactions with Organic Compounds*; Elsevier, New York, 1984.

Restaino, L.; Frampton, E. W.; Hemphill, J. B.; Palnikar, P. Efficacy of Ozonated Water against Various Food Related Microorganism. *Appl. Environ. Microbiol.* **1995,** *61* (9), 3471–3475.

Rice, R. G.; Overbeck, P.; Larson, K. A. Costs of Ozone in Small Drinking Water Systems. *Proc. Small Drinking Water and Wastewater Systems*; NSF International: Ann Arbor, MI, 2000; p 27.

Rodgers, S. L.; Cash, J. N.; Siddiq, M.; Ryser, E. T. A Comparison of Different Chemical Sanitizers for Inactivating *Escherichia coli* O157:H7 and *Listeria monocytogenes* in Solution and on Apples, Lettuce, Strawberries, and Cantaloupe. *J. Food Prot.* **2004**, *67* (4), 721–731.

Rodoni, L.; Casadei, N.; Concellon, A.; Alicia, A. R. C.; Vicente, A. R. Effect of Short Term Ozone Treatments on Tomato (*Solanum Lycopersicum* L.) Fruit Quality and Cell Wall Degradation. *J. Agric. Food Chem.* **2010**, *58* (1), 594–599.

Rombach, C. *Mastering the Fundamentals of Ozone: Monitoring & Control*, 2009. Retrieved from http://pacificozone.com/wp-content/uploads/2014/04/app_1264803650.pdf (accessed May 31, 2016).

Salama, A.; Salama, M. *Inventors of Portable In Situ Ozone Generator and Use Thereof for Purifying Water*, 2016. www.ozomax.com. Retrieved from accessed June 23, 2016).

Sandermann, H.; Ernst, D.; Heller, W.; Langebartels, C. Ozone: An Abiotic Elicitor of Plant Defense Reactions. *Trends Plant Sci.* **1998**, *3*, 47–50.

Saranraj, P.; Stella, D.; Reetha, D. Microbial Spoilage of Vegetables and Its Control Measures: A Review. *Int. J. Nat. Prod. Sci.* **2012**, *2* (2), 1–12.

Sarig, P.; Zahavi, T.; Zutkhi, Y.; Yannai, S.; Lisker, N.; Ben-Arie, R. Ozone for Control of Post-harvest Decay of Table Grapes Caused by *Rhizopus stolonifera*. *Physiol. Mol. Plant Pathol.* **1996**, *48*, 403–415.

Schmidt, G. *Dissolved Ozone Measurements with EC Sensors: Installation and Calibration according to ASTM D7677-11*. n.d. Retrieved from http://www.europeanpharmaceutical-review.com/wp-content/uploads/White_Paper-Hach-Lange.pdf (accessed June 10, 2016).

Schraudner, M.; Enrst, D.; Langerbartels, C.; Sandermann, H. Biochemical Plant Responses to Ozone III. *Plant Physiol.* **1992**, *99*, 1321–1328.

Selma, M. V.; Ibanez, A. M.; Cantwell, M.; Suslow, T. Reduction by Gaseous Ozone of *Salmonella* and Microbial Flora Associated with Fresh Cut Cantaloupe. *Food Microbiol.* **2008**, *25* (4), 558–565.

Sen, C. K.; Packer, L. Thiol Homeostasis and Supplements in Physical Exercise. *Am. J. Clin. Nutr.* **2000**, *72*, 653–669.

Shammas, N. K.; Wang, L. K. Ozonation. In *Handbook of Environmental Engineering—Physicochemical Treatment Processes*; Wang, L. K., Hung, Y. T., Shammas, N. K., Eds.; Humana Press, Inc.: Totowa, NJ, 2005; vol 3, pp 315–355.

Sharpe, D.; Fan, L.; McRae, K.; Walker, B.; MacKay, R.; Doucette, C. Effects of Ozone Treatment on *Botrytis cinerea* and *Sclerotinia sclerotiorum* in Relation to Horticultural Product Quality. *J. Food Sci.* **2009**, *74* (6), 250–257.

Siemens, V. Uber die Electrostatische Induktion und die Verzogerung des Storms in Flaschen-drahten. *Poggendorff's Ann.* **1857**, *102*, 120–124.

Singh, N.; Singh, R. K.; Bhunia, A. K.; Stroshine, R. L. Efficacy of Chlorine Dioxide, Ozone, and Thyme Essential Oil or a Sequential Washing in Killing *Escherichia coli* O157:H7 on Lettuce and Baby Carrots. *Lebensm-Wiss. Technol.—Food Sci. Technol.* **2002**, *35* (8), 720–729.

Skog, L. J.; Chu, C. L. Effect of Ozone on Qualities of Fruits and Vegetables in Cold Storage. *Can. J. Plant Sci.* **2001**, *81*, 773–778.

Smilanick, J. L.; Crisosto, C.; Mlikota, F. Postharvest Use of Ozone on Fresh Fruit. *Perishables Handling Quarterly Issue No. 99*, 1999. Retrieved from http://ucanr.edu/datastore-Files/234-75.pdf (accessed June 29, 2016).

Smilanick, J. L.; Margosan, D. M.; Gabler, F. M. Impact of Ozonated Water on the Quality and Shelf Life of Fresh Citrus Fruit, Stone Fruit and Table Grapes. *Ozone-Sci. Eng.* **2002,** *24* (5), 343–356.

Smith, W. *Principles of Ozone Generation. Watertec Engineering Pty Ltd Information Sheet.* n.d. Retrieved from http://watertecengineering.com/TZ000002%20Principles%20of%20 Ozone%20Generation.pdf (accessed May 20, 2016).

Smith-McCollum, B. *Mastering the Fundamentals of Ozone: Feed Gas Preparation.* 2008. Retrieved from http://www.wqpmag.com/mastering-fundamentals-ozone-feed-gas-prepa-ration (accessed May 27, 2016).

Staehelin, J.; Hoignè, J. Decomposition of Ozone in Water in the Presence of Organic Solutes Acting as Promoters and Inhibitors of Radical Chain Reactions. *Environ. Sci. Technol.* **1985,** *19,* 1206–1213.

Tabakoğlu, N. Effects of Ozone Gas on the Microbiological and Chemical Quality of Mulberry Fruit (*Morus nigra* L.). *M.Sc. Thesis.* Pamukkale University Institute of Science Food Engineering, Turkey, 2016 (in Turkish).

Tizaoui, C.; Bickley, R. I.; Slater, M. J.; Wang, W. J.; Ward, D. B.; Al-Jaberi, A. A Comparison of Novel Ozone Based Systems and Photocatalysis for the Removal of Water Pollutants. *Desalination* **2008,** *227,* 57–71.

Toivonen, P. M. A.; Brummell, D. A. Biochemical Bases of Appearance and Texture Changes in Fresh-Cut Fruit and Vegetables. *Postharvest Biol. Technol.* **2008,** *48,* 1–14.

Tuffi, R.; Lovino, R.; Canese, S.; Cafiero, L. M.; Vitali, F. Effects of Exposure to Gaseous Ozone and Negative Air Ions on Control of Epiphytic Flora and the Development of *Botrytis cinerea* and *Penicillium expansum* During Cold Storage of Strawberries and Tomatoes. *Italian J. Food Sci.* **2012,** 102–114.

Tzortzakis, N.; Borland, A.; Singleton, I.; Barnes, J. Impact of Atmospheric Ozone-Enrich-ment on Quality-Related Attributes of Tomato Fruit. *Postharvest Biol Technol.* **2007b,** *45,* 317–325.

Tzortzakis, N.; Singleton, I.; Barnes, J. Deployment of Low-Level Ozone-Enrichment for the Preservation of Chilled Fresh Produce. *Postharvest Biol. Technol.* **2007a,** *43,* 261–270.

Tzortzakis, N.; Singleton, I.; Barnes, J. Impact of Low-Level Atmospheric Ozone-Enrich-ment on Black Spot and Anthracnose Rot of Tomato Fruit. *Postharvest Biol. Technol.* **2008,** *47* (1), 1–9.

Venta, M. B.; Broche, S. S. C.; Torres, I. F.; Perez, M. G.; Lorenzo, E. V.; Rodriguez, Y. R.; Cepero, S. M. Ozone Application for Postharvest Disinfection of Tomatoes. *Ozone-Sci. Eng.* **2010,** *32* (5), 361–371.

Vurma, M.; Pandit, R. B.; Sastry, S. K.; Yousef, A. E. Inactivation of *Escherichia coli* O157:H7 and Natural Microbiota on Spinach Leaves Using Gaseous Ozone During Vacuum Cooling and Simulated Transportation. *J. Food Prot.* **2009,** *72* (7), 1538–1546.

Wei, K. J.; Zhou, H. D.; Zhou, T.; Gong, J. H. Comparison of Aqueous Ozone and Chlorine as Sanitizers in the Food Processing Industry: Impact on Fresh Agricultural Produce Quality. *Ozone—Sci. Eng.* **2007,** *29* (2), 113–120.

Whangchai, K.; Saengnil, K.; Singkamanee, C.; Uthaibutra, J. Effect of Electrolyzed Oxidizing Water and Continuous Ozone Exposure on the Control of *Penicillium digitatum* on Tangerine cv. 'Sai Nam Pung' During Storage. *Crop Prot.* **2010,** *29,* 386–389.

White, G. C. *Ozone.* In *Handbook of Chlorination and Alternative Disinfectants,* 4th ed.; Van Nostrand Reinhold: New York, 1999; pp 1203–1261.

Wlodek, L. Beneficial and Harmful Effects of Thiols. *Pol. J. Pharmacol.* **2002,** *54* (3), 215–223.

Wu, J. G.; Luan, T. G.; Lan, C. Y.; Lo, W. H.; Chan, G. Y. S. Efficacy Evaluation of Low-Concentration of Ozonated Water in Removal of Residual Diazinon, Parathion, Methyl Parathion and Cypermethrin on Vegetable. *J. Food Eng.* **2007,** *79,* 803–809.

Yaseen, T.; Ricelli, A.; Turan, B.; Albanese, P.; D'onghia, A. M. Ozone for Post-harvest Treatment of Apple Fruits. *Phytopathol. Mediterr.* **2015,** *54* (1), 94–103.

Zambre, S. S.; Venkatesh, K. V.; Shah, N. G. Tomato Redness for Assessing Ozone Treatment to Extend the Shelf Life. *J. Food Eng.* **2010,** *96* (3), 463–468.

Zhang, L. K.; Lu, Z. X.; Yu, Z. F.; Gao, X. Preservation of Freshcut Celery by Treatment of Ozonated Water. *Food Contr.* **2005,** *16* (3), 279–283.

ADVANCES IN EDIBLE COATINGS AND FILMS FOR FRESH FRUITS AND VEGETABLES

EVENING STONE MARBOH* and ALOK KUMAR GUPTA

ICAR-National Research Centre on Litchi, Muzaffarpur, Bihar, India

*Corresponding author. E-mail: esmarboh@gmail.com

ABSTRACT

Postharvest loss has been one of the main global challenges toward ensuring food security for a growing population. Fruits and vegetables being living entities, continue to respire even after harvest. In the last few years, there has been a growing interest toward development of new edible packaging materials for preserving product shelf life. Starches, cellulose derivatives, chitosan, gums, and lipids are among the well-known edible coatings used which act in a manner similar to modified atmosphere storage. Recently, emphasis has also been paid toward use of plant derived compounds as they provide low-cost raw materials which are highly beneficial in the large-scale application of edible packaging. Several nanoparticles have been recognized as possible additives to enhance polymer coatings performance. The advent of nanotechnology has made it possible to engineer these nanostructures in edible coatings for achieving desirable functional properties whose efficacy can be further enhanced through electrospraying, renowned for a series of superiorities over the traditional spraying technique. Although considerable efforts have been dedicated to understand and modify film properties, issues on sensory acceptance pertaining to their impact on taste, flavor and color, still remains to be a major challenge. Much emphasis needs to be paid toward understanding the interactions among active ingredients and coating materials when developing new edible film and coating applications.

9.1   INTRODUCTION

In recent years, there has been a growing interest on the use of edible coating for fresh fruits and vegetables. With the growing demands and consumers interests in safe, fresh, and quality food, the role of food industry sector in developing new and effective methods for preserving food quality and extending shelf-life also increases. A lot of efforts have been made to develop natural edible coatings which are safe for human consumption. Though controlled atmosphere storage and modified atmosphere storage have been traditionally in vogue for preserving the quality of fresh produce, edible coatings being relatively cheaper and easy to produce can provide an alternative means in retaining quality of stored product through modification and control of the internal atmosphere of the individual product. The application of edible films and coatings to prolong the shelf-life, maintaining quality and improving appeal of fresh food and vegetables, is a recent approach. This functional aspect of edible films and coatings can further be improved by including antioxidant and antimicrobial agents in their formulation (Bonilla et al., 2012). Fresh fruits and vegetables, being living entities, undergo metabolic activity even after harvesting. Through a better understanding of the respiration process, several techniques have been developed in extending their shelf-life. In nature, fruits and vegetables are coated by a cuticle which consists of a layer of fatty-acid-related substances, such as waxes and resins, having low permeability to water (Baldwin, 1994). There is increasing public interest in development of edible natural biodegradable coatings to replace the commercial synthetic waxes for preserving quality of stored fruit and vegetables. Starches, cellulose derivatives, chitosan, gums, and lipids are among edible coatings used to coat fresh or processed fruits and vegetables for extending their shelf-life (Rhim and Ng, 2007; Hansen and Plackett, 2008; Razavi et al., 2015). These coatings control the moisture transport and permeability of gases similar to modified atmosphere storage. The purpose of edible coatings for fruits and vegetables is essentially to mimic or enhance their natural barrier or to replace it in case when washing and handling have partially removed or altered it (Baldwin, 1994). Coatings could also help to maintain the structural integrity of coated commodities, thereby improving the mechanical handling properties and retaining volatile flavor compounds. In addition, edible films can act as carriers for antioxidant, antimicrobial, and additives by preventing food oxidation, development of off-flavors, nutritional losses, and food spoilage (Gomez-Estaca et al., 2009; Ramos et al., 2016). One of the important advantages offered by edible coatings is the ability to be consumed together with the product. This approach delivers a single solution for two major postharvest problems related to physiological disorder and disease incidence on stored fruits and vegetables.

9.2 DEFINITION AND HISTORICAL BACKGROUND

According to Guilbert et al. (1995), edible coatings may be defined as a thin layer of material covering the surface of the food and which can be eaten along with the whole product. Although the terms "edible films" and "edible coatings" are sometimes used interchangeably, there is a difference in that films are preformed separately and then applied to food surface or sealed into edible pouches, whereas coatings are formed directly onto food surfaces. An edible coating is also defined as a thin cover of edible material applied to the food surface either as additional layer or as substitution for natural protective coatings, providing barrier to moisture, oxygen, and solute movement from the food (Avena-Bustillos et al., 1997; McHugh and Senesi, 2000). Others defined edible film as a thin layer of edible material formed on a product surface as a coating or placed on or between food components (Krochta and de Mulder-Johnston, 1997). Thus, an edible film refers to a thin skin, which has been preformed while an edible coating is a suspension or an emulsion, which is applied directly to the food surface and later forms a film (Souza et al., 2010). Edible coatings and films do not replace traditional packaging materials but provide an additional option to be used in food preservation. The material employed in preparation of edible films and coatings should be generally recognized as safe (GRAS) because they will be consumed along with the product (Park et al., 1994; Krochta and de Mulder-Johnston, 1997), approved by Food and Drug Administration (FDA), and must conform to the regulations of food products (Guilbert et al., 1996). An ideal coating should be one that can extend storage life of fresh fruits and vegetables without hampering their quality.

The application of edible films and coatings to improve shelf-life of fruits and vegetables dates back to 19 centuries ago involving use of wax coatings. In fact, coating of citrus fruits (oranges and lemons) with wax to retard desiccation was practiced in China in the 12th and 13th centuries (Hardenburg, 1967). Although the Chinese did not realize the role of edible coatings in minimizing the rate of respiratory gas exchange, wax-coated fruits were found to store longer than nonwaxed fruits. In the 1930s, hot-melt paraffin waxes became commercially available as edible coatings for fresh fruits such as apples and pears. Lowings and Cutts (1982) reported an edible coating material containing a mixture of sucrose fatty acid esters, sodium carboxymethyl cellulose, and mono- and diglycerides that is nonphytotoxic, tasteless, odorless, and effective in preserving fruits and vegetables. Since the 1980s, this coating material became commercially available under the trade names "TAL Pro-long" and "Semperfresh" (Chu, 1986; Santerre et al., 1989). Till then, several attempts have been made to evaluate the efficacy of other coating materials which can modify internal gas composition of products during storage. Currently, edible coatings are widely used on whole fruits like apple, pear, orange, lemon, and grapefruit, with the aim of reducing water loss, improvement of appearance, incorporation of fungicides or growth regulators, and creation of barrier for gas exchange between the commodity and the external atmosphere. The first documented use of edible coatings was based on a carrageenan coating on fresh-cut fruits to reduce shrinkage, leakage, and deterioration of flavor of grapefruit halves (Bryan, 1972).

9.3 ADVANTAGES OF EDIBLE COATINGS

The potentials advantages of edible coatings (Olivas and Barbosa-Canovas, 2005) are listed below:

- Improves product appeal by imparting gloss to the fruit surface.
- Reduce moisture and weight loss and maintains product freshness and firmness.
- Restrict gas exchange and check respiration and ethylene production, thereby delaying product deterioration and senescence.
- Retard the enzymatic oxidation and protect product browning and texture softening during storage.
- Protects the product from physical damage caused by mechanical impact, pressure, vibrations, and other mechanical factors.
- Prevents fruits and vegetables against chilling injuries and storage disorders.
- Prevents loss of natural volatile flavor compounds and color components from fresh product and the acquisition of different odors.
- Serves as carrier for postharvest chemical treatments.
- Acts as carriers of functional components like antimicrobial and antioxidant agents, nutraceuticals, and color and flavor ingredients for reducing microbial loads, delaying oxidation and discoloration, and improving the overall quality.

9.4 PHYSIOLOGICAL MECHANISM OF EDIBLE COATINGS ON FRUITS AND VEGETABLES

Fruits and vegetables continue to respire even after harvest and use up all the oxygen within the produce. The presence of edible coating restricts the gas exchange between the product which depletes oxygen surrounding the produce and accumulates carbon dioxide within the produce. Eventually, the produce will shift to partial anaerobic respiration that requires less oxygen, thereby disrupting ethylene biosynthesis (Park et al., 1994; Guilbert et al., 1996; McHugh and Senesi, 2000). The extent to which edible coatings influence the internal oxygen and carbon dioxide composition and reduction in weight loss is dependent upon the natural barrier on the produce and the type and amount of coating applied. Thus, extension of the shelf-life of fresh fruits and vegetables demands an in-depth knowledge on fruit and vegetable physiology. In general, the higher the product's respiration rate, the shorter is its storage life. Thus, shelf-life of fruits and vegetables can be possibly enhanced by minimizing their respiration rate using approaches that modify or control internal gas atmosphere of the produce. In most products, when there is restricted air flow, the equilibrium is shifted toward a decrease in internal oxygen and an increase in carbon dioxide. With the decrease in oxygen concentration, the respiration also decreases until a critical oxygen concentration called extinction point is reached. Beyond this point as oxygen concentrations decline, the respiration rate increases, aerobic respiration is blocked, and anaerobic fermentation begins. On the contrary, an increase in the external carbon dioxide concentration affects the respiratory pathway, resulting in a decrease in the respiration rate. This is the basis of the technology based on edible coatings which could also be applied to obtain similar effects as controlled or modified atmospheres storage. Thus, surface coatings can delay the ripening of fruits and vegetables by modifying the internal gas atmospheres and inhibiting ethylene biosynthesis and action.

9.5 REQUIREMENT OF AN IDEAL EDIBLE FILMS AND COATINGS

An ideal edible film and coatings should possess the following characteristics (Arvanitoyannis and Gorris, 1999):

- The coating should be water-resistant so that it remains intact and covers a product adequately, when applied.

- Controls water migration both in and out of the produce to maintain desired moisture content.
- Provide semipermeability to maintain internal equilibrium of gases. A minimum of 1–3% oxygen is required around a commodity to avoid a shift from aerobic to anaerobic respiration.
- It should not contain toxic, allergic, and nondigestible components.
- Have good surface adhesion with uniform coverage.
- Provide structural stability and prevent mechanical damage during transportation, handling, and display.
- It should be easily emulsifiable, nonsticky, and have efficient drying performance.
- It should never interfere with the quality of fresh produce nor impart undesirable odor and flavor.
- Provide biochemical and microbial surface stability while protecting against contamination, pest infestation, microbe proliferation, and other types of decay.
- Maintain or enhance esthetics and sensory attributes of product.
- Serve as carrier for desirable additives such as flavor, fragrance, colors, nutrients, and vitamins.
- It should be easily manufactured and economically viable.

9.6   CLASSIFICATION OF EDIBLE COATINGS

A wide array of compounds can be used in the formulation of edible coatings ranging from hydrophobic group such as lipid based or waxes to hydrophilic group or hydrocolloids, for example, polysaccharides based, protein based, or combination of both groups to improve function of edible coating. Protein, polysaccharides, and lipids are the most common film-forming materials for edible coatings. Edible films and coatings may be classified according to the kind of material from which they are derived. Each chemical class has its inherent properties, advantages, and limitations for being used as films.

9.6.1   POLYSACCHARIDE-BASED COATINGS

Polysaccharides being present in most commercially available formulations are the most widely used components found in edible coatings for fruits and vegetables (Krochta and de Mulder-Johnston, 1997). Though they have high hydrophobicity and water vapor permeability (WVP) in comparison to commercial plastic films, the oxygen and moisture barrier properties of these substances can be sufficient enough to protect fresh-cut fruit and vegetables from dehydration and in some cases retard their respiration rate. These coatings may be used to retard ripening and increase shelf-life of coated produce without creating severe anaerobic conditions (Baldwin et al., 1995). Polysaccharides that have been used for coating applications in fresh-cut fruits and vegetables include cellulose and derivatives, starch, alginates, gums, chitosan, pectin, carrageenan, and some mucilage compounds.

9.6.1.1   CELLULOSE AND DERIVATIVES

Cellulose is the most abundant natural polymer on earth. It is highly crystalline, fibrous, and insoluble. Cellulose derivatives such as methylcellulose (MC), hydroxypropylmethyl-cellulose, and the ionic carboxymethylcellulose are commonly found in the formulation of edible coatings, especially in commercial products. However, cellulose derivative films are poor water vapor barriers because of the inherent hydrophilic nature of polysaccharides and they possess poor mechanical properties (Gennadios et al., 1997). Despite this, application of cellulose enhanced the storage life of cut apples (Baldwin et al., 1996), decreased surface discoloration of apple slices (Rojas-Grau et al., 2008) and fresh-cut pear wedges (Olivas et al., 2007), and maintained visual quality and flavor of fresh-cut mangoes (Plotto et al., 2004).

9.6.1.2   STARCH

As one of the most abundant natural polysaccharides, starch has been widely used as food hydrocolloid (Narayan, 1994) because it is inexpensive, abundant, biodegradable, and easy to use. Starch consists of two kinds of molecules, namely, amylose (20–30%) and amylopectin (70–80%), which was primarily derived from cereal grains and tubers like corn, wheat, potato, tapioca, and rice. The presence of amylose was accountable for the film-forming capacity of starch (Romero-Bastida et al., 2005). Starch-based edible polymer exhibits physical characteristics similar to plastic polymer in which they were odorless, tasteless, colorless, nontoxic, biologically absorbable, semipermeable to carbon dioxide, and resistant to passage of oxygen (Myllarinen et al., 2002). Garcia et al. (1998) found a significant effect of a starch-based coating on the color, weight loss, firmness, and shelf-life of coated strawberries.

9.6.1.3   PULLULAN

Pullulan is a polysaccharide produced by the fungus *Aureobasidium pullulans*. Purified pullulan in the form of white powder is tasteless and odorless and well soluble in water (Milind et al., 1992). The application of pullulan-based edible coating could have delayed the aging process of kiwi fruits (Xu et al., 2001) and reduced mass losses of strawberries (Diab et al., 2001).

9.6.1.4   CHITOSAN

Chitosan is a natural polysaccharide commercially produced from exoskeletons of crustacean, such as crab, shrimp, and crawfish. This is the second most abundant natural and nontoxic polymer in nature after cellulose. It is environment friendly because of its biodegradability, biocompatibility, antimicrobial activity, nontoxicity, and versatile chemical and physical properties. This coating material has excellent film-forming properties, broad antimicrobial activity, and compatibility with other substances, such as vitamins, minerals, and antimicrobial agents (Park and Zhao, 2004; Chien et al., 2007; Ribeiro et al., 2007). When applied on fruit and vegetables, chitosan-based coatings effectively delayed ripening and decreased respiration rates. It is highly effective in maintaining the quality and extending shelf-life of sliced mango (Chien et al., 2007) and reducing the microbial load of stored strawberries (Park et al., 2006). Han et al. (2004) also proposed chitosan for extending the shelf-life of sliced apples and fresh strawberries.

9.6.1.5   ALGINATES

Alginates are sodium salts of alginic acid extracted from brown seaweeds of the *Phaephyceae* class. They possess good film-forming properties and produce uniform and transparent films but have poor water resistance. Alginate coatings are good oxygen barriers and retard lipid oxidation in fruits and vegetables. They are effective in reducing weight loss and natural microflora counts in minimally processed carrots (Amanatidou et al., 2000). In fresh mushrooms, calcium alginate coatings reduce shrinkage, moisture and weight loss, oxidative rancidity, and oil absorption and improve the overall appeal (Hershko and Nussinovitch, 1998).

9.6.1.6 CARRAGEENANS

Carrageenan is a complex mixture of several polysaccharides extracted from red seaweeds. Thermoreversible carrageenan gels can be used as food coatings to retard produce's moisture loss by acting as a sacrificing agent (Kester and Fennema, 1986). Carrageenan has been applied to a variety of foods as carrier of antimicrobials besides reducing moisture loss and oxidation (Nieto, 2009) and has been studied for flavor encapsulation (Fabra et al., 2009). In apples, carrageenan-based coatings retard moisture loss, oxidation, or disintegration of the flesh (Bryan, 1972; Lee et al., 2003). In combination with ascorbic acid, carrageenan-based coatings delivered positive sensory results with reduction in microbial loads on processed apple slices (Lee et al., 2003).

9.6.1.7 GUMS

Gums encompassing exudate gums (arabic, ghatti, tragacanth, and karaya), seed gums (locust bean and guar), and microbial fermentation gums (xanthan and gellan) are structurally complex heteropolysaccharides. In edible film-forming preparations, guar gum was used as a water binder, stabilizer, and viscosity builder. Gum arabic, owing to its solubility in hot or cold water, was the least viscous among hydrocolloid gums. Xanthan gum being readily dispersed in water could produce film with high consistency in both hot and cold systems. A blend of guar gum, gum arabic, and xanthan gum produced uniform coatings with improved adhesion in wet batters (Subhas and Pathik, 2014) by which they are commonly used as an encapsulation coating material (Pegg and Shahidi, 2007).

9.6.1.8 PECTIN

Edible pectin film is formed by evaporating water from pectin gel. Pectin from fruit purees can be used to extend the shelf-life of fresh-cut fruits and vegetables. Although pectinate coatings are not adequate moisture barriers, they can serve as a sacrificing agent when moisture evaporates from their gel matrix retard, thus preventing water loss and dehydration of the coated products. Pectin coatings have been studied for their ability to check moisture loss, lipid migration, and improving handling and appearance of foods (Kester and Fennema, 1986). In avocados, pectin-based coating was effective in controlling the spread and severity of stem end rot caused by *Lasiodiplodia theobromae* (Maftoonazad and Ramaswamy, 2008). McHugh and Senesi (2000) significantly reduced moisture loss and browning rates of fresh-cut apples using apple puree-based coating containing various concentrations of fatty acids, fatty alcohols, beeswax, vegetable oil, and high methoxyl pectin. An edible coating containing a mixture of apple puree and alginate was also applied to preserve the quality of apple slices (Rojas-Grau et al., 2008).

9.6.2 LIPID-BASED COATINGS

Edible lipid materials as protective coatings that have been utilized to prevent moisture loss and impart surface gloss to the produce. Unlike other macromolecules, lipid and resin compounds are not biopolymers and lack a large number of repeating units connected by covalent bonds to form cohesive, self-supporting film structures. Owing to their relatively low polarity, they must be incorporated into composite films to provide a moisture barrier. The other disadvantages of employing lipids in edible packaging materials are attributed to development of waxy taste and texture, greasy surface, and potential rancidity.

9.6.2.1 WAXES

Waxes are esters of a long-chain fatty acid with a long-chain alcohol. They are substantially more resistant to diffusion of water than most lipid or nonlipid edible films, owing to very low level of polar groups (Kester and Fennema, 1986). Both natural waxes, such as carnauba wax, candelilla wax, rice-bran wax, beeswax, and synthetic waxes, like paraffin and petroleum wax, have been used as protective coatings, alone or in combination with other ingredients (Baldwin, 2007). The most effective lipid substances used in edible coatings are paraffin wax and beeswax. Paraffin wax is derived from distillate fraction of crude petroleum and consists of a mixture of solid hydrocarbon resulting from ethylene catalytic polymerization. It is permitted for the use on raw fruits and vegetables and cheese. Carnauba wax is derived from the palm tree leaves (*Copernicia cerifera*) has very high melting point and is used as an additive to other waxes to increase toughness and luster. Beeswax is made by honeybees, whereas candelilla wax is derived from the candelilla plant. Wax coatings are applied on fruits like citrus and apples and vegetables such as tomatoes, asparagus, beans, beets, carrots, celery, eggplant, peppers, potatoes, radishes, cucumbers, squash, and turnips to restrict gas exchange and moisture loss and to improve the surface appearance (Corbo et al., 2015).

9.6.2.2 RESIN

Edible resins, such as shellac, terpene resin, and wood rosin, are used to impart gloss to food commodities. Shellac has been used extensively as edible coating for confectionary and fresh produce and as an enteric coating for pharmaceuticals (Baldwin, 2007). Resin coatings are effective at reducing water loss but are the least permeable to gases; thus, fruit can suffer undesirable anaerobic respiration and off-flavor development. Rosins obtained from the oleoresins of the pine tree are residues left after distillation of volatiles from the crude resin. Primarily, resin and rosins were used to impart high gloss. Citrus fruits coated with shellac and wood resin-based coatings generally have lower internal oxygen, higher internal carbon dioxide, and higher ethanol content than wax-coated fruits (Hagenmaier and Baker, 1995).

9.6.3 PROTEIN-BASED COATINGS

Proteins derived from animal sources, such as casein and whey protein (WP), or plant sources like zein, wheat gluten, soy protein, and peanut protein are commonly used in the formulation of edible coatings (Gennadios et al., 1994). Like polysaccharides, protein edible films and coatings also exhibit excellent gas and lipid-barrier properties, particularly at low relative humidity, and provide mechanical strength with good structural integrity. However, owing to their hydrophilicity, protein films and coatings exhibit relatively poor water-barrier characteristics (Kester and Fennema, 1986; Gennadios et al., 1994; Sothornvit and Krochta, 2001). Proteins are generally categorized into two groups, that is, fibrous proteins and globular proteins. Fibrous proteins, comprising casein, WP, collagen, gelatin, and keratin, are water insoluble and obtained from animal tissues whereas globular proteins such as wheat gluten, soy protein, peanut protein, corn zein, and cotton seed protein are derived from plants and are soluble in water or aqueous solutions of acids, bases, or salts (Gennadios et al., 1994). This diverse biological origin and molecular characteristics dictates the utility of particular proteins to form coatings and the characteristics of the coatings formed.

9.6.3.1 CORN ZEIN

Corn zein is the most important protein in corn comprising a group of prolamins (alcohol-soluble proteins). Zein was commercially and successfully used as finishing agent in imparting surface gloss and as barrier of oxygen, lipid, and moisture for nuts, candies, confectionery products, and other foods (Krochta and de Mulder-Johnston, 1997). Zein films of moderate gas barrier reduce moisture loss and respiration rates of vegetables (Park et al., 1994). When applied on vegetable and fruits, zein-based coatings provide an adhesive and stable coating and retard the ripening of tomatoes, maintain color and firmness of broccoli florets, reduce the growth of *Listeria monocytogenes* on cooked sweet corn, and maintain the gloss and other qualities of apples (Lin and Zhao, 2007).

9.6.3.2 SOY PROTEIN

Soy protein is commercially available in soy flour (50–59% protein), soy concentrate (65–72%), or isolate (>90%) (Park et al., 2002). These films have good potential to carry flavoring, antimicrobial, and antioxidant compounds. Edible films based on soy protein can be produced either by two ways: surface film formation on heated soymilk or film formation from solutions of soy protein isolates (SPIs) (Gennadios and Weller, 1990). SPI films may find application as microencapsulating agents of flavors and medications or in coatings of fruits and vegetables (Petersen et al., 1999). Eswaranandam et al. (2006) extended the shelf-life of fresh-cut cantaloupe melon using a soy protein coating containing malic and lactic acids.

9.6.3.3 COLLAGEN AND GELATIN

Collagen is a hydrophilic protein rich in glycine, hydroxyproline, and proline; thus, it swells in polar liquids with high solubility parameters. Collagen as a major constituent of skin, tendon, and connective tissues is most prevalent and widely distributed fibrous protein in the animals. It is the basic raw material for gelatin production. Gelatin is produced via either partial acid or alkaline hydrolysis of collagen. Gelatin coatings can reduce oxygen, moisture, and oil migration or carry bioactive ingredients (Krochta and de Mulder-Johnston, 1997). Gelatin is widely used as an encapsulating agent in hard and soft gel capsules for low-moisture or oil-based food ingredients, dietary supplements, and pharmaceuticals (Baldwin, 2007).

9.6.3.4 MILK PROTEIN

The two main classes of milk proteins are caseins (80%) and WPs (20%). Casein-based films possessed excellent oxygen barrier and good moisture barriers for water-soluble pouches, fresh produce, dried fruits, and frozen foods. Furthermore, casein film is highly potent as carriers of flavor, nutrients, or bioactive ingredients (Gennadios et al., 1994). According to Krochta et al. (1990), casein-based emulsion films when emulsified with lipid-based materials were more effective than pure caseinate films in controlling moisture loss of fruits and vegetables. The main drawback of casein is its relatively high price. WPs remain soluble after casein is precipitated at pH 4.6 during the cheese-making process. WPs are commercially available as WP concentrates (25–80% protein) and WP isolates (>90% protein). Like caseinate films, WPs produce transparent, flavorless, and flexible films. They possess excellent oxygen barrier properties comparable to the synthetic polymer films and also are good grease barriers (Lin and Krochta, 2003). WP films and coatings have been used as protective barriers against oxygen uptake and development of rancidity in roasted peanuts (Mate et al., 1996). Sonti et al. (2003) observed a delay in browning and texture decay of apple cubes coated with WP concentrate and WP isolate. A reduction in browning of cut apples and potatoes treated with edible coating containing sour whey flour was also reported by Shon and Haque (2007).

9.6.4 BILAYER COATINGS AND EMULSIONS

Polysaccharides and proteins polymer, despite their hydrophilic nature, are good film formers with excellent oxygen, aroma, and lipids barriers at low relative humidity. In fact, each individual coating material has some unique, but limited, functions. The integration of proteins, polysaccharides, or lipids together can improve functionality of the coating; in fact, they are more effective when used in a combination (Lin and Zhao, 2007). Because of the presence of two or more components, the characteristics of the film are enhanced by individual contribution of each component. Therefore, protein, lipid, or polysaccharide–lipid composite films are more effective than stand-alone films (Kester and Fennema, 1986). Baldwin (1994) proposed that composite and bilayer coatings are the edible coatings of the future. The improved moisture-barrier properties of composite coatings have made them promising candidates for coating of fresh-cut fruits and vegetables. According to Lin and Zhao (2007), lipid generally forms an additional layer over the polysaccharide or protein layer in bilayer composite film and coatings, while the lipid in the emulsion composite layer is dispersed and entrapped in the matrix of protein or polysaccharide. Bilayer coatings combine the greaseless feel and gas permeability properties of polysaccharide coatings with the good water barrier characteristics of lipid coatings (Wong et al., 1994). Such coating is highly effective in reducing gas exchange and weight loss of fresh produce. Tanada-Palmu and Grosso (2005) reported that wheat gluten–lipid (beeswax, stearic acid, and palmitic acid) composite coatings significantly retained firmness and reduced weight loss of fresh strawberries. In fresh-cut apples, milk protein–lipid (acetylated monoglyceride) bilayer coating effectively check moisture loss and oxidative browning up to 3 days (Baldwin et al., 1995). Coating comprising double layers of polysaccharides (cellulose, carrageenan, pectin, or alginate) and acetylated monoglyceride maintains apple quality (Wong et al., 1994). Coatings prepared from WP isolate or concentrate and beeswax or carnauba wax was also effective in inhibiting browning of apple slices (Perez-Gago et al., 2003).

9.6.5   HERBAL EXTRACT

The search for naturally derived biodegradable coatings materials has enabled researchers to explore the potential applications of plant-derived compounds in the formulation of edible coatings. Such coatings are prepared by mixing extracts of different herbs such as thyme (*Thymus vulgaris*), cinnamon (*Cinnamon cassia*), oregano (*Origanum vulgare*), lemon grass (*Cymbopogon citratus*), neem (*Azadirictica indica*), and *Aloe vera* (Chauhan et al., 2014; Padmaja and John Don Bosco, 2014; Zinoviadou et al., 2009). These herbal extracts possess antimicrobial, antioxidant, nutraceutical, and therapeutic properties. Films containing lemon grass, cinnamon oil, and oregano have lower oxygen permeability (OP) (Rojas-Grau et al., 2006, 2007). The common herbs used in edible coatings are described hereunder.

9.6.5.1   ALOE VERA

At present, *Aloe vera* extract is used in edible coatings at large scale for increasing the shelf-life of fruits and vegetables (Martinez-Romero et al., 2006). The main component of *Aloe vera* extract is "Aloin and Aloe-emodin." *Aloe vera* gel-based edible coatings are good moisture and gas barrier. Valverde et al. (2005) and Martinez-Romero et al. (2006) proposed *Aloe vera* gel-based edible coatings for preventing moisture loss, reducing texture loss, controlling respiratory rate, and reducing microbial proliferation in table grapes and sweet cherries. In addition, Martinez-Romero et al. (2006) maintained the initial taste, aroma, or flavors sweet cherries during storage using *Aloe vera*-based coating. *Aloe vera* gel has also been used to extend the shelf-life of grapes up to 40 days (Chauhan et al., 2014).

9.6.5.2 NEEM

Neem is a medicinal plant possessing excellent antimicrobial properties. Its main active components include azadirachtin and nimbidin. Neem oil and extract are also used as an antimicrobial agent in edible coatings. Neem extract is used in many biodegradable edible coatings for fruits and vegetables to enhance their shelf-life (Rao, 1990). Chauhan et al. (2014) extends the storage life of apple for 45 days based on neem extract.

9.6.5.3 TULSI

Tulsi also called "Queen of Herbs" is known for its therapeutic potentials (Kumar et al., 2011). The extract and oil of tulsi have many beneficial properties such as antibacterial, antifungal, antiviral, anti-oxidative, and insecticidal and it also acts as anti-oxidative agents. Therefore, it was used in shelf-life enhancement of fruits and vegetables (Kayastha, 2014).

9.6.5.4 CINNAMON

This spice possesses good antimicrobial, anti-oxidative, and inhibitory properties against pathogenic microbes. It is a natural food additive used as flavoring agent and preservative which prevent the growth of bacteria and fungi (Roller and Seedhar, 2002). Cinnamon as an antimicrobial agent has been found to reduce microbial load and increased the storage life of fresh cut apples (Muthuswamy et al., 2008) and kiwifruit (Roller and Seedhar, 2002).

9.7 APPLICATIONS OF EDIBLE COATINGS FOR FRUITS AND VEGETABLES

Edible coatings are naturally synthesized. Generally, it is used to improve appearance and preservation of fruits and vegetables. The main advantages of edible coating are its edibility, nontoxic nature, and cost-effectiveness as compared to other synthetic coating (Prasad and Batra, 2015). As carriers of active compounds that can be released in a controlled way, edible coatings have good potentials in improving the shelf-life and quality of minimally processed fruits and vegetables. The ability to consume along with the product, incorporation of additional nutrients, enhancement of sensory characteristics, and inclusion of quality-enhancing antimicrobials provides additional benefits. The main potential of edible coatings in maintaining quality of fresh fruits and vegetables can be attributed to its unique functional properties that have rendered its application in diversified ways.

9.7.1 BARRIER

Edible coating functions as barrier that protecting the product from exposure to the environment. This property has extended its commercial applications to protect fruits and vegetables from pathogenic microbial contamination. The important functional barriers of edible coatings include barrier to moisture, oxygen and other gases, fats, and oils. These barriers can be applied to ready-to-eat food and fresh produce such as fruits and vegetables. The extent of the barrier performance is influenced by the chemical properties of the material used, environmental conditions, such as temperature and relative humility and product handling (Krochta et al., 1994).

9.7.1.1 MOISTURE BARRIER

Fresh-cut products lack a cuticle and subepidermal layers, thereby exposing the internal tissues. Thus, water loss occurs at a rapid rate which eventually leads to loss of turgor and crispness (Shackel et al., 1991). Watada et al. (1996) attributed the moisture barrier property of edible coatings in preventing moisture loss to maintenance of a moisture-saturated environment throughout the surface of cut produce. Application of an emulsion containing calcium caseinate and an acetylated monoglyceride successfully reduced water loss in apples (Avena-Bustillos et al., 1997). Fresh-cut pineapple coated with alginate has improved shelf-life with higher juice retention (Montero-Calderon et al., 2008). Moisture loss was significantly reduced in fresh-cut apples after applying wraps made from apple puree containing lipids (McHugh and Senesi, 2000). Han et al. (2004) applied chitosan-based coating containing calcium and reported a 24% reduction in the drip loss of frozen-thawed raspberries with improved firmness in comparison to uncoated fruits.

9.7.1.2   GAS BARRIER

Edible coatings are also used as a protective barrier to restrict gas exchange between the tissues and the surrounding atmosphere, leading to a decrease in respiration and stress-mediated deteriorative response (Lin and Zhao, 2007). The coating basically creates a modified atmosphere inside each coated piece (Rojas-Grau et al., 2008). However, consideration must be taken to avoid extremely impermeable coatings that may induce anaerobic conditions which eventually lead to a decrease in the production of characteristic aroma volatile compounds (Mattheis and Fellman, 2000). Lee et al. (2003) reported a reduction of the initial respiration rate of fresh-cut Fuji apples coatèd with WP concentrate which was attributed to the calcium ions contained in the film-forming solution and to the oxygen barrier properties inherent to the film. Wong et al. (1994) obtained a large reduction in the rates of gas evolution-coated cut apples with a bilayer of acetylated monoglyceride and ascorbate buffer containing calcium ions. The production of ethanol and acetaldehyde formation increases from the second week of storage in apple wedges (Rojas-Grau et al., 2008). Hence, selection of an edible coating material with appropriate gas permeability as well as the control of environmental conditions such as temperature and relative humidity is of prime importance as this affects coating permeability and produce respiration (Lin and Zhao, 2007).

9.7.2   CARRIER

One of the significant applications of edible coatings is the ability of incorporating active ingredients such as antioxidants, antimicrobials, additives, enhancers, and nutraceuticals into the matrix. These ingredients should be incorporated at a predetermined level so as to avoid interference with the mechanical and barrier properties of the coating (Kester and Fennema, 1986; Guilbert and Gontard, 1995).

9.7.2.1 ANTIOXIDANTS

Processing operations can induce undesirable changes in color and appearance of fresh-cut fruit products during storage and marketing. The phenomena are usually caused by an enzyme polyphenol oxidase, which in the presence of oxygen converts phenolic compounds into dark-colored pigments. Application of antioxidant treatments is the most common strategy to control browning of fresh-cut fruits. Ascorbic acid is most extensively used to avoid enzymatic browning of fruit. Perez-Gago et al. (2006) reported a substantial reduction in browning of fresh-cut apples treated with WP concentrate–beeswax coating containing ascorbic acid, cysteine, or 4-hexylresorcinol. Oms-Oliu et al. (2008) maintained quality of fresh-cut pears wedges without browning by using coating formulations containing *N*-acetylcysteine and glutathione. Application of alginate edible coating in conjunction with anti-browning agents such as ascorbic and citric acid preserves the color of fresh-cut mangoes cubes (Robles-Sanchez et al., 2013).

9.7.2.2 ANTIMICROBIAL

Fresh-cut fruits are more perishable than their corresponding whole uncut commodities due to wounding during preparation (Brecht, 1995). The physical and chemical barrier provided by the epidermis, which prevents the development of microbes on the fruit surface, are removed during processing. Several categories of antimicrobials can be potentially incorporated into edible films and coatings, including organic acids (acetic, benzoic, lactic, propionic, sorbic), fatty acid esters (glyceryl monolaurate), and polypeptides (lysozyme, peroxidase, lactoferrin, nisin). The incorporation of antimicrobials direct into edible coating is receiving considerable attention as a means of extending the shelf-life and maintaining product quality and safety. This can be achieved by adding antimicrobial agents into the melt form of polymer or by adding into the wet polymer solution. Organic acids and plant essential oils are the main antimicrobial agents incorporated into edible coatings for fresh-cut fruits. Lee et al. (2003) extended the shelf-life of apple slices coated with a carrageenan-based layer containing ascorbic acid, citric acid, and oxalic acid by at least 2 weeks at 3°C. Krasaekoopt and Mabumrung (2008) observed that MC coating incorporated with 1.5% and 2% (w/v) chitosan preserved quality of fresh-cut cantaloupe. Chitosan-based coatings were shown to protect highly perishable fruits like strawberries, raspberries, grapes, and fresh-cutgreen pepper from fungal decay (Vargas et al., 2006; Park et al., 2006; Raymond et al., 2012). The application of chitosan edible coating on fresh-cut broccoli (Moreira et al., 2011) significantly reduced mesophilic and psychrotrophic counts and inhibited the growth of total coliform. Raybaudi-Massilia et al. (2008) reported that the incorporation of 0.3% (v/v) palmarosa oil into the alginate coating inhibited the growth of the native microbiota and reduced the population of inoculated *Salmonella enteritidis* in fresh-cut melon. Chiu and Lai (2010) reported the efficacy of antimicrobial edible coatings based on tapioca starch/decolorized hsian-tsao leaf gum matrix with various green tea extracts to preserve various types of salads. Azarakhsh et al. (2014) reported a significant reduction in yeast and mold counts in fresh-cut pineapple treated with alginate-based edible coating formulation incorporated with 0.3% (w/v) lemongrass.

9.7.2.3 PROBIOTICS

Probiotics are microbial cell preparations or components of microbial cells that have a beneficial effect on the health and well-being of the host (Salminen et al., 1999). This definition implies that probiotics do not necessarily need to be viable. Probiotic adhesion to host tissues is thought to be a key determinant for probiotic efficacy as it facilitates the host–microbial interactions such as the effects of microbes on the immune system of the host. The most commonly used microorganisms are bifidobacteria, lactic acid bacteria, and certain yeast. Tapia et al. (2007) developed the first edible films for probiotic coatings on fresh-cut apple and papaya and observed that both fruits were successfully coated with alginate or gellan film-forming solutions containing viable bifidobacteria. Based on probiotic immobilization technique, Moayednia et al. (2009) observed that immobilization of *Lactobacillus acidophilus* in alginate matrix of strawberry coating effectively protected bacteria against the low temperature storage. Recently Moayednia et al. (2010) immobilized *L. acidophilus* and *Bifidobacterium lactis* in calcium alginate and coated strawberries by using 2% (w/v) concentration of sodium alginate solution.

9.7.2.4   NUTRACEUTICALS

Edible films and coatings are also an excellent vehicle to enhance the nutritional value of fruits and vegetables by carrying basic nutrients that lack or are present in low amounts in fruits and vegetables. Chien et al. (2007) maintained the ascorbic acid content of sliced red pitayas (dragon-fruit) coated with low-molecular weight chitosan. Tapia et al. (2008) reported that the addition of ascorbic to the alginate edible coating helped to preserve the natural ascorbic acid content in fresh-cut papaya, thus helping to maintain its nutritional quality throughout storage. Hernandez-Munoz et al. (2006) indicated that chitosan-coated strawberries retained more calcium gluconate than strawberries dipped into calcium solutions. Likewise, Han et al. (2004) noted a high calcium gluconate or vitamin E content in fresh and frozen strawberries and red raspberries treated with chitosan-based coatings.

9.7.2.5   ENHANCERS

Application of edible coatings can enhance sensory attributes, including visual appeal (e.g., color, glossiness) and tactile features (e.g., surface smoothness, nongreasy/sticky surface). Shellac and wax coating as lipid-based coatings have ordinarily been used to provide high-gloss finish on fruits. Texture enhancers such as calcium chloride can be incorporated into the formulation of edible coatings to better maintain quality during storage of fresh-cut produce since it plays a role in crosslinking polymers, hence minimize softening phenomena. Perez-Gago et al. (2006) maintained firmness of fresh-cut apple pieces through the incorporation of 1% calcium chloride within a WP concentrate coating formulation. Oms-Oliu et al. (2008) reported that calcium chloride, as a crosslinker of polysaccharide chains (alginate, gellan, and pectin), retained firmness of fresh-cut melon during storage. Similar results were observed by Rojas-Grau et al. (2008) where apple wedges coated with alginate or gellan edible coatings and calcium chloride solution had their initial firmness maintained throughout refrigerated storage. Hernandez-Munoz et al. (2006) observed an increase in firmness of strawberries treated with 1% chitosan coating formulation containing calcium gluconate. Lee et al. (2003) indicated that incorporating 1% calcium chloride within a WP concentrate coating formulation helped to maintain firmness of fresh-cut apple pieces. Likewise, Olivas et al. (2007) maintained firmness of fresh-cut Gala apples dipped in calcium chloride solution and subsequently treated with alginate-based coating. Ribeiro et al. (2007)

checked firmness loss of fresh strawberries treated with calcium-enriched carrageenan coating compared to the noncoated fruit.

9.7.2.6 FLAVORS

Encapsulation of flavors by the use of edible coatings represents an effective method for adding flavors to foods, which allows controlled flavor loss and controlled release. Mourtzinos et al. (2008) reported that matrices incorporated with thymol and geraniol in β-cyclodextrin and modified starch prevent the oxidation of thymol and geraniol, which remain intact in temperatures at which free monoterpenes were oxidized. Application of flavor precursors that react with food components to produce flavoring compounds is another approach to increase flavor of coated foods (Olivas et al., 2007). In apple wedges, application of alginate–calcium coatings as a holding matrix for linoleic acid and isoleusine increased the concentration of aroma compounds which can be attributed to the metabolization of linoleic acid and isoleucine by the fruit-producing compounds such as hexanal *trans*-2-hexenal, 2-methyl-1-butanol, and 2-methyl-butyl acetate (Olivas et al., 2012).

9.8 NANOTECHNOLOGY: A NOVEL APPLICATION IN EDIBLE FOOD PACKAGING

In today's competitive market, new frontier technology is essential to develop new products and new processes, with the goal of enhancing the product performance, shelf-life, freshness, safety, and quality, so as to keep pace with consumers demand for fresh, authentic, and flavorful food. The concept of nanotechnology was proposed by Richard Feyman in 1959 at a meeting of the American Physical Society (Khademhosseini and Langer, 2006). Nanotechnology, the science of very small materials, is poised to have a big impact in food production and packaging. Nanotechnology has been provisionally defined as relating to materials, systems, and processes and it encompasses a range of technologies that operate at the scale of the building blocks of biological and manufactured materials, the nanoscale (Heena Jalal et al., 2013). The first commercial applications of nanotechnology in food sector are nanofood packaging (Roach, 2006). The commercialization of nanocomposite materials was started in the late 1980s by Toyota. However, research on use of nanocomposites for food packaging started in the 1990s mostly based on montmorillonite (MMT) clay as the nanocomponent in a number of polymers like nylon, starch, polyethylene, and polyvinyl chloride (Heena Jalal et al., 2013).

Nanotechnology and its application in food science have recently been studied. Some of the novel nanofood applications involve the use of nanoparticles, such as micelles, liposomes, nanoemulsions, and biopolymeric nanoparticles aimed at ensuring food safety. The main purpose of nanofood packaging is to improve product's shelf-life by enhancing the barrier functions against UV light exposure and exchange of moisture and gas between the product and surroundings (Lagaron et al., 2005; Sorrentino et al., 2007). Further, they could serve as carrier to deliver flavors and colors and release nanoscale enzymes, antimicrobials, antioxidants, flavors, fragrances, or nutraceuticals into the product to improve its shelf-life, taste, or smell (LaCoste et al., 2005; Lopez-Rubio et al., 2006; Weiss et al., 2006). Several nanoparticles have been identified as possible additives to promote polymer performance. Considering the availability and cost factors, the potential role of cheaper and easily available coating materials, for example, layered inorganic solid such as clay and silicates, should be given attention (Azeredo et al., 2009). MMT, a hydrated alumina–silicate-layered clay and most widely studied type of clay fillers, possesses the ability to minimize OP when incorporated into pectins (Mangiacapra et al., 2005). An antimicrobial nanocomposite film is highly desirable considering the structural integrity and barrier properties imparted by the matrix, and the antimicrobial properties of the impregnated natural antimicrobial agents (Rhim and Ng, 2007). Nanocomposites food packaging based on silver is well known for exhibiting strong toxicity toward a wide range of microorganisms (Liau et al., 1997). Moreover, silver nanoparticles (Ag-NPs) may contribute to shelf-life extension of fruits and vegetables as they absorb and decompose ethylene (Hu and Fu, 2003). Li et al. (2009) reported that a nanocomposite polyethylene film incorporated with Ag-NPs can retard the senescence of Chinese jujube. An et al. (2008) reported the effectiveness of coating containing Ag-NPs in decreasing microbial growth and increasing shelf-life of asparagus. The significance of chitosan nanoparticles in imparting higher antimicrobial effect combining with other antimicrobial agents has also been studied by many researchers (Ramani et al., 2014; Song and Jang, 2014; Madureira et al., 2015).

9.9 PREPARATION OF EDIBLE FILMS AND COATINGS

Films can be formed via several processes, depending on the starting material, and can be obtained by extrusion, co-extrusion, spreading, casting, roll coating, drum coating, and coating or laminating techniques. The film-forming mechanisms of biopolymers include intermolecular forces such as covalent bonds (e.g., disulfide bonds and cross-linking) and electrostatic, hydrophobic, or ionic interactions. Lipid and wax films can be formed through solidification of the melted material (Fennema and Kester, 1991). Biopolymers in solution can form films by changing the conditions of the solution. The application of heat, salt, or changing the pH may alter conditions in the solution, which eventually influence biopolymers aggregation. This process is called coacervation (Krochta et al., 1994; Debeaufort et al., 1998). Gelation or thermal coagulation consists of heating the macromolecules involving denaturation, gelling, and precipitation. Traditionally, the methods used for the production of edible films have been divided in two main groups: wet and dry processes. Both these processes begin by dissolving the ingredients in a solvent and then removing the liquid phase by drying (Peressini et al., 2003). The composition of the films contains a substance that forms a tough and solid matrix, through the interactions between the molecules undergoes chemical or physical treatments. The method selected for the production of active edible films could affect and modify the final properties of the material.

Wet process such as mold casting, drawdown bar, and compression molding that are used to form edible films needs solvents for the solution and dispersion of the polymer onto a flat surface; this is followed by drying in controlled conditions for the removal of the solvent and the formation of the film. It is a high-energy-consuming procedure, adequate for laboratories but not for the industrial scale up. Casting is a simple method for the production of edible films, but it is a batch procedure used on a very small scale. Nevertheless, a continuous casting method (knife coating or tape casting) can be used on the industrial scale because the film-forming suspension is prepared on continuous carrier tapes with effective control of the thickness (De Moraes et al., 2013). Because the final product should be edible and biodegradable, only water and ethanol or their combinations are suitable solvents (Campos et al., 2011). It is known that materials with thermoplastic behavior can be processed into films by the application of different thermal or mechanical processing techniques. The most common dry processing methods of edible films include extrusion, injection, blow molding, and heat pressing. The extrusion process is based on the thermoplastic properties of polymers when plasticized and heated above their glass transition temperature under low water content conditions. The different methods of preparation of edible films and coatings are described below.

9.9.1 COMPRESSION MOLDING

In compression molding, protein–plasticizer mixtures are placed in an open mold. Subsequently, a plunger applies pressure forcing the material to assume the desired form. This occurs relatively quickly at low moisture contents, high temperature, and pressure to transform the blends into visco-elastic melts. At the same time under such conditions, protein gets denatured (Guerrero and La Caba, 2010). Through subsequent cooling, the product attains a form that is determined by its covalent, ionic, and hydrogen bonding as well as hydrophobic and hydrophilic interactions (Balny et al., 2002).

9.9.2 SOLVENT CASTING

This is the preferred method for forming edible protein films for research. Various types of equipment are available for solvent casting of films, from simple casting plates to more advanced batch and continuous lab coaters. In this method, films are formed by manually spreading dilute film solutions (usually 5–10% solids) of protein and plasticizer into level petri dishes, followed by drying under ambient conditions or controlled relative humidity. In case of proteins, a larger film can be obtained using sophisticated equipment which mechanically spread the solution to a fixed thickness. Kozempel and Tomasula (2004) developed a continuous process for solvent casting of protein films using casein-based films. The physical properties of the final film such as film morphology, appearance, and barrier and mechanical properties are highly affected by the drying method (Perez-Gago and Krochta, 2000).

9.9.3 EXTRUSION

Till date, extrusion is the most commonly used process for polymer production (Robertson, 2012). It is a continuous process where the raw materials are constantly introduced into a hopper feeding a horizontal barrel. This process can involve various operations such as heating, shearing, mixing, compressing, melting, and shaping. As indicated by the name of this section, this is where the raw material changes into extrudate under the rising pressure and temperature. Knowledge about optimal ingredient ratio and processing parameters is also vital for film formation by extrusion (Sothornvit et al., 2007). The extrusion process can proceed under various conditions using different screw configurations, lengths, diameters, speeds, temperature profiles, feeding rates, and the addition of ingredients at the beginning and during the process (Hernandez-Izquierdo and Krochta, 2008). Extrusion method of film production has certain advantages over solvent casting. Comparatively, extrusion is faster and requires less energy, due to the fact that more concentrated film solutions can be fed into the extruder. Besides, this method does not require solvent addition and a time of evaporation making it attractive for industrial processes (Guan and Hanna, 2006). Recent technical developments, especially in twin-screw extrusion, have increased the application of this technique for edible film production (Liu et al., 2009). Other processing techniques, such as injection molding, film-blown die, and thermopressing (Lopez et al., 2014), are combined with extrusion to produce the final edible films.

9.9.4 SPINNING

This processing technique is most commonly used by the textile industry to form fibers (Rampon et al., 1999). For formation of flat films, this method was modified by Frinnault et al. (1997) replacing the spinneret with a plate die. Film formation involved the extrusion of the polymer solutions into a coagulating bath and then collected onto a roller. This modified process has been used to prepare films from casein and soy protein. Using the spinning method, Rampon et al. (1999) hypothesized that soy protein chains could be oriented similar to textile fibers. They further theorized that the elongation values of the films are dependent on the extent of orientation of protein chains within the edible film. Despite the successful effort in forming films, they could not observe an increase in the orientation of the protein chains. They suggested that shear could possibly play a role in orientation since the modified film formation process used a pressure lower than that commonly used textile processing.

9.10 APPLICATION METHODS OF EDIBLE COATING

Edible films or coating can provide either clear or milky coatings, but consumers generally favor transparent, clear coatings. The simplest way to apply a film is directly from solution. Depending on concentration of coating solution, the product will absorb an appropriate amount of coating material necessary to form the desired layer, which when dried forms a protective layer at the food surface. In most cases, some plasticizers like glycerol, mannitol, sorbitol, and sucrose need to be added to coating solution to prevent film brittleness; otherwise, mobility of various components will increase by orders of magnitude, resulting in mass flow instead of diffusion. Coatings should have good adhesion to rough surfaces (Hershko et al., 1996) though application of a uniform film or coating layer to cut fruit and vegetable surfaces is generally difficult. This can be improved by adding surfactants to solution to reduce surface tension. This strategy will also reduce the superficial and in turn reduce water loss (Roth and Loncin, 1984). Several approaches have been employed for depositing the film directly on food surface and include dipping, brushing, spraying, and creating standalone film from solution or through thermoformation (Gontard and Guilbert, 1994). Edible coatings can also be applied by methods such as panning, fluidized bed, dipping, and spraying. All these techniques exhibit several advantages and disadvantages and their performance depends principally on the characteristics of the foods to be coated and the physical properties of the coating.

9.10.1   FLUIDIZED-BED PROCESSING

This technique was originally applied in the pharmaceutical industry, which then extended to a wide variety of food products of high markup, such as functional ingredients and additives including processing aids (leavening agents and enzymes), preservatives (acids and salts), fortifiers (vitamins and minerals), flavors (natural and synthetic), and spices (Dewettinck and Huyghebaert, 1999; Chen et al., 2009). Typically, in this method, a coating material, either in the form of a solution or suspension, is sprayed through a set of nozzles onto the surface of fluidized powders to form a shell-type structure. Fluidized beds are categorized by three different configurations: top spray, bottom spray, and rotating-fluidized bed; but the conventional top-spray method has a greater possibility of success in the food industry compared to the other methods (Dewettinck and Huyghebaert, 1999). Fluidized-bed coating is used in the food industry to produce a large variety of encapsulated food ingredients and additives, such as puffed wheat, nuts, and peanuts.

9.10.2   PANNING

The panning process consists of depositing the product to be coated into a large, rotating bowl, referred to as the "pan." This method is specially designed for extruded products as they can be produced in a round or oval form and in different sizes which are relatively easy to coat. Other products, particularly small items such as nuts and raisins, are also coated by panning (Geschwindner and Drouven, 2009; Talbot, 2009).

9.10.3   DIPPING

Dipping method of applying coatings to fruits and vegetables is used when the coating solution is highly viscous (Dhanapal et al., 2012). It is performed by introducing the product in a coating solution between 5 and 30 s under controlled conditions of density and surface tension. This method is advantageous for products with complex and rough surface that require a total uniform coating. The effectiveness of coating and fruit acceptability are greatly affected by the temperature of coating emulsion and the dipping time. However, dipping method has its own inherent disadvantages such as coating dilution and dilution of the natural waxy layer of the food surface, microorganism growth, and coating solution contamination (Lin and Zhao, 2007). On the contrary, this method usually forms thick coating which may pose problems with product respiration and storage characteristics (Grant and Burns, 1994; Martin-Belloso et al., 2009).

9.10.4   SPRAYING

According to Debeaufort and Voilley (2009), spray coating is the most commonly used technique for applying food coatings. Spraying method is used when the viscosity of coating solution is low. In this method, a uniform thick or thin layer of coating over a food surface can be obtained. Different from other systems, spray coating can work with large surface areas making it possible to deposit various kinds of aqueous solutions or suspensions, such as liquefied lipids or chocolate (Debeaufort and Voilley, 2009). Spray applications are also suitable when dual or more successive applications are required, for example, to make a gel layer with alginate and calcium chloride solutions (Cutter, 2006; Dangaran et al., 2009). The main advantages of this method are its uniform coating, thickness control, and the possibility of multilayer applications, such as using alternating sodium alginate and calcium chloride solutions (Martin-Belloso et al., 2009; Ustunol, 2009). Moreover, spraying systems avoid contamination of coating solution, allow coating solution temperature control, and can facilitate automation of continuous production. Fruit-based salads coated through spraying of tapioca starch with green tea extracts showed reduce growth of aerobic microorganisms and yeasts (Chiu and Lai, 2010). Chitosan solution applied as a preharvest spray or postharvest coating reduced decay in table grapes and affected the content of total phenolic compounds and the activities of anti-oxidative enzymes of the product (Meng et al., 2008).

9.10.5   BRUSHING

This method is used for highly perishable products, such as fresh beans and strawberries, where moisture loss is an issue. A thin coating onto the surface of the product is obtained which could act as semipermeable membranes reducing gas transfer rates and creating new packaging materials to extend food shelf-life.

9.10.6   ELECTROSPRAYING

The development of edible nano- or microencapsulation matrices to protect the biologically active compounds against adverse conditions (Dube et al., 2010) faces some limitation related to stability of bioactive compounds, when exposed to high temperatures and use of organic solvents (Lopez-Rubio and Lagaron, 2012). Electrospraying or e-spraying has recently emerged as an alternative for the generation of polymeric particles incorpo-rating bioactive agents (Bock et al., 2012) with application in therapeutics, cosmetics, and the food industry (Jaworek and Sobczyk, 2008). Beginning in the paint industry, electrostatic spraying has a series of superiorities over the traditional spraying technique. In electrospraying, microdroplets of size down to 20 µm are generated giving the potential of very thin and even coatings (Khan et al., 2012). The charged droplet follows a trajectory to the nearest ground surface which, under electrostatic attraction, leads to a high transfer efficiency (80%) which might reduce the processing cost (Luo et al., 2012; Maski and Durairaj, 2010; Oh et al., 2008). Electrospraying has been successful for lipid-based materials (Gorty and Barringer, 2011; Luo et al., 2012; Marthina and Barringer, 2012). Ganesh et al. (2012) pointed out that electrostatic spray of food-grade acids and plant extracts is more effec-tive compared to conventional spray in decontaminating *Escherichia coli* O157:H7 on spinach and iceberg lettuce. The selection of an appropriate coating method not only impacts the preservation effect of the coatings formed on the food items but also determines the production cost and process efficiency.

9.11 ANALYTICAL TECHNIQUES FOR MEASURING EDIBLE COATINGS APPLIED ON FRUITS AND VEGETABLES

The applicability of edible films and coatings varied according to the types of product to be coated. In contrast, the coating material with different properties influences its overall function and compatibility with a product. It also influences the mouth feel of the material when consumed by customers. The development and analysis of edible films must be conducted to select the optimum coating for a specific product (Park, 1999). Since some polymer characteristics are very difficult to measure on the applied products, the examination of edible films alone can be made to provide a notion about their behavior once applied to processed fruits and vegetables. Thus, analysis of edibles films must be conducted under conditions similar to the environments induced on the coated fruits and vegetables. Since one of the main functions of edible coatings is to act as a protective barrier to environmental moisture, gases, flavors, aromas, or oils (Han, 2014), other properties that are often determined on stand-alone films are WVP, OP, carbon dioxide permeability, and flavor permeability. Aroma and oil permeability are also very important for many foods but have generally received less attention. In other cases, other properties of interest are water solubility, gloss, and color.

9.11.1 INTERNAL GAS COMPOSITION OF COATED PRODUCE

The internal gas modification in coated fruits and vegetables gives an indication of the potentialities of coating in controlling the internal gas atmosphere of produce. Different coatings have different permeability characteristics that affect the quality and shelf-life of a preserved product. Permeability testing can be used to evaluate the types of foods that can be packaged by a material and it can also be used to predict the shelf-life of the packaged product (Park and Zhao, 2004; Prommakool et al., 2011). An edible film with good gas barrier properties can be used to prolong the shelf-life of the product. For example, a material with high oxygen barrier helps to reduce lipid oxidation and nutrition loss if used to coat peanuts (Han et al., 2008). Therefore, it is important to understand the gas permeabilities of edible films to predict the shelf-life of a product and the end-use application of the materials (Siracusa et al., 2008). Saltveit (1982) and Park (1999) proposed a procedure for extracting and analyzing internal gas samples.

9.11.2 OXYGEN TRANSMISSION RATE

The oxygen transmission rate (OTR) of a packaging material plays a vital role in influencing the shelf-life of oxygen sensitive products such as fresh fruits, vegetables, and salads. Most researches use the ASTM D3985-05 (2010) method to measure the OTR of films at 23°C and 50% relative humidity (Du et al., 2008; Prommakool et al., 2011). According to ASTM D3985-05, oxygen permeance (PO) is the ratio of the OTR to the difference between the partial pressure of oxygen on the two sides of the film. The SI unit of PO is the mol/(m^2 s Pa). These conditions include the environmental temperature and relative humidity. In general, higher the crystallinity of the polymeric structures, the lower the OTR of the materials. Also, the higher the amorphous region of the material, the higher is the OTR (Lacroix, 2009).

9.11.3 CARBON DIOXIDE TRANSMISSION RATE

The carbon dioxide transmission rate CO_2TR is measured using a technique similar to that of the OP. Knowing the CO_2TR is important as the respiration rate of a product affects the quality of the food. The rate of carbon dioxide loss from the package provides one of the clues to prolonging the shelf-life of these products (Siracusa et al., 2008).

9.11.4 WATER VAPOR TRANSMISSION RATE

The WVP of a packaging material refers to the rate of moisture transfer between the storage environment and the internal environment of the package. Lower the WVP, higher is the capability of the coatings to extend the product shelf-life (Bravin et al., 2006). The water vapor transmission of a film sample can be determined by the ASTM E96-92 method (ASTM, 1992). In this method, the water vapor transmission rate (WVTR) of the package can be assessed by the change in weight over time, which is influence by dehydration and adsorption phenomena that occur within and outside the coating materials. Henrique et al. (2007) indicated that WVTR of edible films could be directly related to the quantity of –OH group on the molecule. It also influences the properties of the polymeric structure like crystallinity and cohesiveness and the molecular interactions between the polymer chains. For example, increasing the cross-linking between polymeric chains and reducing the impact of solvents and plasticizers are reported to improve the water vapor barrier of protein-based films (Sabato et al., 2001).

9.11.5 COATING THICKNESS

The thickness of coatings directly relates the permeability function of the coatings to water and gases through coating gas resistance parameters (Hagenmaier and Baker, 1995) and through the solid concentration of the coating solutions (Park et al., 1994). Coatings thickness exceeding a critical limit can cause undesirable build-up of anaerobic fermentation. The solution properties such as density, viscosity, and surface tension, as well as surface withdrawal speed from the coating solution, influence coating thickness (Cisneros-Zevallos and Krochta, 2003). The coating thickness can be determined destructively by measuring the film thickness of the peeled coatings using a micrometer or undestructively. Viscosity measurement to understand the visco-elasticity properties of edible film solutions has been used to determine how much wet film thickness should be applied to a casting plate before drying in a film processing operation. In addition, viscosity also helps to understand the thoroughness of a blended emulsion for both Newtonian and non-Newtonian fluids. Recently, confocal Raman microspectrometry (CRM), surface enhanced Raman scattering (SERS), and Fourier transform (FT)-Raman spectrometer have been employed to evaluate coating thickness directly from the surface of coated foods (McAnally et al., 2003; Hsu et al., 2005). Coating thickness varies with viscosity, concentration, density, and draining time of the biopolymer solution and relates to the square root of viscosity and the inverse square root of draining time, which agrees with the theoretical approach for flat plate dip-coating in low-capillary-number Newtonian liquids (Cisneros-Zevallos and Krochta, 2003).

9.11.6 WETTABILITY

The effectiveness of fruit and vegetables edible coatings primarily depends on controlling the coating solution wettability which affects a film's coating thickness (Park, 1999). The coating should be able to wet and spread uniformly over product surface and, upon drying, should have suitable adhesion, cohesion, and durability to function effectively. Coating involves the wetting of the product to be coated by the coating solution, which may penetrate into the fruit skin (Hershko et al., 1996; Krochta and de Mulder-Johnston, 1997), followed by possible adhesion between these two commodities. Choi et al. (2002) discussed the theoretical background of the wettability measurement. They investigated wettability of chitosan-coating solutions on "Fuji" apple skin using the Du Nouy ring method and the sessile-drop method described by Harkins and Jordan (1930).

9.11.7 SURFACE CHARACTERISTICS OF COATINGS

Various analytical instruments are used to determine surface characteristics of coatings on fruits and vegetables. In apple, the surface morphologies of the applied coating were observed using a scanning electron microscope (Choi et al., 2002; Fornes et al., 2005). Leica stereomicroscope was used to study the uniformity of the coatings and its adherence to the fresh-cut apples (Rojas-Grau et al., 2007). X-ray diffraction can be used to determine the morphology as well as the details of the chemical components of polymeric materials (Garcia et al., 2000; Yoo and Krochta, 2008). Zhong and Xia (2008) reported that X-ray diffraction helped to conceptualize the interaction and molecular miscibility among the major components of a chitosan/cassava starch/gelatin film. Surface gloss can be measured using a Micro-TRI-Gloss meter in accordance with American Society for Testing and Materials method D523.11 (Trezza and Krochta, 2001) at 20°, 60°, and 85° angles from normal to the coating surface. In addition, CRM, SERS, and FT-Raman spectrometer can also be utilized to analyze coating structure and surface characteristics of the polymeric materials (Hsu et al., 2005).

9.11.8 MECHANICAL STRENGTH

Texture profile analysis describing the structure of materials provides the stress or strain of samples and records the force changes as time progress

(Rosenthal, 1999). Tensile strength (TS) and elongation at break are two commonly used test methods for knowing the mechanical strength films and sheets. The film tensile properties can be described in terms of Young's modulus indicating that the mechanical strength of a film, TS, reflecting the pulling force per film cross-sectional area required to break the film, and elongation at break (E), which gives the degree to which the film can stretch before breaking (Porat et al., 2004). ASTM D882-00 (2010) is the standard test method for assessing tensile properties of thin films. Both rheology and texture profile analyses also describe the physical properties of polymers and establish the procedure for handling them during food processing. Edible films are sensitive to environmental conditions such as relative humidity and temperature. During analytical testing, edible films could exchange moisture with the environment which can affect the mechanical properties of the film and alter its properties. A small amount of water can act as a plasticizer within the polymeric structure but brittleness may results when moisture content reduces (Rossman, 2009). Other standardized mechanical examinations include compression strength, puncture strength, stiffness, tearing strength, burst strength, abrasion resistance, adhesion force, and folding endurance (Krochta, 2002; Dhall, 2013).

9.11.9   THERMAL PROPERTIES

When a polymer is heated from a low to a high temperature, it goes through a series of transition zones that could be described as rigid, thermoelastic, and thermoplastic (Mathew et al., 2006). In each of these zones, the material will exhibit certain properties that are characteristic of that material and of each zone. The glass transition temperature (T_g) and melt temperature (T_m) are used to described this elastic state of the polymers as the properties of edible components are highly affected by temperature changes. Differential scanning calorimetry and thermo-gravimetric analysis are usually combined to gain a better understanding of the thermal characteristics of a sample. Dynamic mechanical analysis is a more versatile tool for testing the thermal properties of a sample when compared with the texture analyzer and differential scanning calorimetry methods. This is so because it provides a controlled system of both temperature and stress/strain to a detected sample (McFarland, 2001).

9.12   FACTORS AFFECTING FUNCTIONALITIES OF EDIBLE FILMS

9.12.1   TYPE OF RAW MATERIAL

The raw materials used in film solutions are classified, based on their solubility, into hydrophilic (water-soluble) and hydrophobic materials (water insoluble but alcohol soluble). The difference in soluble properties of these raw materials influences the amount of energy needed to obtain dried films and their application on foods. Carbohydrates such as alginate, carregeenan, pectin, starch, cellulose, and cellulose derivatives provide a strong matrix-free standing film, but these films are poor water barrier properties because of the hydrophilic nature of raw materials used (Kester and Fennema, 1986). Proteins provide good gas barrier but poor water vapor barrier properties. However, some protein films such as corn zein films exhibit better water resistance than other protein films because zein contains high amount of hydrophobic side chain amino acid. Lipid films, made from hydrophobic materials such as wax and fatty acid, show excellent water vapor barriers but poor mechanical properties.

9.12.2   POLYMER CHEMISTRY

The regular structure molecule is more diffusible than the irregular stereochemical structure whereas branched molecules may provide a greater cohesive strength than nonbranched molecules. A lower molecular weight fraction shows a greater cohesion and a greater change in cohesion with temperature changes. In highly polar polymers such as protein and protein-based cellulosic, self-adhesion by diffusion is not significant due to the minimal flexibility and fixed order of the macromolecule. This is caused by the internal molecular forces holding the polymer chains. Cellulosics have a backbone with a rigid ring structure chain whereas proteins tend to form helical chain structure (Banker, 1966).

9.12.3   pH

pH plays an important role in protein films made from water-soluble materials, such as SPI and WP isolate, as the solubility of these proteins depends on their isoelectric point (pI). During the dissolution of macromolecular substances, the cohesive forces between the solute macromolecules are neutralized by unions with the solvent molecules (Banker, 1966). The greater the degree of dissolution and the more extensively the chain is charged, the greater is the uncoiling of the chain. The maximum protein solubility is obtained at pH away from its pI. But to produce an edible film at extreme pH, the sensory property must also be considered along with other film properties. Gennadios et al. (1994) studied the effect of pH on SPI film and found that highly acidic (pH < 1) or alkaline conditions (pH > 12) inhibit the formation of SPI film. Kinsella and Phillip (1989) reported that films formed near the pI of major proteins are more condensed and stronger.

9.12.4 DRYING TEMPERATURE

Production of films based on casting method involves drying of a complex colloidal matrix containing the polymer, solvent, and plasticizer. In the process, the polymer matrix is exposed to high temperatures which affect interaction forces at structural level. High temperature during edible film processing will cause browning in polysaccharide-based materials and oxidation in lipid-based films and affect protein degradation and the dehydrations of protein polymeric chains (Hernandez-Izquierdo and Krochta, 2008). Water loss during the drying period alters the conformation of the proteins. This results in the facilitation of adhesion between polymer films and the substrate (Banker, 1966). High temperature (70–100°C) affects the forming of rigid structures in protein solutions because of protein denaturation (Perez-Gago et al., 1999). Excessive heat or an excessive solvent evaporation rate during processes may produce noncohesive films (Guilbert et al., 1996). Water-soluble proteins such as soy protein and WP need a higher temperature and longer time for film formation than films from alcohol-soluble protein such as corn zein or wheat gluten. The higher drying temperature of water-solubility-based films may limit a film's use.

9.12.5 CONCENTRATION

The concentration of a solution could affect the mobility of polymeric materials and their ability to form films, the self-adhesion of high polymers, and the rate of matrix forming in film preparations. Besides, the polymer concentration in film solution can also influence the formation of the polymer matrix. For instance, there is probably less protein–protein interaction at a lower protein concentration, while at higher protein concentrations, self-diffusion is promoted resulting in inferior properties. At the optimum concentration of film solutions, an intermediate viscosity could be obtained which result in the highest cohesive strength. The production of films with WP isolate requires a relatively high protein concentration (>8%) in the film-forming solution so that the formation of S–S bridges occurs whereas the films produced from muscle fish protein prepared with 1.5–2% showed stronger films than other concentrations (Sothornvit and Krochta, 2005).

9.12.6   RELATIVE HUMIDITY

The adsorption of water and vapor by dried materials is generally assumed to involve the binding of water molecules to specific hydrophilic sites, such as carboxylic, amino, and hydroxy residues, in addition to backbone peptide groups. At high relative humidity, multimolecular adsorption in protein occurs through swelling and conformational changes in the macromolecular structure. The relationship between equilibrium relative humidity and film water content has been assessed by measuring water sorption isotherms. In addition, an understanding of water sorption properties is necessary to tailor film applications. Pochat-Bohatier et al. (2006) reported an increase in gas permeability at high relative humidity owing to swelling of the polymer matrix which allows chemical interactions to take place between amino acids and the gas.

9.12.7   FILM ADDITIVE

Film additives are incorporated into edible films with the purpose of improving its mechanical, protective, sensory, or nutritional properties. In protein-based films, addition of plasticizers improves the flexibility and capacity for processing of polymers by lowering the glass transition temperature (T_g). These substances reduce the tension of the deformation, hardness, density, viscosity, and electrostatic charge of a polymer. It also affects the degree of crystallinity, optical clarity, electric conductivity, fire behavior, and resistance of polymer to biological degradation (Vieira et al., 2011). The compatibility between polymer and plasticizer plays a critical role in plasticization adjudged by polarity, hydrogen bonding, dielectric constant, and solubility parameters (Choi et al., 2004). The addition of plasticizer aids in producing flexible and stronger films. This can also be achieved by reducing the intermolecular bonds between the polymer chains, and thus the overall cohesion to facilitate elongation of the films, and reduces its glass transition temperature. Such property is manifested by a reduction in the barrier properties to gases, vapors, and film solutes (Banker, 1966).

9.13 CHALLENGES AND OPPORTUNITIES OF EDIBLE COATINGS

The potential application of edible coating in fresh fruits and vegetables has delivered significant impact in enhancing protection and quality preservation of fresh products. Some novel applications of the promising edible films and coatings have been proposed as that also serves as alternatives to existing technologies. The advent of nanotechnology has paved way for engineering the nanostructure of packaging materials to achieve desirable functional properties as per their designed functions. Nevertheless, edible packaging still has to overcome several issues to achieve significant commercial application. Though edible coatings have been successfully applied to fresh fruits and vegetables, there are also reports indicating its adverse effect on product quality such as development of off-flavor disorders (Ben-Yehoshua, 1969). Hence, for edible packaging to function both as a packaging and food component, it must fulfill rigorous requirements and challenges.

9.13.1 REGULATION

The application of edible coating has delivered promising potential in tackling the quality concerns of fresh produce of fruits and vegetables. However, issues pertaining to its commercial applications need to be addressed to meet the legislation needs of the country whether imported, exported, or domestically consumed. In general, edible coating and its components should be at minimum of food grade without toxic effects meeting all the required regulations regarding food products (Guilbert and Gontard, 1995) and must be generally recognized as safe (GRAS) for intended use or sanctioned by the United States FDA Code of Federal Regulations or the U.S. Pharmacopoeia/National Formulary. Their use must be in accordance with good manufacturing practice and within any limitations specified by the FDA (Krochta and de Mulder-Johnston, 1997). The current classification and regulations of food ingredients can be found at FDA's everything added to Food in the United States database. Different countries and regional single market such as European Community have their own specific regulations on food manufacturing and food additives used in food products. In the USA, federal FDA regulates almost all final food products and food additives that are used in food matrices and U.S. Department of Agriculture regulates "organic" labeling of fresh produces, dairy products, or others. The ingredients of edible packaging must be declared on the label under regulation of the Federal Food, Drug, and Cosmetic Act (21 USC 343). Edible packaging made from common allergens (i.e., milk, eggs, peanuts, tree nuts, fish, shellfish, soy, wheat) must be clearly indicated to provide information to consumers who are sensitive to particular food components according to Food Allergen Labeling and Consumer Protection Act of 2004. The European Union regulations for food and food packaging have made it mandatory for assessing the introduction of new nanotechnology under specific safety standards and testing procedures (Halliday, 2007). The perspective on nanotechnology has been provided on the FDA Web site (http://www.fda.gov/nanotechnology/). One important fact to remember is that the FDA regulates products, not technologies.

9.13.2 SENSORY IMPLICATIONS

The development of undesirable sensory properties on the coated products is one of the potential adverse effects of the use of edible coating. Off-flavors owing to the coating materials itself or anaerobic respiration resulting from excess inhibition of oxygen and carbon dioxide exchange may develop. In addition, the appearance of nonuniform or sticky surfaces on the product may affect its appeal and consumers' acceptance (Zhao and McDaniel, 2005). Adjusting the thickness or permeability of the coating according to the variety and storage and marketing temperatures is highly imperative. Park et al. (1994) reported the development of alcoholic off-flavors fermentation in tomatoes coated with 2.6-mm zein film owing to low oxygen and high carbon dioxide concentration. Smith et al. (1987) attributed the occurrence of core flush, flesh breakdown, accumulation of ethanol, and off-flavors disorders in processed fruits to use of edible coatings. Burt (2004) reported that incorporation of antimicrobial agent especially essential oils into edible coatings could impart unacceptable sensorial modifications in foods. The inclusion of antibrowning compounds, involving high concentrations of sulfur-containing compounds such as N-acetylcysteine and glutathione, as dipping agents into edible coatings can yield an unpleasant odor (Iyidogan and Bayindirli, 2004; Rojas-Grau et al., 2006). There is also the possibility of imparting bitter taste, astringent, or off-flavor when nutraceutical compounds are incorporated into edible coatings (Drewnowski and

Gomez-Carneros, 2000). Hence, while developing edible films and coating, quality parameters such as color change, firmness loss, ethanol fermentation, decay ratio, and weight loss of edible coated fruits and vegetables must be considered. Since edible coatings are consumed along with the coated products, the incorporation of antimicrobials, antioxidants, and nutraceutical compounds into its formulation should not affect consumer acceptance.

9.13.3 CONSUMER ACCEPTANCE

The acceptability of edible coatings by consumers is significantly influenced by the film sensory properties, safety, marketing, and cultural and religious restrictions regarding the use of new materials. As consumers are becoming increasingly aware of labels on food products, edible packaging materials must possess neutral sensorial properties without toxicity effects and must be compatible with edible-packaged food to be consumed. Although FDA requires food manufacturer to list ingredients and allergens on labels, the application of new edible coatings might present issues related to allergies or cultural and religious restrictions for sensitive groups of consumers. Other safety issues related to the potential microflora changes of the coated food products must also be addressed. Marketing factors such as price, special instructions required for opening, cooking, consuming the packaged foods, or disposing of the packaging and consumer reluctance to use new materials may also affect the commercialization of edible packaging.

9.14 FEASIBILITY OF COMMERCIALIZED SYSTEMS

Till date, the production of edible films is mainly done at the laboratory scale which is expensive compared to synthetic films. Research on cost reduction and production in larger scales are necessary to promote the feasibility of commercialized edible packaging. This depends on the extent of investment for film production or coating equipment, complexity of the production process, manufacturer resistance to the use of new materials, and potential conflicts with conventional food packaging systems (Han and Gennadios, 2005). The demand for products with long shelf-life for trading purpose presents another barrier. Edible coating materials are being inherently susceptible to biodegradation; thus, their protective functions are comparatively stable only for a short duration than conventional packaging.

Therefore, the stability and safety of edible coating under specific storage conditions require further investigation.

9.14.1   LACK OF KNOWLEDGE AND MACHINERY

Basic information on film-coating formulation, properties, methods of application to fruit or vegetable surface, and demonstration of effectiveness are lacking. Tremendous research is required in the area of applications of edible coatings and films on fresh and fresh fruits and vegetables. Moreover, as development of edible coatings and films is still in infancy stage, enterprises do not have the technology needed to apply the system (Dhall, 2013).

9.15   FUTURE THRUST

The development of new technologies to improve the delivery properties of edible films and coatings is a major challenge for future research. At present, most studies on the application of coating technology to fruit and vegetable products are still limited. At the same time, it is also expected that future use of coatings applications will spread through all kind of products, whether fresh or treated, in a manner which is more environment friendly. Hence, future development of edible film and coatings that could be applied on a broad spectrum of foods, besides adding value while increasing their shelf-life, should be largely focused. Emphasis needs to be diverted toward development of effective delivery coatings with bioactive components to extend the shelf-life and control the safety and quality of food products. Such coatings will release the novel bioactive components like vitamins, nutraceuticals, enzymes, pre-, or probiotics into the food matrix throughout storage period. Taking into account the current trends in consumer preferences and the requirements imposed by food regulation policies, these bioactive ingredients will be preferably natural components. In this regard, the role of nanotechnology in edible packaging needs prime attention. Interest on production of coatings with improved incorporation and controlled release of active compounds has paved way for a new generation of edible coatings based on nanotechnological solutions such as nanoencapsulation and multi-layered systems. At present, nanotechnologies are being used to enhance the nutritional aspects of food by means of nanoscale additives and nutrients and nanosized delivery systems for bioactive compounds. Nanotechnology has demonstrated great potential in the fabrication of the whole food-packaging sector by imparting desirable changes. However, important safety concerns about nanotechnology applications to food materials must be addressed. Because of their tiny dimensions, it is reasonable to assume that migration of nanostructures from packaging materials into food may occur. Hence, significant research is still required to generate scientific data and evaluate the potential health effects of nanotechnology products, as well as the environmental safety of their use. Besides, consumer information, involvement, and education are indispensable for the success of nanotechnology.

Edible films and coatings have gained popularity, but so far, on a laboratory scale. Hence, a great effort is needed to scale up this technology at commercial level with the purpose of providing more realistic information to commercialize fresh-cut products coated with edible films or coatings. There are numerous advantages' offers by edible packaging over traditional synthetic polymers. However, the cost of raw materials and manufacturing represents other hurdles in the large-scale application of edible packaging. Owing to biodegradability, edible packaging cannot be used for long-term storage. Thus, development of coatings with longer durability is another area of focus. In today's scenario, it is quite difficult to deliver edible packaging at the price of traditional packaging. Hence, utilization of naturally available materials or waste by products from food industry, for generating edible films and coatings which also solve dual problem of cost and global pollution, needs emphasis.

9.16 CONCLUSION

Edible films and coatings have delivered promising application in the food industry. They have shown encouraging results in enhancing the shelf-life, quality, and safety of perishable food products such as fresh and minimally processed fruits and vegetables. Although considerable efforts have been made to understand and modify film properties, it still remains to be a major challenge. Issues on sensory acceptance pertaining to their impact on taste, flavor, and color, which are of prime importance in food industry, must be addressed. Owing to high cost involved in procurement of raw materials, attention must be paid to harness cheap, natural, and easy available materials. Promising and novel technologies must be commercially tested at large scale to yield better results and improve consumer acceptance. Coatings should be designed to provide highly specific functional performances which are required by different food types to further enhance their performance. Considerations on understanding the interactions among active ingredients and coating materials when developing new edible film and coating applications must be emphasized. Such information will be of great utility in developing new coating with improved and acceptable functionalities.

KEYWORDS

- fresh fruit
- quality
- shelf life
- coating
- formulation
- antimicrobial
- edible film

REFERENCES

Amanatidou, A.; Slump, R. A.; Gorris, L. G. M.; Smid, E. J. High Oxygen and High Carbon Dioxide Modified Atmospheres for Shelf Life Extension of Minimally Processed Carrots. *J. Food Sci.* **2000,** *65,* 61–66.

An, J.; Zhang, M.; Wang, S.; Tang, J. Physical, Chemical and Microbiological Changes in Stored Green Asparagus Spears as Affected by Coating of Silver Nanoparticles-PVP. *LWT—Food Sci. Technol.* **2008,** *41* (6), 1100.

Arvanitoyannis, I.; Gorris, L. G. M. Edible and Biodegradable Polymeric Materials for Food Packaging or Coating. In *Processing Foods: Quality Optimization and Process Assessment*; Oliveira, J. C., Ed.; CRC Press: Boca Raton, FL, 1999; pp 357–371.

ASTM D3985-05. Standard Test Method for Oxygen Gas Transmission Rate Through Plastic Film and Sheeting Using a Coulometric Sensor, ASTM International, West Conshohocken, PA, 2010.

ASTM D882-00. Standard Test Methods for Tensile Properties of Thin Plastic Sheeting (Annual Book of ASTM Standards). Philadelphia: ASTM, 2010.

ASTM. Designation E 96-92: Standard Test Methods for Water Vapor Transmission of Materials. In *Annual Book of ASTM Standards*, Philadelphia, PA, American Society for Testing Materials, 1992; pp 398-405.

Avena-Bustillos, R. J.; Krochta, J. M.; Saltveit, M. E. Water Vapor Resistance of Red Delicious Apples and Celery Sticks Coated with Edible Caseinate-Acetylated Monoglyceride Films. *J. Food Sci.* **1997,** *62,* 351–354.

Azarakhsh, N.; Osman, A.; Ghazali, H. M.; Tan, C. P.; Mohd Adzahan, N. Lemongrass Essential Oil Incorporated into Alginate-Based Edible Coating for Shelf-Life Extension and Quality Retention of Fresh-Cut Pineapple. *Postharvest Biol. Technol.* **2014,** *88,* 1–7.

Azeredo, H. M. C.; Mattoso, L. H. C.; Wood, D.; Williams, T. G.; Avena-Bustillos, R. J.; McHugh, T. H. Nanocomposite Edible Films from Mango Puree Reinforced with Cellulose Nanofibers. *J. Food Sci.* **2009**, *74* (5), 31–35.

Baldwin, E. A. Edible Coatings for Fresh Fruits and Vegetables: Past, Present, and Future. In *Edible Coatings and Films to Improve Food Quality*; Krochta, J. M., Baldwin, E. A., Nisperos-Carriedo, M. O., Eds.; Technomic Publishing Company, Inc.: Lancaster, PA, 1994; pp 25–64.

Baldwin, E. A. *Surface Treatments and Edible Coatings in Food Preservation*. In *Handbook of Food Preservation*; Rahman, M. S., Ed.; CRC Press: Boca Raton, FL, 2007; vol 21, pp 78–508.

Baldwin, E. A.; Nisperos-Carriedo, M. O.; Chen, X.; Hagenmaier, R. D. Improving Storage Life of Cut Apple and Potato with Edible Coating. *Postharvest Biol. Technol.* **1996**, *9*, 151–163.

Baldwin, E. A.; Nisperos-Carriedo, M. O.; Show, P. E.; Burns, J. K. Effect of Coatings and Prolonged Storage Conditions on Fresh Orange Flavor Volatiles, Degrees Brix, and Ascorbic Acid Levels. *J. Agric. Food Chem.* **1995**, *43*, 1321–1331.

Balny, C.; Masson, P.; Heremans, K. High Pressure Effects on Biological Macromolecules: From Structural Changes to Alteration of Cellular Processes. *Biochim. Biophys. Acta* **2002**, *1595*, 3–10.

Banker, G. S. Film Coating Theory and Practice. *Pharm. Sci.* **1966**, *55* (1), 81–89.

Ben-Yehoshua, S. Gas Exchange, Transportation, and the Commercial Deterioration in Storage of Orange Fruits. *J. Am. Soc. Horticult. Sci.* **1969**, *94*, 524–528.

Bock, N.; Dargaville, T. R.; Woodruff, M. A. Electrospraying of Polymers with Therapeutic Molecules: State of the Art. *Prog. Polym. Sci.* **2012**, *37* (11), 1510–1551.

Bonilla, J.; Atarés, L.; Vargas, M.; Chiralt, A. Edible Films and Coatings to Prevent the Detrimental Effect of Oxygen on Food Quality: Possibilities and Limitations. *J. Food Eng.* **2012**, *110*, 208–213.

Bravin, B.; Perssini, D.; Sensidoni, A. Development and Application of Polysaccharide–Lipid Edible Coating to Extend Shelf-Life of Dry Bakery Products. *J. Food Eng.* **2006**, *76*, 280–290.

Brecht, J. K. Physiology of Lightly Processed Fruits and Vegetables in Lightly Processed Fruits and Vegetables. *Hort. Sci.* **1995**, *30*, 18.

Bryan, D. S. Prepared Citrus Fruit Halves and Method of Making the Same. *U.S. Patent 3,707,383*, December 26, 1972.

Burt, S. Essential Oils: Their Antibacterial Properties and Potential Applications in Foods: A Review. *Int. J. Food Microbiol.* **2004**, *94*, 223–253.

Campos, C. A.; Gerschenson, L. N.; Flores, S. K. Development of Edible Films and Coatings with Antimicrobial Activity. *Food Bioprocess. Technol.* **2011**, *4*, 849–875.

Chauhan, S.; Gupta, K. C; Agrawal, M. Efficacy of Natural Extracts on the Storage Quality of Apple. *Int. J. Curr. Microbiol. Appl. Sci.* **2014**, *3* (3), 706–711.

Chen, Y.; Yang, J.; Mujumdar, A.; Dave, R. Fluidized Bed Film Coating of Cohesive Geldart Group C Powders. *Powder Technol.* **2009**, *189* (3), 466–480.

Chien, P. J.; Sheu, F.; Lin, H. R. Coating Citrus (*Murcott tangor*) Fruit with Low Molecular Weight Chitosan Increases Postharvest Quality and Shelf Life. *Food Chem.* **2007**, *100*, 1160–1164.

Chiu, P.; Lai, L. Antimicrobial Activities of Tapioca Starch Decolorized Hsian-Tsao Leaf Gum Coatings Containing Green Tea Extracts in Fruit-Based Salads, Romaine Heart Sand Pork Slices. *Int. J. Food Microbiol.* **2010,** *139* (1–2), 23–30.

Choi, J. S.; Lim, S. T.; Choi, H. J.; Mohanty, A. K.; Drzal, L. T.; Misra, M. Preparation and Characterization of Plasticized Cellulose Acetate Biocompostite with Natural Fiber. *J. Mater. Sci.* **2004,** *39* (21), 6631–6633.

Choi, W. Y.; Park, H. J.; Ahn, D. J.; Lee, J.; Lee, C. Y. Wettability of Chitosan Coating Solution on 'Fuji' Apple Skin. *J. Food Sci.* **2002,** *67*, 2668–2672.

Chu, C. L. Postsorage Application of TAL Prolong on Apples from Controlled Atmosphere Storage. *HortScience* **1986,** *21*, 267–268.

Cisneros-Zevallos, L.; Krochta, J. M. Dependence of Coating Thickness on Viscosity of Coating Solution Applied to Fruits and Vegetables by Dipping Method. *J. Food Sci.* **2003,** *68*, 503–510.

Corbo, M. R.; Campaniello, D.; Speranza, B.; Bevilacqua, B.; Sinigaglia, M. Non-conventional Tools to Preserve and Prolong the Quality of Minimally-Processed Fruits and Vegetables. *Coatings* **2015,** *5*, 931–961.

Cutter, C. N. Opportunities for Bio-based Packaging Technologies to Improve the Quality and Safety of Fresh and Further Processed Muscle Foods. *Meat Sci.* **2006,** *74* (1), 131–142.

Dangaran, K.; Tomasula, P. M.; Qi, P. Structure and Formation of Protein-Based Edible Films and Coatings. In *Edible Films and Coatings for Food Applications*; Embuscado, M. E., Huber, K. C., Eds.; Springer: New York, 2009; pp 25–56.

De Moraes, J. O.; Scheibe, A. S.; Sereno, A.; Laurindo, J. B. Scale-Up of the Production of Cassava Starch Based Films Using Tape-Casting. *J. Food Eng.* **2013,** *119* (4), 800–808.

Debeaufort, F.; Quezada-Gallo, J. A.; Voilley, A. Edible Films and Coatings: Tomorrow's Packagings: A Review. *Crit. Rev. Food Sci.* **1998,** *38*, 299–313.

Debeaufort, F.; Voilley, A. Lipid-Based Edible Films and Coatings. In *Edible Films and Coatings for Food Applications*; Embuscado, M. E., Huber, K. C., Eds.; Springer: New York, 2009; pp 135–168.

Dewettinck, K.; Huyghebaert, A. Fluidized Bed Coating in Food Technology. *Trends Food Sci. Technol.* **1999,** *10* (4–5), 163–168.

Dhall, R. K. Advances in Edible Coatings for Fresh Fruits and Vegetables: A Review. *Crit. Rev. Food Sci. Nutr.* **2013,** *53*, 435–450.

Dhanapal, A.; Sasikala, P.; Rajamani, L.; Kavitha, V.; Yazhini, G. Edible Films from Polysaccharides. *Food Sci. Qual. Manage.* **2012,** *3*, 9–18.

Diab, T.; Biliaderis, C. G.; Gerasopoulos, D. Physicochemical Properties and Application of Pullulan Edible Films and Coatings in Fruit Preservation. *J. Sci. Food Agric.* **2001,** *81*, 988–1000.

Drewnowski, A.; Gomez-Carneros, C. Bitter Taste, Phytonutrients and the Consumer: A Review. *Am. J. Clin. Nutr.* **2000,** *72*, 1424–1435.

Du, W. X.; Olsen, C. W.; Avena-Bustillos, R. J.; Mchugh, T. H.; Levin, C. E.; Friedman, M. Antibacterial Activity against *E. coli* 0157:H7, Physical Properties, and Storage Stability of Novel Carvacrol Containing Edible Tomato Film. *J. Food Sci.* **2008,** *73* (3), 378–383.

Dube, A.; Ng, K.; Nicolazzo, J. A.; Larson, I. Effective Use of Reducing Agents and Nanoparticle Encapsulation in Stabilizing Catechins in Alkaline Solution. *Food Chem.* **2010,** *122* (3), 662–667.

Eswaranandam, S.; Hettiarachchy, N. S.; Meullenet, J. F. Effect of Malic and Lactic Acid Incorporated Soy Protein Coatings on the Sensory Attributes of Whole Apple and Fresh-Cut Cantaloupe. *J. Food Sci.* **2006**, *71*, 307–313.

Fabra, M. J.; Hambleton, A.; Talens, P.; Debeaufort, F.; Chiralt, A.; Voilley, A. Influence of Interactions on Water and Aroma Permeabilities of ι-Carrageenan–Oleic Acid–Beeswax Films Used for Flavor Encapsulation. *Carbohydr. Polym.* **2009**, *76* (2), 325–332.

Fennema, O.; Kester, J. J. Resistance of Lipid Films to Transmission of Water Vapor and Oxygen. In *Water Relationships in Food*; Levine, H., Slade, L., Eds.; Springer: New York, 1991; pp 703–719.

Fornes, F.; Almela, V.; Abad, M.; Agusti, M. Low Concentration of Chitosan Coating Reduce Water Spot Incidence and Delay Peel Pigmentation of Clementine Mandarin Fruit. *J. Sci. Food Agric.* **2005**, *85*, 1105–1112.

Frinnault, A.; Gallant, D. J.; Bouchet, B.; Dumont, J. P. Preparation of Casein Films by a Modified Wet Spinning Process. *J. Food Sci.* **1997**, *62* (4), 744–747.

Ganesh, V.; Hettiarachchy, N. S.; Griffis, C. L.; Martin, E. M.; Ricke, S. C. Electrostatic Spraying of Food-Grade Organic and Inorganic Acids and Plant Extracts to Decontaminate *Escherichia coli* O157:H7 on Spinach and Iceberg Lettuce. *J. Food Sci.* **2012**, *77* (7), 391–396.

Garcia, D.; Guo, R.; Bhalla, A. S. Growth and Properties of $BA_{0.9}Sr_{0.1}TiO_3$ Single Crystal Fibers. *Mater. Lett.* **2000**, *42*, 136–141.

Garcia, M. A.; Martino, M. N.; Zaritzky, N. E. Plasticized Starch-Based Coatings to Improve Strawberry (*Fragaria* × *ananassa*) Quality and Stability. *J. Agric. Food Chem.* **1998**, *46*, 3758–3767.

Gennadios, A.; Hanna, M. A.; Kurth, L. B. Application of Edible Coatings on Meat, Poultry and Seafoods: A Review. *Leb. Wiss. Technol.* **1997**, *30*, 337–350.

Gennadios, A.; McHugh, T. H.; Weller, G. L.; Krochta, J. M. Edible Coatings and Films Based on Proteins. In *Edible Coatings and Films to Improve Food Quality*; Krochta, J. M., Baldwin, E. A., Nisperos-Carriedo, M. O.; Eds. Technomic Publishing Co.: Lancaster, 1994; pp 201–277.

Gennadios, A.; Weller, C. L. Edible Films and Coatings from Wheat and Corn Proteins. *Food Technol.* **1990**, *44*, 63–69.

Geschwindner, G.; Drouven, H. Manufacturing Processes: Chocolate Panning and Inclusions. In *Science and Technology of Enrobed and Filled Chocolate, Confectionery and Bakery Products*; Talbot, G., Ed.; Woodhead Publishing Limited: Cambridge, 2009; pp 397–413.

Gomez-Estaca, J.; Bravo, L.; Gomez-Guillen, M. C.; Aleman, A.; Montero, P. Antioxidant Properties of Tuna-Skin and Bovine-Hide Gelatin Films Induced by the Addition of Oregano and Rosemary Extracts. *Food Chem.* **2009**, *112*, 18–25.

Gontard, N.; Guilbert, S. Biopackaging Technology and Properties of Edible and/or Biodegradable Material of Agricultural Origin. In *Food Packaging and Preservations*; Mathlouthi, M., Ed.; Blackie Academic and Professional: Glasgow, 1994; pp 159–181.

Gorty, A. V.; Barringer, S. A. Electrohydrodynamic Spraying of Chocolate. *J. Food Process. Preserv.* **2011**, *35* (4), 542–549.

Grant, L. A.; Burns, J. Application of coatings. In *Edible Coatings and Films to Improve Food Quality*; Krochta, J. M., Baldwin, E. A., Nisperos-Carriedo, M. O., Ed.; Technomic Publishing Co.: Lancaster, 1994; pp 189–200.

Guan, J.; Hanna, M. A. Functional Properties of Extruded Foams Composites of Starch Acetate and Corn Cob Fiber. *Ind. Crop Prod.* **2004**, *19*, 255–269.

Guerrero, P.; La Caba, K. D. Thermal and Mechanical Properties of Soy Protein Films Processed at Different pH by Compression. *J. Food Eng.* **2010,** *100,* 261–269.

Guilbert, S.; Gontard, N. Technology and Applications of Edible Protective Films. In *VII Biotechnology and Food Research—New Shelf-Life Technologies and Safety Assessments. VTT Symposium 148*; Ahvinainen, R., Ohlsson, T., Matilla-Sandholm, T., Eds.; VTT: Helsinki, Finland, 1995; pp 49–60.

Guilbert, S.; Gontard, N.; Cuq, B. Technology and Applications of Edible Protective Films. *Packag. Technol. Sci.* **1995,** *8,* 339–346.

Guilbert, S.; Gontard, N.; Gorris, L. G. M. Prolongation of the Shelf Life of Perishable Food Products Using Biodegradable Films and Coatings. *LWT—Food Sci. Technol.* **1996,** *29,* 10–17.

Hagenmaier, R. D.; Baker, R. A. Layered Coatings to Control Weight Loss and Preserve Gloss of Citrus Fruit. *HortScience* **1995,** *30,* 296–298.

Halliday, J. EU Parliament Votes for Tougher Additives Regulation. *Food Navigator.com Europe,* 12 July 2007.

Han, C.; Zhao, Y.; Leonard, S. W.; Traber, M. G. Edible Coatings to Improve Storability and Enhance Nutritional Value of Fresh and Frozen Strawberries (*Fragaria ananassa*) and Raspberries (*Rubus ideaus*). *Postharvest Biol. Technol.* **2004,** *33,* 67–78.

Han, J. H. *Innovations in Food Packaging*; Elsevier, Academic Press: Cambridge, MA, 2014; pp 213–255.

Han, J. H.; Gennadios, A. *Edible Films and Coatings*: A Review. In *Innovations in Food Packaging*; Han, J. H., Ed.; Elsevier: London, 2005; vol 15, pp 239–262.

Han, J. H.; Wang, H. M.; Min, S.; Krochta, J. M. Coating of Peanuts with Edible Whey Protein Film Containing Alpha-Tocopherol and Ascorbyl Palmitate. *J. Food Sci.* **2008,** *73* (8), 349–355.

Hansen, N. M.; Plackett, D. Sustainable Films and Coatings from Hemicelluloses: A Review. *Biomacromolecules* **2008,** *9,* 1493–1505.

Hardenburg, R. E. Wax and Related Coatings for Horticultural Products—A Bibliography. *Agric. Res. Bull.* **1967,** *51,* 1–15.

Harkins, W. D.; Jordan, H. F. A Method for the Determination of Surface and Interfacial Tension from the Maximum Full on a Ring. *J. Am. Chem. Soc.* **1930,** *52,* 1751–1772.

Heena Jalal, H.; Salahuddin, M.; Gazalli, H. Nanotechnology in Food Packaging. *Int. J. Food Nutr. Saf.* **2013,** *3* (3), 111–118.

Henrique, C. M.; Teoofilo, R. F.; Sabino, L.; Ferreira, M. M. C.; Cereda, M. P. Classification of Cassava Starch Films by Physicochemical Properties and Water Vapor Permeability Quantification by FT-IR and PLS. *J. Food Sci.* **2007,** *72* (4), 184–189.

Hernandez-Izquierdo, V. M.; Krochta, J. M. Thermoplastic Processing of Proteins for Film Formation—A Review. *J. Food Chem.* **2008,** *73,* 30–39.

Hernandez-Munoz, P.; Almenar, E.; Ocio, M. J.; Gavara, R. Effect of Calcium Dips and Chitosan Coatings on Postharvest Life of Strawberries (*Fragaria ananassa*). *Postharvest Biol. Technol.* **2006,** *39,* 247–253.

Hershko, V.; Klein, H.; Nuvissovitch, A. Relationships between Edible Coating and Garlic Skin. *J. Food Sci.* **1996,** *61,* 769–777.

Hershko, V.; Nussinovitch, A. Relationships between Hydrocolloid Coating and Mushroom Structure. *J. Agric. Food Chem.* **1998,** *46,* 2988–2997.

Hsu, B. L.; Weng, Y. M.; Liao, Y. H.; Chen, W. Structural Investigation of Edible Zein Films/ Coatings and Directly Determining their Thickness by FT-Raman Spectroscopy. *J. Agric. Food Chem.* **2005,** *53,* 5089–5095.

Hu, A. W.; Fu, Z. H. Nanotechnology and its Application in Packaging and Packaging Machinery. *Packaging Eng.* **2003,** *24,* 22.

Iyidogan, N. F.; Bayindirli, A. Effect of L-Cysteine, Kojic Acid and 4-Hexylresorcinol Combination on Inhibition of Enzymatic Browning in Amasya Apple Juice. *J. Food Eng.* **2004,** *62,* 299–304.

Jaworek, A.; Soczyk, A. T. Electrospraying Route to Nanotechnology: An Overview. *J. Electrostat.* **2008,** *66* (3.4), 197–219.

Kayastha, B. L. Queen of Herbs Tulsi (*Ocimum sanctum*) Removes Impurities from Water and Plays Disinfectant Role. *J. Med. Plants Stud.* **2014,** *2* (2), 1–8.

Kester, J. J.; Fennema, O. R. Edible Films and Coatings: A Review. *Food Technol.* **1986,** *40,* 47–59.

Khademhosseini, A.; Langer, R. Nanobiotechnology: Drug Delivery and Tissue Engineering. *Chem. Eng. Prog.* **2006,** *102* (2), 38.

Khan, M. K. I.; Schutyser, M. A. I.; Karin Schroen, K.; Remko Boom, R. The Potential of Electrospraying for Hydrophobic Film Coating on Foods. *J. Food Eng.* **2012,** *108* (3), 410–416.

Kinsella, J. E.; Phillips, L. G. Film Properties of Modified Protein. In *Food Protein*; Kinsella, J. E., Soucie, W. G., Eds.; The American Oil Chemists Society: Champaign IL, 1989; pp 78–99.

Kozempel, M.; Tomasula, P. M. Development of a Continuous Process to Make Casein Films. *J. Agric. Food Chem.* **2004,** *52* (5), 1190–1195.

Krasaekoopt, W.; Mabumrung, J. Microbiological Evaluation of Edible Coated Fresh-Cut Cantaloupe. *Kasetsart J. Nat. Sci.* **2008,** *42,* 552–557.

Krochta, J. M. Proteins as raw materials for films and coatings: Definitions, Current Status, and Opportunities. In *Protein-Based Films and Coatings*; Gennadios, A., Ed.; CRC Press: Boca Raton, FL, 2002; pp 1–41.

Krochta, J. M.; Baldwin, E. A.; Nisperos-Carriedo, M. *Edible Coating and Films to Improve Food Quality*; Technomic Pub Co.: Lancaster, PA, 1994.

Krochta, J. M.; de Mulder-Johnston, C. D. Edible and Biodegradable Polymer Films: Challenges and Opportunities. *Food Technol.* **1997,** *51,* 61–74.

Krochta, J. M.; Pavlath, A. E.; Goodman, N. Edible Films from Casein–Lipid Emulsions for Lightly Processed Fruits and Vegetables. In *Engineering and Food, Preservation Processes and Related Techniques*; Spiess, W. E., Schubert, H., Eds.; Elsevier Science Publishers: New York, NY, 1990; pp 329–340.

Kumar, V.; Andola, H. C.; Lohani, H.; Chauhan, N. Pharmacological Review on *Ocimum sanctum* Linnaeus: A Queen of Herbs. *J. Pharm. Res.* **2011,** *4,* 366–368.

LaCoste, A.; Schaich, K.; Zumbrunnen, D.; Yam, K. Advanced Controlled Release Packaging through Smart Blending. *Packag. Technol. Sci.* **2005,** *18,* 77–87.

Lacroix, M. Mechanical and Permeability Properties of Edible Films and Coatings for Food and Pharmaceutical Application. In *Edible films and Coatings for Food Applications*; Embuscado, M. E., Huber, K. C., Eds.; Springer Science Business Media, LLC: New York, NY, 2009; pp 367–390.

Lagaron, J.; Cabedo, L.; Cava, D.; Feijoo, J.; Gavara, R.; Gimenez, E. Improved Packaging Food Quality and Safety. Part 2: Nano-composites. *Food Addit. Contam.* **2005,** *22,* 994–998.

Lee, J. Y.; Park, H. J.; Lee, C. Y.; Choi, W. Y. Extending Shelf-Life of Minimally Processed Apples with Edible Coatings and Antibrowning Agents. *Lebens. Wissen. Technol.* **2003,** *36,* 323–329.

Li, H.; Li, F.; Wang, L.; Sheng, J.; Xin, Z.; Zhao, L.; Xiao, H.; Zheng, Y.; Hu, Q. Effect of Nano-Packing on Preservation Quality of Chinese Jujube (*Ziziphus jujuba* Mill. var. Inermis (Bunge) Rehd). *Food Chem.* **2009,** *114* (2), 547.

Liau, S. Y.; Read, D. C.; Pugh, W. J.; Furr, J. R.; Russell, A. D. Interaction of Silver Nitrate with Readily Identifiable Groups: Relationship to the Antibacterial Action of Silver Ions. *Lett. Appl. Microbiol.* **1997,** *25,* 279–283.

Lin, D.; Zhao, Y. Innovations in the Development and Application of Edible Coatings for Fresh and Minimally Processed Fruits and Vegetables. *Compr. Rev. Food Sci. Food Saf.* **2007,** *6,* 60–75.

Lin, S. Y.; Krochta, J. M. Plasticizer Effect on Grease Barrier Properties of Whey Protein Concentrate Coatings on Paperboard. *J. Food Sci.* **2003,** *68,* 229–233.

Liu, H.; Xie, F.; Yu, L.; Chen, L.; Li, L. Thermal Processing of Starch-Based Polymers. *Progr. Polym. Sci.* **2009,** *34,* 1348–1368.

Lopez, O. V.; Garcia, M. A.; Villar, M. A.; Gentili, A.; Rodriguez, M. S.; Albertengo, L. Thermo-Compression of Biodegradable Thermoplastic Corn Starch Films Containing Chitin and Chitosan. *LWT—Food Sci. Technol.* **2014,** *57,* 106–115.

Lopez-Rubio, A.; Gavara, R.; Lagaron, J. Bioactive Packaging: Turning Foods into Healthier Foods through Biomaterials. *Trends Food Sci. Technol.* **2006,** *17,* 567–575.

Lopez-Rubio, A.; Lagaron, J. M. Whey Protein Capsules Obtained through Electrospraying for the Encapsulation of Bioactives. *Innov. Food Sci. Emerg. Technol.* **2012,** *13,* 200–206.

Lowings, P. H.; Cutts, D. F. The preservation of fresh fruits and vegetables. In *Proceedings of the Institute of Food Science and Technology Annual Symposium,* Nottingham, UK, July 1982; p 52.

Luo, C. J.; Loh, S.; Stride, E.; Edirisinghe, M. Electrospraying and Electrospinning of Chocolate Suspensions. *Food Bioprocess Technol.* **2012,** *5* (6), 2285–2300.

Madureira, A. R.; Pereira, A.; Castro, P. M.; Pintado, M. Production of Antimicrobial Chitosan Nanoparticles against Food Pathogens. *J. Food Eng.* **2015,** *167,* 210–216.

Maftoonazad, N.; Ramaswamy, H. S. Effect of Pectin-Based Coating on the Kinetics of Quality Change Associated with Stored Avocados. *J. Food Process. Preserv.* **2008,** *32,* 621–643.

Mangiacapra, P.; Gorrasi, G.; Sorrentino, A.; Vittoria, V. Biodegradable Nanocomposites Obtained by Ball Milling of Pectin and Montmorillonites. *Carbohydr. Polym.* **2005,** *64,* 516.

Marthina, K.; Barringer, S. A. Confectionery Coating with an Electro-hydrodynamic (EHD) System. *J. Food Sci.* **2012,** *77* (1), 26–31.

Martin-Belloso, O.; Rojas-Grau, M. A.; Soliva-Fortuny, R. Delivery of Flavor and Active Ingredients Using Edible Films and Coatings. In *Edible Films and Coatings for Food Applications;* Embuscado, M. E., Huber, K. C., Eds.; Springer: New York, 2009; pp 295–313.

Martinez-Romero, D.; Alburquerque, N.; Valverde, J. M.; Guillen, F.; Castillo, S.; Valero, D.; Serrano, M. Postharvest Sweet Cherry Quality and Safety Maintenance by *Aloe vera* Treatment: A New Edible Coating. *Postharvest Biol. Technol.* **2006,** *39,* 93–100.

Maski, D.; Durairaj, D. Effects of Electrode Voltage, Liquid Flow Rate, and Liquid Properties on Spray Chargeability of an Air-Assisted Electrostatic-Induction Spray Charging System. *J. Electrostat.* **2010**, *68* (2), 152–158.

Mate, J. I.; Saltweil, M. E.; Krochta, J. M. Peanut and Walnut Rancidity: Effects of Oxygen Concentration and Relative Humidity. *J. Food Sci.* **1996**, *61*, 465–472.

Mathew, S.; Brahmakumar, M.; Abraham, T. E. Microstructureal Imaging and Characterization of the Mechanical, Chemical, Thermal, and Swelling Properties of Starch–Chitosan Blend Films. *Biopolymers* **2006**, *82*, 176–187.

Mattheis, J.; Fellman, J. K. Impacts of Modified Atmosphere Packaging and Controlled Atmospheres on Aroma, Flavor, and Quality of Horticultural Commodities. *HortTechnology* **2000**, *10*, 507–510.

McAnally, G. D.; Everall, N. J.; Chalmers, J. M.; Smith, W. E. Analysis of Thin Film Coatings on Poly(Ethylene Terephthalate) by Confocal Raman Microscopy and Surface-Enhanced Raman Scattering. *Appl. Spectrosc.* **2003**, *57*, 44–49.

McFarland, M. J. *Biosolids Engineering*; McGraw-Hill Companies, Inc.: New York, NJ, 2011; vol B-2, p 2.12.

McHugh, T. H.; Senesi, E. Apple Wraps: A Novel Method to Improve the Quality and Extend the Shelf Life of Fresh-Cut Apples. *J. Food Sci.* **2000**, *65*, 480–485.

Meng, X.; Li, B.; Liu, J.; Tian, S. Physiological Responses and Quality Attributes of Table Grape Fruit to Chitosan Preharvest Spray and Postharvest Coating during Storage. *Food Chem.* **2008**, *106* (2), 501–508.

Milind, S. D.; Rale, V. B.; Lynch, J. M. *Aureobasidium pullulans* in Applied Microbiology. A Status Report. *Enzyme Microbiol. Technol.* **1992**, *14*, 514–527.

Moayednia, N.; Ehsani, M. R.; Emamdjomeh, Z.; Asadi, M. M.; Mizani, M.; Mazaheri, A. F. The Effect of Sodium Alginate Concentrations on Viability of Immobilized *Lactobacillus acidophilus* in Fruit Alginate Coating during Refrigerator Storage. *Aust. J. Basic Appl. Sci.* **2009**, *3* (4), 3213–3226.

Moayednia, N.; Ehsani, M. R.; Emamdjomeh, Z.; Asadi, M. M.; Mizani, M.; Mazaheri, A. F. Effect of Refrigeration on Viability of Immobilized Probiotic Bacteria in Alginate Coat of Strawberry. *World Appl. Sci. J.* **2010**, *10* (4), 472–476.

Montero-Calderon, M.; Rojas-Grau, M. A.; Martin-Belloso, O. Effect of Packaging Conditions on Quality and Shelf-Life of Fresh-Cut Pineapple (*Ananas comosus*). *Postharvest Biol. Technol.* **2008**, *50*, 182–189.

Moreira, M. R.; Roura, S. I.; Ponce, A. Effectiveness of Chitosan Edible Coatings to Improve Microbiological and Sensory Quality of Fresh Cut Broccoli. *Food Sci. Technol.* **2011**, *44*, 2335.

Mourtzinos, I.; Kalogeropoulos, N.; Papadakis, S.; Konstantinou, K.; Karathanos, V. Encapsulation of Nutraceutical Monoterpenes in Cyclodextrin and Modified Starch. *J. Food Sci.* **2008**, *73* (1), 89–94.

Muthuswamy, S.; Rupasinghe, H. P. V.; Stratton, G. W. Antimicrobial Effect of Cinnamon Bark Extract on *Escherichia coli* O157:H7, *Listeria innocua* and Fresh-Cut Apple Slices. *J. Food Saf.* **2008**, *28* (4), 534–549.

Myllarinen, P.; Buleon, A.; Lahtinen, R.; Forssell, P. The Crystallinity of Amylase and Amylopectin Films. *Carbohydr. Polym.* **2002**, *48*, 41–48.

Narayan, R. Polymeric Materials from Agricultural Feedstocks. In *Polymers from Agricultural Coproducts*; Fishman, M. L., Friedman, R. B., Huang, S. J., Eds.; American Chemical Society: Washington, DC, 1994; pp 2–28.

Nieto, M. Structure and Function of Polysaccharide Gum-Based Edible Films and Coatings. In *Edible Films and Coatings for Food Applications*; Embuscado, M. E., Huber, K. C., Eds.; Springer: London/New York, 2009; vol 3, pp 57–112.

Oh, H.; Kim, K.; Kim, S. Characterization of Deposition Patterns Produced by Twin-Nozzle Electrospray. *J. Aerosol Sci.* **2008**, *39* (9), 801–813.

Olivas, G. I.; Barbosa-Canovas, G. V. Edible Coatings for Fresh-Cut Fruits. *Crit. Rev. Food Sci. Nutr.* **2005**, *45*, 657–670.

Olivas, G.; Maya, I.; Espino-Diaz, J.; Molina-Corral, J.; Olivas-Dorantes, C.; Sepulveda, D. Metabolization of Linoleic Acid and Isoleucine for Aroma Production in Fresh-Cut 'Golden Delicious' Apples Using Alginate Coatings as the Holding Matrix. *Technical Research Abstract Institute of Food Technologists IFT Annual Meeting*, Las Vegas, NV, EU, June 25–28, 2012.

Olivas, G. I.; Mattinson, D. S.; Barbosa-Canovas, G. Alginate Coatings for Preservation of Minimally Processed 'Gala' Apples. *Postharvest Biol. Technol.* **2007**, *45*, 89–96.

Oms-Oliu, G.; Soliva-Fortuny, R.; Martin-Belloso, O. Edible Coatings with Antibrowning Agents to Maintain Sensory Quality and Antioxidant Properties of Fresh-Cut Pears. *Postharvest Biol. Technol.* **2008**, *50*, 87–94.

Padmaja, N.; John Don Bosco, S. Preservation of Jujube fruits by Edible Aloe Vera Gel Coating to Maintain Quality and Safety. *Ind. J. Sci. Res. Technol.* **2014**, *2* (3), 79–88.

Park, H. J. Development of Advanced Edible Coatings for Fruits. *Trends Food Sci. Technol.* **1999**, *10*, 254–260.

Park, H. J.; Chinnan, M. S.; Shewfelt, R. L. Edible Corn-Zein Film Coatings to Extend Storage Life of Tomatoes. *J. Food Process. Pres.* **1994**, *18*, 317–331.

Park, S. I.; Stan, S. D.; Daeschel, M. A.; Zhao, Y. Antifungal Coatings on Fresh Strawberries (*Fragaria ananassa*) to Control Mold Growth during Cold Storage. *J. Food Sci. B* **2006**, *70*, 202–207.

Park, S. I.; Zhao, Y. Incorporation of a High Concentration of Mineral or Vitamin into Chitosan-Based Films. *J. Agric. Food Chem.* **2004**, *52*, 1933–1939.

Park, S. K.; Hettiarachchy, N. S.; Ju, Z. Y.; Gennadios, A. Formation and Properties of Soy Protein Films and Coatings. In *Protein-Based Films and Coatings*; Gennadios, A., Ed.; CRC Press: New York, **2002**, *4*, 123–138.

Pegg, R. B.; Shahidi, F. Encapsulation, Stabilization and Controlled Release of Food Ingredients and Bioactives. In *Handbook of Food Preservation*; Rahman, M. S., Ed.; CRC Press: Boca Raton, FL, 2007; vol 22, pp 509–570.

Peressini, D.; Bravin, B.; Lapasin, R.; Rizzotti, C.; Sensidoni, A. Starch–Methylcellulose Based Edible Films: Rheological Properties of Film-Forming Dispersions. *J. Food Eng.* **2003**, *59* (1), 25–32.

Perez-Gago, M. B.; Krochta, J. M. Drying Temperature Effect on Water Vapor Permeability and Mechanical Properties of Whey Protein–Lipid Emulsion Films. *J. Agric. Food Chem.* **2000**, *8* (7), 2687–2692.

Perez-Gago, M. B.; Nadaud, P.; Krochta, J. M. Water Vapor Permeability, Solubility and Tensile Properties of Heat-Denatured versus Native Whey Protein Films. *J. Food Sci.* **1999**, *64* (6), 1034–1037.

Perez-Gago, M. B.; Serra, M.; Alonso, M.; Mateos, M.; del Rio, M. A. Effect of Solid Content and Lipid Content of Whey Protein Isolate–Beeswax Edible Coatings on Color Change of Fresh Cut Apples. *J. Food Sci.* **2003**, *68* (7), 2186–2191.

Perez-Gago, M. B.; Serra, M.; del Rio, M. A. Color Change of Fresh-Cut Apples Coated with Whey Protein Concentrate-Based Edible Coatings. *Postharvest Biol. Technol.* **2006,** *39,* 84–92.

Petersen, K.; Nielsen, V. P.; Bertelsen, G.; Lawther, M.; Olsen, M. B.; et al. Potential of Biobased Materials for Food Packaging. *Trends Food Sci. Technol.* **1999,** *10* (2), 52–68.

Plotto, A.; Goodner, K. L.; Baldwin, E. A. Effect of Polysaccharide Coating on Quality of Fresh Cut Mangoes (*Mangifera indica*). *Proc. Fl. State Hort. Soc.* **2004,** *117,* 382–388.

Pochat-Bohatier, C.; Sanchez, J.; Gontard, N. Influence of Relative Humidity on Carbon Dioxide Sorption in Wheat Gluten Films. *J. Food Eng.* **2006,** *77* (4), 983–991.

Porat, R.; Weiss, B.; Cohen, L.; Daus, A.; Aharoni, N. Reduction of Postharvest Rind Disorders in Citrus Fruit by Modified Atmosphere Packaging. *Postharvest Biol. Technol.* **2004,** *33,* 35–43.

Prasad, N.; Batra, E. Edible Coating (the Future of Packaging): Cheapest and Alternative Source to Extend the Post-Harvest Changes: A Review. *Asian J. Biochem. Pharm. Res.* **2015,** *3* (5), 2231–2560.

Prommakool, A.; Sajjaanantakul, T.; Janjarasskul, T.; Krochta, J. M. Whey Protein–Okra Polysaccharide Fraction Blend Edible Films: Tensile Properties, Water Vapor Permeability and Oxygen Permeability. *J. Sci. Food Agric.* **2011,** *91* (2), 362–369.

Ramani, M.; Ponnusamy, S.; Muthamizhchelvan, C.; Marsili, E. Amino Acid-Mediated Synthesis of Zinc Oxide Nanostructures and Evaluation of their Facet-Dependent Antimicrobial Activity. *Colloids Surf., B: Biointerfaces* **2014,** *117,* 233–239.

Ramos, M.; Jimenez, A.; Garrigos, M. C. Carvacrol-Based Films: Usage and Potential in Antimicrobial Packaging. In *Antimicrobial Food Packaging*; Barros-Velazquez, J., Ed.; Academic Press: San Diego, CA, 2016; pp 329–338.

Rampon, V.; Robert, P.; Nicolas, N.; Dufour, E. Protein Structure and Network Orientation in Edible Films Prepared by Spinning Process. *J. Food Sci.* **1999,** *64* (2), 313–316.

Rao, R. V. Natural Decay Resistance of Neem. *Wood J. Indian Acad. Wood Sci.* **1990,** *21* (1), 19–21.

Raybaudi-Massilia, R. M.; Mosqueda-Melgar, J.; Martin-Belloso, O. Edible Alginate-Based Coating as Carrier of Antimicrobials to Improve Shelf-Life and Safety of Fresh-Cut Melon. *Int. J. Food Microbiol.* **2008,** *121,* 313–327.

Raymond, L. V.; Zhang, M.; Roknul Azam, S. M. Effect of Chitosan Coating on Physical and Microbial Characteristics of Fresh Cut Green Peppers (*Capsicum annuum* L.). *Pak. J. Nutr.* **2012,** *11,* 806.

Razavi, S. M. A.; Mohammad Amini, A.; Zahedi, Y. Characterization of a New Biodegradable Edible Film Based on Sage Seed Gum: Influence of Plasticiser Type and Concentration. *Food Hydrocolloids* **2015,** *43,* 290–298.

Rhim, J. W.; Ng, P. K. W. Natural Biopolymer-Based Nanocomposite Films for Packaging Applications. *Crit. Rev. Food Sci. Nutr.* **2007,** *47,* 411–433.

Ribeiro, C.; Vicente, A. A.; Teixeira, J. A.; Miranda, C. Optimization of Edible Coating Composition to Retard Strawberry Fruit Senescence. *Postharvest Biol. Technol.* **2007,** *44* (1), 63–70.

Roach, S. Most Companies Will Have to Wait Years for Nanotech's Benefits. *Foodproductiondaily.com*, 21st August 2006.

Robertson, G. L. *Food Packaging: Principles and Practice*, 3rd ed.; CRC Press: Boca Raton, FL, 2012; p. 131–164.

Robles-Sanchez, R. M.; Rojas-Grau, M. A.; Odriozola-Serrano, I.; Gonzalez-Aguilar, G.; Martin-Belloso, O. Influence of Alginate-Based Edible Coating as Carrier of Antibrowning Agents on Bioactive Compounds and Antioxidant Activity in Fresh-Cut Kent Mangoes. *LWT—Food Sci. Technol.* **2013**, *50* (1), 240–246.

Rojas-Grau, M. A.; Avena-Bustillos, R. J.; Friedman, M.; Henika, P. R.; Martin-Belloso, O.; McHugh, T. H. Mechanical, Barrier, and Antimicrobial Properties of Apple Puree Edible Films Containing Plant Essential Oils. *J. Agric. Food Chem.* **2006**, *54*, 9262–9267.

Rojas-Grau, M. A.; Avena-Bustillos, R. J.; Olsen, C.; Friedman, M.; Henika, P. R.; Martin-Belloso, O.; Pan, Z.; McHugh, T. H. Effects of Plant Essential Oil Compounds on Mechanical, Barrier and Antimicrobial Properties of Alginate–Apple Puree Edible Films. *J. Food Eng.* **2007**, *81*, 634–641.

Rojas-Grau, M. A.; Tapia, M. S.; Martin-Belloso, O. Using Polysaccharide-Based Edible Coatings to Maintain Quality of Fresh-Cut Fuji Apples. *LWT—Food Sci. Technol.* **2008**, *41*, 139–147.

Roller, S.; Seedhar, P. Carvacrol and Cinnamic Acid Inhibit Microbial Growth in Fresh-Cut Melon and Kiwifruit at 4 and 8°C. *Lett. Appl. Microbiol.* **2002**, *35*, 390–394.

Romero-Bastida, C. A.; Bello-Perez, L. A.; Garcia, M. A.; Martino, M. N.; Solorza-Feria, J.; Zaritzky, N. E. Physicochemical and Microstructural Characterization of Films Prepared by Thermal and Cold Gelatinization from Non-conventional Sources of Starches. *Carbohydr. Polym.* **2005**, *60* (2), 235–244.

Rosenthal, A. J. Food *Texture: Measurement and Perception*; Aspen Publishers, Inc.: Gaithersburg, MD, 1999; p 274.

Rossman, J. M. Edible Films and Coatings for Food Applications. In *Edible Films and Coatings for Food Applications*; Embuscado, M. E., Huber, K. C., Eds.; Springer Science Business Media, LLC: New York, NY, 2009; pp 367–390.

Roth, I.; Loncin, M. Superficial Activity of Water. In *Engineering and Food*; McKenna, B. M., Ed. Elsevier Applied Sciences Publishers: London, 1984; vol I, p 433.

Sabato, S. F.; Ouattara, B.; Yu, H.; D'Aprano, G.; Lacroix, M. Mechanical and Barrier Properties of Cross-Linked Soy and Whey Protein Based Films. *J. Agric. Food Chem.* **2001**, *49*, 1397–1403.

Salminen, S.; Ouwehand, A.; Benno, Y.; Lee, Y. Probiotics: How Should They Be Defined? *Trends Food Sci. Technol.* **1999**, *10* (3), 107–110.

Saltveit, M. Procedures for Extracting and Analyzing Internal Gas Samples from Plant Tissues by Gas Chromatograph. *HortScience* **1982**, *17*, 878–881.

Santerre, C. R.; Leach, T. F.; Cash, J. N. The Influence of the Sucrose Polyester, Semperfresh™, on the Storage of Michigan Grown 'McIntosh' and 'Golden Delicious' Apples. *J. Food Process. Preserv.* **1989**, *13*, 293–305.

Shackel, K. A.; Greve, C.; Labavitch, J. M.; Ahmadi, H. Cell Turgor Changes Associated with Ripening in Tomato Pericarp Tissue. *Plant Physiol.* **1991**, *97*, 814–816.

Shon, J.; Haque, Z. U. Efficacy of Sour Whey as a Shelf-Life Enhancer: Use in Antioxidative Edible Coatings of Cut Vegetables and Fruit. *J. Food Quality* **2007**, *30*, 581–593.

Siracusa, V.; Rocculi, P.; Romani, S.; Rosa, M. D. Biodegradable Polymers for Food Packaging: A Review. *Trends Food Sci. Technol.* **2008**, *19*, 634–643.

Smith, S.; Geeson, J.; Stow, J. Production of Modified Atmospheres in Deciduous Fruits by the Use of Films and Coatings. *Hortic. Sci.* **1987**, *22*, 772–776.

Song, J.; Jang, J. Antimicrobial Polymer Nanostructures: Synthetic Route, Mechanism of Action and Perspective. *Adv. Colloid Interface Sci.* **2014**, *203*, 37–50.

Sonti, S.; Prinyawiwatkul, W.; Gillespie, J. M.; McWatters, K. H.; Bhale, S. D. Probit Analysis of Consumer Perception of Fresh-Cut Fruits and Vegetables and Edible Coating. In *IFT Annual Meeting Technical Program Abstracts*, Chicago, USA, 104D-26, 2003.

Sorrentino, A.; Gorrasi, G.; Vittoria, V. Potential Perspectives of Bio-nanocomposites for Food Packaging Applications. *Trends Food Sci. Technol.* **2007,** *18,* 84–95.

Sothornvit, R.; Krochta, J. M. Plasticizer Effect on Mechanical Properties of Beta-Lactoglobulin Films. *J. Food Eng.* **2001,** *50* (3), 149–155.

Sothornvit, R.; Krochta, J. M. Plasticizers in Edible Films and Coatings. In *Innovations in Food Packaging*; Han, J. H., Ed.; Academic Press: London, UK, 2005; pp 403–433.

Sothornvit, R.; Olsen, C. W.; McHugh, T. H.; Krochta, J. M. Tensile Properties of Compression-Molded Whey Protein Sheets: Determination of Molding Condition and Glycerol-Content Effects and Comparison with Solution-Cast Films. *J. Food Eng.* **2007,** *78,* 855–860.

Souza, B. W. S.; Cerqueira, M. A.; Teixeira, J. A.; Vicente, A. A. The Use of Electric Fields for Edible Coatings and Films Development and Production: A Review. *Food Eng. Rev.* **2010,** *2,* 244–255.

Subhas, C. S.; Pathik, M. S. Edible Polymers: Challenges and Opportunities. *J. Polym.* **2014,** *2014,* 1–13.

Talbot, G. Product Design and Shelf-Life Issues: Moisture and Ethanol Migration. In *Science and Technology of Enrobed and Filled Chocolate, Confectionery and Bakery Products*; Talbot, G. Ed.; Woodhead Publishing Ltd.: Cambridge, UK, 2009; pp 211–232.

Tanada-Palmu, P. S.; Grosso, C. R. F. Effect of Edible Wheat Gluten-Based Films and Coatings on Refrigerated Strawberry (*Fragaria ananassa*) Quality. *Postharvest Biol. Technol.* **2005,** *36,* 199–208.

Tapia, M. S.; Rojas-Grau, M. A.; Carmona, A.; Rodriguez, F. J.; Fortuny, S.; Martin-Belloso, O. Use of Alginate and Gellan-Based Coatings for Improving Barrier, Texture and Nutritional Properties of Fresh-Cut Papaya. *Food Hydrocolloid* **2008,** *22,* 14936.

Tapia, M. S.; Rojas-Grau, M. A.; Rodriguez, F. J.; Ramirez, J.; Carmona, A.; Martin-Belloso, O. Alginate and Gellan-Based Edible Films for Probiotic Coatings on Freshcut Fruits. *J. Food Sci.* **2007,** *72,* 190.

Trezza, T. A.; Krochta, J. M. Specular Reflection, Gloss, Roughness and Surface Heterogeneity of Biopolymer Coatings. *J. Appl. Polym. Sci.* **2001,** *79,* 2221–2229.

Ustunol, Z. Edible Films and Coatings for Meat and Poultry. In *Edible Films and Coatings for Food Applications*; Embuscado, M. E., Huber, K. C., Eds.; Springer: New York, 2009; pp 245–268.

Valverde, J. M.; Valero, D.; Martínez-Romero, D.; Guillen, F. N.; Castillo, S.; Serrano, M. Novel Edible Coating Based on *Aloe vera* Gel to Maintain Table Grape Quality and Safety. *J. Agric. Food Chem.* **2005,** *53,* 7807–7813.

Vargas, M.; Albors, A.; Chiralt, A.; Gonzalez-Martinez, C. Quality of Cold-Stored Strawberries as Affected by Chitosan–Oleic Acid Edible Coatings. *Postharvest Biol. Technol.* **2006,** *41,* 164–171.

Vieira, M. G. A.; Altenhofen, D.; Silva, M.; Oliveira, D. S. L.; Beppu, M. M. Natural-Based Plasticizers and Biopolymer Films: A review. *Eur. Polym. J.* **2011,** *47* (3), 254–263.

Watada, A. E.; Ko, N. P.; Minott, D. A. Factors Affecting Quality of Fresh-Cut Horticultural Products. *Postharvest Biol. Technol.* **1996,** *9,* 115–125.

Weiss, J.; Takhistov, P.; McClements, J. Functional Materials in Food Nanotechnology. *J. Food Sci.* **2006,** *71,* 107–116.

Wong, D. W. S.; Tillin, S. J.; Hudson, J. S.; Pavlath, A. E. Gas Exchange in Cut Apples with Bilayer Coatings. *J. Agric. Food Chem.* **1994,** *42,* 2278–2285.

Xu, S.; Chen, X.; Sun, D. Preservation of Kiwifruit Coated with an Edible Film at Ambient Temperature. *J. Food Eng.* **2001,** *50,* 211–216.

Yoo, S.; Krochta, J. M. Whey Protein–Polysaccharide Blended Edible Film Formation and Barrier, Tensile, Thermal and Transparency Properties. *J. Sci. Food Agric.* **2011,** *91,* 2628–2636.

Zhao, Y.; McDaniel, M. Sensory Quality of Foods Associated with Edible Film and Coating Systems and Shelf-Life Extension. In *Innovations in Food Packaging*; Han, J. H., Ed.; Elsevier Academic Press: San Diego, CA, 2005; pp 434–453.

Zhong, Q. P.; Xia, W. S. Physicochemical Properties of Edible and Preservative Films from Chitosan/Cassava Starch/Gelatin Blend Plasticized with Glycerol. *Food Technol. Biotechnol.* **2008,** *46,* 262–269.

Zinoviadou, K. G.; Koutsoumanis, K. P.; Biliaderis, C. G. Physico-Chemical Properties of Whey Protein Is Late Films Containing Oregano Oil and Their Antimicrobial Action against Spoilage Flora of Fresh Beef. *Meat Sci.* **2009,** *82* (3), 338–345.

CHAPTER 10

POSTHARVEST TREATMENTS TO REDUCE BROWNING IN MINIMALLY PROCESSED PRODUCTS

ALEMWATI PONGENER[1*], M. B. DARSHAN[2], and AABON W. YANTHAN[3]

[1]*ICAR-National Research Centre on Litchi, Muzaffarpur, Bihar, India*

[2]*ICAR-AICRP on PHET, University of Agricultural Sciences, GKVK, Bengaluru, Karnataka, India*

[3]*ICAR Research Complex for North Eastern Hill Region, Nagaland Centre, Jharnapani, Nagaland, India*

Corresponding author. E-mail: alemwati@gmail.com

ABSTRACT

Consumer demand for minimally processed products is continually on the rise due to the fresh-like property and convenience they offer. Minimal processing imposes mechanical injury to the fruit and vegetable tissues and accelerates the process of deterioration. One of the major problems in minimal processing is the browning in cut surface of fruit and vegetables. Control of browning in minimally processed products assumes greater significance because product appearance and color are important factors in the decision to purchase and consume. This chapter covers the ways and means to reduce and slow down the process of browning in minimally processed fruit and vegetables.

10.1 INTRODUCTION

Higher education level and increasing income in developing countries have resulted in more and more demand for nutritious and safe food.

Recommendations for balanced diet always include consumption of fresh fruit and vegetables. Although conventionally processed food products represent an increasingly growing industry, it is quite clear that consumers perceive fresh foods as healthier than heat-treated/processed foods. Over recent years, therefore, per capita consumption of fresh fruit and vegetables has significantly increased over that of processed foods. A growing focus on health and increased preference for convenience have led to fast growth in demand and sales of fresh, minimally processed, and ready-to-eat fruits, vegetables, and salads.

Minimal processing refers to any operation performed on fresh fruit and vegetable with the aim of increasing the functionality without significantly changing the fresh-like properties. IFPA (1999) defines minimally processed product as "any fruit and vegetable, or any combination thereof, which has been physically altered from its original form but has remained in its fresh condition." Also called fresh-cut or ready-to-eat, minimally processed fruit and vegetables are free from additives and only need minimal or no further processing prior to consumption (Sun et al., 2005). Operations for minimal processing can include washing, peeling, cutting, slicing, trimming, shredding, dicing, coring, deseeding, etc., depending on structure of raw materials and intended use (Artés and Allende, 2005; Pasha et al., 2014). Traditional processing techniques such as dehydration, freezing, salting, etc. are not applied in minimal processing. They can be referred to as less processed or less preserved foods and are therefore less stable.

The industry for minimally processed fruits and vegetables has grown rapidly in recent years, especially in the USA and Europe (Siddiqui et al., 2011). Minimal processing allows for convenience in consumption as end users forgo the processing already performed. A minimally processed produce should stay fresh and offer convenience without compromising on sensory and nutritional quality, for a period long enough for transportation and distribution to consumers. Minimal processing, therefore, aims at use of mild but reliable processing and treatments to achieve fresh-like quality and safe product with high nutritional quality. The most important aspects of minimally processed products are that consumers expect are color, texture, flavor, and smell. Wound response is triggered due to peeling or cutting of fruit and vegetables. Therefore, deteriorative processes take place in addition to dehydration and microbial spoilage. After cutting, fruit and vegetables are washed to remove intracellular residues and microorganisms.

10.2 PROBLEMS IN MINIMAL PROCESSING

Minimal processing imposes mechanical injury to the fruit and vegetable tissues and accelerates the process of deterioration. The natural shield of fruit is removed during the steps of minimal processing, which increases the product's susceptibility to microbial spoilage. Therefore, one of the major challenges in minimal processing is the reduced shelf-life due to rapid deterioration. Deterioration of minimally processed fruit and vegetables is therefore due to physiological aging, biochemical changes, and microbial spoilage. Spoilage changes in minimally processed fruit and vegetables are represented by off-flavors, discoloration, softening, and water loss (Willocx et al., 1994). Some of the major changes in minimally processed products are increased metabolic activity, enzymatic browning, and presence of microorganisms and pathogens in the tissues. Deterioration in minimally processed products is irreversible and can only be slowed down to a certain extent through optimal processing, packaging, storage temperature, and use of antibrowning inhibitors.

Infliction of physical injury during minimal processing, especially cutting or cubing, disrupts cell membrane and brings substrates into contact with enzymes leading to increased deterioration (Laurila et al., 1998). Physical wounding induces stress-related responses in products which can include increase in rate of respiration, ethylene evolution, and activity of enzymes phenyl ammonia lyase (PAL), peroxidase (POD), catalases (CAT), polyphenol oxidase (PPO), etc. Higher respiration rate and ethylene evolution are associated with acceleration of senescence in fruit and vegetables. Increase in metabolic activity as a result of minimal processing has been reported in melons, papaya, apples, guava, pineapple, potato, etc. Production of ethylene can adversely affect the quality and loss in color of green vegetables such as spinach and broccoli. This is because ethylene evolution is related to degradation of chlorophyll via acceleration of chlorophyllase enzyme.

Browning in cut tissue is mainly an enzymatic process (Ioannou and Ghoul, 2013; Massantini and Mencarelli, 2007). Cutting increases the surface area of tissues which increases the exposure to oxygen. Browning occurs when metabolites, mostly phenols, get oxidized under the activity of enzymes like PPO and POD.

The very fact that minimally processed products are ready to eat and consumed without heating makes them highly probable agents of food-borne diseases (FBD). Cut fruit and vegetables release exudates that provide an excellent medium for growth of microorganisms, predominantly fungi and bacteria. Most of the FBD are caused by consumption of food contaminated with *Salmonella* sp., *Staphylococcus aureus*, *Clostridium* sp., *Bacillus cereus*, etc. Minimal processing procedures must therefore aim at reduction or elimination of surface contamination.

10.3   BROWNING IN MINIMALLY PROCESSED PRODUCTS

Product appearance is an important criterion and color is a key factor involved in decision to purchase. Minimal processing of fruit and vegetable results in increase of surface area (cut surface). This also implies increased exposure to air or oxygen. More importantly, cutting operation also disturbs the compartmentation and brings substrates into contact with oxidative enzymes. Cutting fruit and vegetables increases the activity of PAL, which catalyzes the biosynthesis of phenylpropanoids. When the products of phenylpropanoids metabolism, such as phenols, are oxidized in reactions catalyzed by enzymes importantly PPO and POD, the cut surface turns brown. Therefore, enzymatic browning requires four different components—oxygen, an enzyme, copper, and a substrate. PPO is a term generally used to refer to a group of enzymes that catalyze the oxidation of phenolic compounds producing a brown color on the cut surfaces of fruit and vegetables. The presence or biosynthesis of ethylene as a result of wounding due to minimal processing can also increase the activity of PAL, thereby accelerating browning reactions. Operations such as peeling and cutting are key steps of minimal processing in fruit and vegetables, during which cell membranes are broken and bring enzymes into contact with appropriate substrates. Phenols are then oxidized to orthoquinones that polymerize to form brown or black pigments, such as melanins.

10.4   MEASURES TO REDUCE BROWNING

Since the factors responsible for enzymatic browning are known, the control measures to reduce browning circles around the concentration of enzyme and substrate, the pH, temperature, and the availability of oxygen. In conventional processing, browning or inactivation of oxidative enzymes can be prevented through heat treatment altogether. But in minimally processed fruit and vegetables, application of heat is avoided so as to prevent cooking and loss of fresh-like characteristics of the product. Several measures are followed for prevention of browning in minimally processed products. These include use of modified atmosphere packaging (MAP), edible films, and chemical treatments. A very important control measure is to maintain pH at levels where the activity of oxidizing enzymes is curtailed. In case of PPO, pH below 4.0 can be very effectively used as long as the product acidity can be tolerated for consumption. Color preservation, like safety, is a vital parameter in minimal processing because most often product appearance relates directly to consumer acceptance, and food color is already known to play a key role in food choice, preference, and acceptability.

10.4.1 RAW MATERIAL AND PROCESSING

One of the important measures to reduce browning during minimal processing is selection of varieties with lesser tendencies to brown. This is because different varieties differ in chemical composition. Even within the same variety, the extent of browning during minimal processing can vary significantly according to pre- and postharvest factors. Agricultural practices, edaphic properties, nutrient management, climate, and harvesting conditions can affect the final quality of minimally processed produce. For instance, in potatoes early harvested produce contains higher phenolic levels and thus, higher degree of browning. Transport and storage of harvested produce are also known to play a role in the rate of browning during minimal processing, especially in potatoes. This is because mechanical shocks during transportation of produce can result in cracks and bruises that can elicit physiological and biochemical responses in both wounded and unwounded distant tissues.

Cutting, peeling, slicing, and trimming are necessary steps in minimal processing of fruit and vegetables. Research has shown that the method of peeling or cutting can have significant effect on the extent and rate of browning in processed product. For example, carrots, potatoes, or apples are normally peeled on a commercial scale mechanically, chemically, or in high-pressure steam peelers. Knife peeling has been however, found to be better in terms of browning inhibition. In potatoes, hand peeling or lye peeling resulted in better quality, while abrasion peeling was undesirable for browning. Dull and blunt knives and blades used for peeling can induce higher respiration compared to use of sharp knives. Wounding, brought about during minimal processing, induces production of secondary products including phenolic compounds.

10.4.2   PHYSICAL METHODS

Physical methods for control of browning in minimally processed fruit and vegetables include reduction of temperature, restriction of oxygen supply, use of MAP and edible coatings, high pressure, treatment with gamma irradiation, ozone, ultraviolet radiation, pulsed-electric fields, etc.

10.4.2.1   REDUCING AVAILABILITY OF OXYGEN

Even in the presence of oxidizing enzyme and substrate, browning can be considerably reduced if the contact with oxygen is cut off or reduced. This principle is commonly used in homes and industry where fresh-cut fruit and vegetables are temporarily immersed in water, brine, or syrup with the sole aim of preventing browning through prevention of contact with oxygen. However, it is important to note that as living tissues, oxygen is a requirement and its supply cannot be completely stopped altogether. Therefore, the concept of MAP and use of edible films have become immensely popular and commercial. In MAP, the atmosphere inside a package is modified in such a way that it is completely different from normal atmospheric gaseous composition.

10.4.2.2   MODIFIED ATMOSPHERE PACKAGING

MAP entails increasing the concentration of CO_2 and decreasing that of oxygen. There are two types of MAP—active and passive MAP. In passive MAP, the in-package atmosphere is modified through the product respiration, whereby oxygen is consumed and CO_2 is liberated. With time, equilibrium is attained that is unique to the product at that particular temperature and packaging film used. In active MAP, the equilibrium gas concentration is achieved by adding oxygen or CO_2 removal so that no time is lost in reaching the state of equilibrium. Fresh-cut fruits and vegetables can tolerate higher extremes of both O_2 and CO_2 but care should to taken to ensure that the limit of tolerance is not exceeded which can lead to anaerobic respiration and development of off-flavors and stimulation of food-borne pathogens. As a supplement to low temperature storage, MAP has been commercially used in several fruits and vegetables, both fresh and minimally processed, as a powerful tool of shelf-life extension. MAP also controls microbial build-up through creation of a barrier due to packaging as well as through the antimicrobial effect of CO_2. Atmospheres with 5–15% CO_2 and 2–8% O_2 have been shown to be effective in maintaining the quality of minimally processed products, the specific atmosphere being dependent of the product. For example, fresh-cut cantaloupes packed in an atmosphere containing 10% CO_2 and 4% O_2 resulted in better retention of color with reduced respiration rate during storage at 10°C (Bai et al., 2001).

10.4.2.3   EDIBLE COATINGS

Fruit and vegetables contain outer epidermis, or peel, which acts as a protective layer to the underlying pulp or tissue. During minimal processing, fruit and vegetables are cut or peeled, thereby rendering them susceptible. Therefore, the use of edible coatings acts as a replacement of the peel. They function as a semipermeable barrier that reduces product respiration, checks water loss, prevents drying, maintain membrane integrity, reduce microbial growth, and checks surface browning (Vargas et al., 2008). The edible coating can be applied in liquid form by dipping, panning, or spraying, while a film wrap is a preformed thin layer of edible material. Most commonly used materials in edible coatings are lipids (oil or paraffin wax, beewax, carnauba wax, vegetable oil, mineral oil, etc.), polysaccharides (alginate, starch, cellulose, gums, pectin, chitosan, etc.), and proteins (casein, gelatin, collagen, albumin, etc.). Edible films are mostly sourced from polysaccharides, particularly starch.

10.4.2.4   TEMPERATURE MANAGEMENT

Temperature is one of the most important factors that decide the storage life and behavior of any fresh or processed product. At low temperature, the metabolic and respiratory rates slow down, and there is reduction in the activity of enzymes. Thus, the shelf-life of processed product increases. It is for this reason that unit operations in minimal processing are carried out in low temperature.

Refrigeration throughout the production and supply chain is an integral part of minimal processing and is making such products available to consumers in the form they expect. Wounding produce during cutting, trimming, or slicing processes of minimal processing can increase the respiration rate by up to 100%, thereby reducing shelf-life. Respiration rate doubles for every 10°C rise in temperature. Thus, low temperature reduces metabolic processes in processed products and minimizes the damaging effects of mechanical injury. Minimally processed products are usually recommended to be kept at a temperature just above freezing, although care should be taken to avoid chilling injury as per the fruit and vegetable used. Fresh-cut produce is usually rinsed in cold water as a common practice to lower the temperature and prevent browning. Although low temperature storage is the norm for storage of minimally processed products, heat treatment is also an effective and popular method integrated into the process to reduce browning. PPO from different plants sources is most active in a temperature range of 20–35°C, and the enzyme activity can be reduced or inhibited at temperatures above 40°C.

10.4.2.5   HIGH-PRESSURE PROCESSING

Inactivation of enzymes has been successfully achieved through application of high-pressure processing (HPP). In addition, HPP has tremendous application in food preservation due to its deleterious effect on microbial flora. One of the main disadvantages of heat preservation is the loss of sensory and nutritional qualities of the product. But HPP nullifies this in which the process is carried out in low temperature and therefore, has minimal effect on the sensory and nutritional properties of the product. However, variation in efficacy of pressure processing on inactivation of enzymes has been found with respect to enzymes from different sources. Greater efficacy can be achieved by combining HPP with other method such as blanching, MAP, refrigeration, etc.

10.4.3   CHEMICAL METHODS

Control of browning in minimally processed fruit and vegetables is achieved through the use of chemicals that can inhibit the activity of oxidizing enzymes, remove the substrates, restrict enzyme–substrate contact, or function as preferred substrate. Earlier, sulfites were popularly used for control of both enzymatic and nonenzymatic browning; but following their generally recognized as safe (GRAS) status revoked due to potential health hazards for sensitive consumers, the use of sulfites has been restricted. The activity of PPO can be effectively inhibited through heat treatment but due to destruction of food quality attributes including fresh-like texture, the use of heat is minimized or restricted to the extent that it should not cease respiration in minimally processed products. Chemicals used in control of browning are popularly referred to as antibrowning agents. They control browning by directly inhibiting enzyme, creating a medium inadequate for browning reaction, or by reacting with the products of enzymatic reaction before formation of brown or dark pigments (Table 10.1).

10.4.3.1   ACID TREATMENT

Maintaining pH below the level optimum for the activity of PPO is an effective means to control enzyme activity and therefore browning. There is very little activity of PPO below pH 4.5, while below pH 3.0, the enzyme is known to be irreversibly inactivated. The pH can be lowered by acid treatment or use of acidulants. Citric acid remains the most widely used acidulant to control browning in minimally processed fruit and vegetables. Antibrowning property of citric acid has been established in minimally processed apples and Chinese water chestnut. However, different products show varied response to different acids with respect to browning control. In addition to this, organic acids have increasingly been tested as alternative disinfectants to sanitize fresh-cut surfaces of fruits and vegetables. They have the ability to retard or prevent the growth of microorganisms due to their ability to maintain pH, homeostasis, disrupts substrate transport, and inhibit metabolic pathways.

10.4.3.2   REDUCING AGENTS

Reducing agents are used to revert back the colorless o-quinones, products of PPO reaction, to o-diphenols. However, the use of reductants has a disadvantage; they are irreversibly oxidized during the reaction. Therefore, the browning protection they confer remains temporary and, once used up, o-quinones may undergo oxidization and polymerization resulting in browning. Important reductants used in the industry include ascorbic acid and cysteine. Ascorbic acid not only has a direct effect on PPO but can also reduce browning through its role in lowering pH. Additionally, acids such as ascorbic acid and citric acid are found naturally in fruits and vegetables, and consumers perceive them as a "natural" product. They are, therefore, popular in use as antibrowning agents. Cysteine or thiol containing compounds are also reducing agents that reduce browning in minimally processed products.

10.4.3.3   ENZYMATIC METHODS

Protease enzymes that can hydrolyze or breakdown PPO are known to be effective against browning in minimally processed products. Once hydrolyzed, the PPO can no longer effect browning reactions as they get inactivated. The mechanism of action of proteases is proteolysis or binding at specific sites required for activation and also the presence of sulfhydryl groups in proteases. Proteases of plant origin, namely ficin, bromelain (from pineapple), or papain (from papaya), have been found effective against enzymatic browning. Papain has been found to work best on apples and ficin on potatoes. Ficin preparation also contains antibrowning agents that are analoques of resorcinol.

10.4.4   OTHER ANTIBROWNING AGENTS

Several other antibrowning agents that find wide usage include ethylenediamine tetraacetic acid (ETDA), cyclodextrins, 4-hexylresorcinol (4-HR), sodium chloride, calcium chloride, etc. ETDA is a chelating agent that inhibits browning by complexing copper from the PPO active site. Common salt, sodium chloride, also inhibits browning in minimally processed fruit and vegetables, especially with decrease in pH. Calcium chloride, possibly through the action of chloride ion, can inhibit browning, in addition to its role in maintaining tissue firmness and texture. Enzymatic inhibitors like 4-HR, isoascorbic acid, N-acetylcysteine, and calcium propionate have been found to reduce browning in minimally processed apple slices. The most widely used is 4-HR as it hinders PPO, especially in combination with other additives. A combination treatment involving 0.01% 4-HR, 0.5% ascorbic acid, and 1% calcium lactate has been found effective in maintaining the surface color of fresh-cut pears (Dong et al., 2000), while combination of 0.001 M 4-HR + 0.5 M isoascorbic acid + 0.05 M calcium propionate + 0.025 M homocysteine maintained minimally processed apples for 4 weeks at 5°C.

Better results in browning inhibition can be achieved through combined application of antibrowning agents. Also, in combination with physical treatments, including mild heat, the efficacy of antibrowning agents can be improved. Table 10.1 details the effect of combined application of different chemicals and methods to prevent browning and improve quality in minimally processed fruits and vegetables.

TABLE 10.1 Combined Application of Different Chemicals and Methods to Prevent Browning in Minimally Processed Fruits and Vegetables.

Fruit/Vegetable	Treatment combination	Physiological effect	Reference
Apple	Ethanol (20–30%) + ascorbic acid (1%)	Prevented browning and softening of apple slices	Yan et al. (2017)
Apple	Carboxymethyl cellulose coating (1% w/v) + CaCl$_2$ (0.5%) + ascorbic acid (2%)	Synergistic control of surface browning through effect on PPO and POX enzymes	Saba and Sogvar (2016)
Ash gourd	Gamma irradiation (2 kGy)	Increased content of α-resorcylic acid that inhibited PPO activity and browning	Tripathi and Variyar (2016)
Lettuce	Dipping in arginine (100 mM) for 5 min	Delayed browning without affecting the taste	Wills and Li (2016)
Mango	Pulsed light (20 pulses at fluence of 0.4 J/cm^2/pulse) + alginate coating (2%) + malic acid (2%)	Reduced *Listeria innocua* and microbial load, and maintained freshness for 14 days	Salinas-Roca et al. (2016)
Melon	Chitosan coating (2%) containing 500 mg/L *trans*-cinnamaldehyde	Reduced activity of G-POD and PPO, thereby browning, maintained firmness, and quality	Carvalho et al. (2016)
Pear	Xanthan gum-based edible coating (2.5 g/L) enriched with cinnamic acid (1 g/L)	Reduces surface browning and improves quality	Sharma and Rao (2015)
Persimmon	Apple pectin-based edible coating containing antioxidants, 2% potassium sorbate, 4% sodium benzoate, and nisin	Browning inhibition and microbial control	Sanchís et al. (2016)
Potato	Sanitizers + passive MAP or Vacuum packaging	Inhibited browning up to 14 days	Beltrán et al. (2005)
Sweet potato	Cassava starch edible coating + ascorbic acid	Prevented enzymatic browning and retained freshness	Ojeda et al. (2014)

10.5 CONCLUSION

Browning in minimally processed fruit and vegetables can be controlled by adopting physical or chemical methods, or combination of both. However, no chemical or physical method can guarantee complete inhibition of browning as the phenomenon is complex. Enzymatic browning contributes to majority of the browning in fresh-cut produce. For manifestation of browning, enzyme needs to act on the substrate in the presence of oxygen. Therefore, strategies to control or reduce browning include ways to inactivate or inhibit enzyme activity and prevent contact with oxygen and enzyme–substrate contact. Combination treatments involving right time of harvest, temperature and relative humidity control, use of optimal atmosphere composition and chemicals offer effective solutions for browning inhibition. Minimally processed products are projected to dominate consumer demand due to increase in purchasing power capacity in emerging economies, and the convenience they bring to people to save time in a busy and fast-paced life. Minimally processed products are highly sensitive to deterioration and if not handled properly, they pose health hazards to consumers. Since color is one of the biggest determinants of quality of minimally processed products, maintaining fresh-like properties of such foods is paramount. Emerging technologies in minimal processing with hurdle principle help develop safer and healthier products to fulfill consumer demands.

KEYWORDS

- minimal processing
- browning
- anti-browning
- fresh produce
- consumers

REFERENCES

Artés, F.; Allende, A. Minimal Fresh Processing of Vegetables, Fruits and Juices. In Emerging Technologies for Food Processing; Sun, D. W., Ed.; Elsevier Academic Press: Cambridge, MA, 2005; pp 677–716.

Bai, J. H.; Saftner, R. A.; Watada, A. E.; Lee, Y. S. Modified Atmosphere Maintains Quality of Fresh Cut Cantaloupe (Cucumis melo L.). J. Food Sci. **2001,** 75, 1–7.

Beltrán, D.; Selma, M. V.; Tudela, J. A.; Gil, M. I. Effect of Different Sanitizers on Microbial and Sensory Quality of Fresh-Cut Potato Strips Stored under Modified Atmosphere or Vacuum Packaging. Postharvest Biol. Technol. **2005,** 37, 37–46.

Carvalho, R. L.; Cabral, M. F.; Germano, T. A.; de Carvalho, W. M.; Brasil, I. M.; Gallao, M. I.; Moura, C. F. H. Chitosan Coating with *Trans*-Cinnamaldehyde Improves Structural Integrity and Antioxidant Metabolism of Fresh-Cut Melon. Postharvest Biol. Technol. **2016,** 113, 29–39.

Dong, X.; Wrolstad, R. E.; Sugar, D. Extending Shelf Life of Fresh-Cut Pears. J. Food Sci. **2000,** 65, 181–186.

IFPA. Fresh-Cut Produce Handling Guidelines, 3rd ed.; Produce Marketing Association: Newark, NJ, 1999.

Ioannou, I.; Ghoul, M. Prevention of Enzymatic Browning in Fruits and Vegetables. Eur. Sci. J. **2013,** 9 (30), 310–341.

Laurila, E.; Kervinen, R.; Ahvenainen, R. The Inhibition of Enzymatic Browning in Minimally Processed Vegetables and Fruits. Postharvest News Inf. **1998,** 9 (4), 53–66.

Massantini, R.; Mencarelli, F. Understanding and Management of Browning in Fresh Whole and Lightly Processed Fruits. Fresh Prod. **2007,** 1 (2), 94–100.

Ojeda, G. A.; Sgroppo, S. C.; Zaritzky, N. E. Application of Edible Coatings in Minimally Processed Sweet Potatoes (*Ipomoea batatas* L.) to Prevent Enzymatic Browning. Int. J. Food Sci. Tecchnol. **2014,** 49 (3), 876–883.

Pasha, I.; Saeed, F.; Sultan, M. T.; Khan, M. R.; Rohi, M. Recent Developments in Minimal Processing: A Tool to Retain Nutritional Quality of Food. Crit. Rev. Food Sci. Nutr. **2014,** 54, 340–351.

Saba, M. K.; Sogvar, O. B. Combination of Carboxymethyl Cellulose-Based Coatings with Calcium and Ascorbic Acid Impacts in Browning and Quality of Fresh-Cut Apples. LWT—Food Sci. Technol. **2016,** 66, 165–171.

Salinas-Roca, B.; Soliva-Fortuny, R.; Welti-Chanes, J.; Martin-Belloso, O. Combined Effect of Pulsed Light, Edible Coating, and Malic Acid Dipping to Improve Fresh-Cut Mango Safety and Quality. Food Cont. **2016,** 66, 190–197.

Sanchís, E.; Gonzalez, S.; Ghidelli, C.; Sheth, C. C.; Mateos, M.; Palou, L.; Perez-gago, M. B. Browning Inhibition and Microbial Control in Fresh-Cut Persimmon (Diospyros kaki Thunb. cv. Rojo Brillante) by Apple Pectin-Based Edible Coatings. Postharvest Biol. Technol. **2016,** 112, 186–193.

Siddiqui, M. W.; Chakraborty, I.; Ayala-Zavala, J. F.; Dhua, R. S. Advances in Minimal Processing of Fruits and Vegetables: A Review. J. Sci. Ind. Res. **2011,** 70, 823–834.

Sharma, S.; Rao, T. V. R. Xanthan Gum Based Edible Coating Enriched with Cinnamic Acid Prevents Browning and Extends the Shelf-Life of Fresh-Cut Pears. LWT—Food Sci. Technol. **2015,** 62, 791–800.

Sun, D. W. Emerging Technologies for Food Processing; Elsevier Academic Press: Cambridge, MA, 2005.

Tripathi, J.; Variyar, P. S. Gamma Irradiation Inhibits Browning in Ready-to-Cook (RTC) Ash Gourd (*Benincasa hispida*) During Storage. Innov. Food Sci. Emerg. Technol. **2016,** 33, 260–267.

Vargas, M.; Pastor, C.; Chiralt, A.; McClements, D. J.; Gonzalez-Martinez, C. Recent Advances in Edible Coatings for Fresh and Minimally Processed Fruits. Crit. Rev. Food Sci. Nutr. **2008,** 48, 496–511.

Willocx, F.; Hendrickx, M.; Tobback, P. The Influence of Temperature and Gas Composition on the Evolution of Microbial and Visual Quality of Minimally Processed Endive. In Minimal Processing of Foods and Process Optimization: An Interface; Singh, R. P., Oliveira, F. A. R., Eds.; CRC Press: Boca Raton, FL, 1994; pp 475–492.

Wills, R. B. H.; Li, Y. Use of Arginine to Inhibit Browning of Fresh Cut Apple and Lettuce. Postharvest Biol. Technol. **2016,** 113, 66–68.

Yan, S.; Luo, Y.; Zhou, B.; Ingram, D. T. Dual Effectiveness of Ascorbic Acid and Ethanol Combined Treatment to Inhibit Browning and Inactivate Pathogens on Fresh-Cut Apples. LWT—Food Sci. Technol. **2017,** 80, 311–320.

CHAPTER 11

ESSENTIAL OILS AND PLANT EXTRACTS AS NATURAL ANTIMICROBIAL AGENTS

A. GARCHA HEREDIA[1†], J. E. DÁVILA-AVICA[1*], C. ZOELLNER[2], L. E. GARCIA-AMEZQUITA[3], N. HEREDIA[1], and S. GARCÍA[1]

[1]*Universidad Autónoma de Nuevo León, Facultad de Ciencias Biológicas, Apdo. Postal 124-F, Ciudad Universitaria, San Nicolás de los Garza, Nuevo León 66451, México*

[2]*Department of Food Science, Cornell University, 306 Stocking Hall, Ithaca, NY 14850, USA*

[3]*Centro de Investigación en Alimentaciyn y Desarrollo, Coordinación de Fisiología y Tecnología de Alimentos de la Zona Templada, Av. Río Conchos S/N Parque Industrial, C.P. 31570 Cd. Cuauhtémoc, Chihuahua, México*

Corresponding author. E-mail: jda011@gmail.com

†*Current address: University of Massachusetts Amherst, Amherst, MA 01003, USA*

ABSTRACT

Pathogenic microorganisms not only cause edible quality deterioration and resulting postharvest losses but also produce toxins. Essential oils and plant extracts have been historically used as a natural source of antimicrobials for managing harmful microbial invasions on food products, particularly fresh fruits and vegetables. These are categorized as "generally recognized as safe" by the United States Food and Drug Administration as they are environmental friendly and leave no residues behind. Synthetic antimicrobials have wide-ranging applications, are more stable and hence are more persistent in biological system inducing consumer health hazards, which creates an aversion for their usage. The occurrence of antimicrobial-resistant microorganisms is widespread, due to their wide range of defense mechanisms, raising difficulties in food preservation. Antimicrobial activity of natural plant extracts and essential oils ascribes to several molecules, favorable for combating antibiotic-resistant microbial populations. In this chapter, source, composition, and activity of natural antimicrobials of plant origin, plant extracts and essential oils are discussed with the aim to identify their possible varied applications to the postharvest fruit and vegetable commodities.

11.1   INTRODUCTION

The need for antimicrobial agents, natural or synthetic, to prevent both food quality degradation and contamination with harmful human pathogens is more apparent now than ever due to demands of an increasing food supply.
In spite of the use of available means of food protection during production and modern storage, molds, mycotoxin and bacterial contamination, and oxidative deterioration still present different challenges for producers along the postharvest chain in all corners of the world (Alvarez et al., 2015; Prakash et al., 2015). Moreover, processed fruit and vegetable products, such as beverages, require either antimicrobial processing or additives to protect both consumers and the product quality. Due to the wide range of defense mechanisms in microorganisms, there is no "magic bullet" in terms of antimicrobial agents. Rather there are important questions to be addressed when designing food preservation systems for specific products (Roller and Board, 2003).

Synthetic preservatives have found wide application and efficacy as antimicrobials and antioxidants in the food industry. For fruits and vegetables, potent fungicides are applied using dipping methods, during production or immediately postharvest. Other common postharvest antimicrobial control agents include chlorine, carbonates and bicarbonates, ammonia, sulfur dioxide, and low-molecular weight amines. While postharvest fungicides and antimicrobial agents in food crops are among the most rigorously tested and regulated chemicals, recently their use has faced concerns over the effects of residues on human health and the environment. For example, it has been suggested that there are human health risks associated with antimicrobial residues, particularly in the diets of children, as they can persist in fatty tissues and organs such as liver and kidneys. On the other hand, the occurrence of antimicrobial-resistant microorganisms is widespread, and there are still research gaps in approved antimicrobial agents for the control of some plant pathogens (Smilanick and Sorenson, 2001).

For application in the food industry, ideal antimicrobials must be available in large volumes and have already established as "generally recognized as safe" (GRAS) by the United States Food and Drug Administration (Calo et al., 2015). Whether consumers are aware of it or not, the source of most antimicrobials has historically been from "natural" sources or products (i.e., plants, bacteria) and has been subjected through approval process. The main "natural" antimicrobial compounds are essential oils (EOs), bacteriocins, organic acids, and naturally occurring polymers (Lucera et al., 2012) and often provide the "structural motifs" for the development of synthetic and commercial products (Morcia et al., 2011). During the last 25 years, many in vitro and in vivo investigations have been made with different plant sources of these natural antimicrobial agents for use against agricultural pests and the number of publications in this area over the past 10 years has quintupled. The environmental and human toxicity advantage of naturally sourced preservatives, in contrast to synthetic, is that they break down more rapidly because they lack the persistent, unnatural structures and seldom contain halogens that contribute to the high stability of synthetic compounds (Regnier et al., 2012). Additionally, their antimicrobial activity is usually due to a variety of compounds (some of which may not be identified), which could be beneficial to combat antibiotic-resistant microbial populations.

This chapter will focus on two categories of natural antimicrobial agents of plant origin, EOs, and plant extracts, in terms of their sources, composition, and demonstrated activity as antimicrobial agents, with the goal of providing insight into their potential for postharvest application to the fruit and vegetable industry.

11.2   PLANT EXTRACTS AS NATURAL ANTIMICROBIAL AGENTS

Plants have been proven to contain a wide array of natural compounds with antimicrobial activity. The antimicrobial effect of plant extracts obtained from spices, herbs, fruits, and vegetables has been well documented in recent years (Burt, 2004; Tajkarimi et al., 2010; Tiwari et al., 2009; Davidson, 1997; Sánchez et al., 2014; Heredia et al., 2005). The antibiotic resistance observed in pathogenic microorganisms has been a factor for the research into plant extracts as a more sustainable alternative. Plant extracts seem to be a promising solution to the increasing antibiotic resistance and may also provide better results than synthetic preservatives (Hayek et al., 2013), promoting synergistic effects between natural antimicrobials and antibiotics at the same time.

Antibiotic resistance has been largely increased due to extensive use of these compounds in foods and produce; this urging the need of new, safe, and more effective antimicrobial agents (Perez-Montaño et al., 2012). Plant-derived compounds could be used in the development of new preservation systems; thus, these compounds could prove to be among the most promising solutions for microbial resistance (Hayek et al., 2013; Perez-Montaño et al., 2012; Castillo et al., 2015). Natural compounds could meet the consumer demand for healthier foods. Plant antimicrobials could also be used to combat infections caused by multidrug-resistant bacteria. Extracts from plants could be used in combination with antibiotics to restore the efficacy of these drugs for resistant pathogenic bacteria (Chinsembu, 2016; Abreu et al., 2012; Celestino et al., 2014).

11.2.1 ANTIMICROBIAL ACTIVITY OF PLANT EXTRACTS

According to Davidson (1997), there are some possible mechanisms of anti-microbial activity; these include membrane-disrupting, causing leakage of the cellular content, interference with active transport or metabolic enzymes, dissipation of cellular energy in ATP form, depression of the internal cellular pH, or disruption of substrate transport by alteration of cell membrane permeability (Sánchez et al., 2010, 2013). Organic acids can also inhibit nicotinamide adenine dinucleotide (NADH) oxidation, thus eliminating supplies of reducing agents to electron transport systems. Other mechanisms reported include the attack of the phospholipid bilayer of the cell membrane, disrupting the cell enzyme systems and compromising the genetic material of a bacterial cell (Burt et al., 2007).

Secondary plant metabolites, particularly resins, isolated from the family Apiaceae, Burseraceae, Anacardiaceae, Palmaceae, Euphorbiaceae, Dracena-ceae, Pinaceae, and Cupressaceae have revealed antifungal, antibacterial, and antiprotozoal activity (Termentzi et al., 2011; Paraschos et al., 2012; Rosas-Taraco et al., 2011; García et al., 2006). Various species of the plant genus *Hypericum*, used in traditional medicine, contain several compounds including hyperenone A, hypercalin B, and hyperphorin, responsible for anti-bacterial activity against antibiotic-resistant *Staphylococcus aureus* and also against *Mycobacterium tuberculosis* (Osman et al., 2012; Shiu et al., 2011). The antimicrobial effect of pomegranate extracts on antibiotic-sensitive and resistant strains of *S. aureus* were reported by Braga et al. (2005). The syner-gistic effect of pomegranate extract (0.1–2.0% (v/v)), with chloramphenicol, gentamicin, ampicillin, tetracycline, and oxacillin, indicated the potential for plant extracts to enhance the activity of these antibiotics. Because plant extracts kill bacteria or inhibit their growth, quorum-sensing, and virulence factors, plant products could become important ingredients in new antimicrobial drugs to combat antibiotic-resistant microorganisms (Castillo et al., 2014; Clatworthy et al., 2007).

For example, a study tested 10 commonly used spices and herbs on 41 foodborne bacterial strains using a series of evaluation methods to identify synergistic combinations in vitro with the goal of guiding more focused in vivo studies. Only three showed promising antibacterial activity across the pathogens (with inhibition zone diameter >11 mm), those being coriander, cumin, and mustard seed. Of the possible combinations, only coriander/cumin seed showed synergism (FICI 0.25–0.5) with 2 log reduction in 24 h (Bag and Chattopadhyay, 2015). Other tested combinations showed additive effects and no antagonistic effects were observed. Subsequent toxicity tests on human colon cell lines did not demonstrate cytotoxicity. These findings demonstrate the potential for further study in food systems but raise questions regarding the desired level of microbial reduction or duration of activity.

11.2.2 POTENTIAL APPLICATIONS OF PLANT EXTRACTS IN FOOD AND PHARMACEUTICAL INDUSTRY

Green tea extract has been used in various food applications such as bread, extra virgin olive oil (Rosenblat et al., 2008), meat, fish (Alghazeer et al., 2008), dehydrated apple products (Lavelli et al., 2010), rice starch products (Wu et al., 2009), and biscuits (Mildner-Szkudlarz et al., 2009). Grape seed extract has also demonstrated antimicrobial activities alone or in combination with other technologies in various food applications such as tomatoes, frankfurters, raw and cooked meat, poultry products, and fish (Bisha et al., 2010; Brannan, 2009; Perumalla and Hettiarachchy, 2011). According to Pina-Pérez et al. (2013), the combined effect of polyphenol-rich cocoa powder with pulsed electric field significantly inactivated the level of *Cronobacter sakazakii* in infant milk formula over 12 h at 8°C storage. García et al. (2005) and Heredia et al. (2005) reported that partial purification of the active fraction suggested that polyphenols may play a role in the antimicrobial activity exhibited by *Haemathoxylon brasiletto* extracts against *Vibrio cholera* and enterohemorrhagic *Escherichia coli* 0157:H7, respectively.

The extract of lime alone or in combination with other edible fruits has been reported to reduce the population of *Campylobacter jejuni* and *Campylobacter coli* in chicken skin by >4.0 log cycles (Valtierra-Rodríguez et al., 2010). Ibrahim et al. (2009) showed that crude chive extract can be effective against the growth of *Salmonella* and could be used in food products to prevent the growth of this pathogen. Also, the inhibitory effect of bell pepper (*Capsicum annum*) extract against *Salmonella typhimurium* in raw beef was reported by Careaga et al. (2003); the authors reported 1.5 mL/100 g of meat as the minimum concentration of the extract to prevent the growth of this bacterium.

Gram-positive bacteria are more susceptible to antimicrobial plant extracts (Deans et al., 1995; Yoda et al., 2004; Shan et al., 2007; Belguith et al., 2009). Plant extracts have been reported to be more effective in culture medium than in food systems (Orue et al., 2013). According to Burt (2004), a higher concentration of EOs is needed to achieve the same effect in food experiments as in vitro (growth media). Poor water solubility of most EOs extracted from thyme, oregano, and clove makes it difficult to incorporate EOs into complex food systems (Ananda Baskaran et al., 2009; Friedman et al., 2004). High levels of these EOs can also affect the original sensory properties of the food products, altering the taste and the organoleptic effects must be considered when using plant extracts (Nazer et al., 2005). Conversely, lower concentrations of these extracts in combination with other natural preservatives may lower the negative sensory effect.

11.3 ESSENTIAL OILS AS NATURAL ANTIMICROBIAL AGENTS

EOs are complex mixtures of aromatic oily liquids synthesized and obtained from vegetal organisms as secondary metabolites and may vary depending on the part of the plant (flowers, stems, leaves, seeds, fruits, roots, wood). The functions of the EOs are diverse, including the defense system of higher plants (Burt, 2004; Avila-Sosa et al., 2012; Teixeira et al., 2013). The amount and chemical composition of the plants oils vary greatly depending on external factors such as seasonal variation, ripeness, geographical region, nutrition, and climate. The EOs usually contain about 20–60 components in different proportions. These components are mixtures of different lipophilic and volatile substances, chemically derived from terpenes and their oxygenated derivatives, terpenoids, which are low-weight aromatic and aliphatic acid esters and phenolic compounds. Normally, just a few components (two to three) are present in high concentration; such molecules tend to determine the biological proprieties of the EO (Hussain et al., 2010; Bakkali et al., 2008; Solórzano-Santos and Miranda-Novales, 2012; Reichling et al., 2009; Fisher and Phillips, 2008).

The composition of EOs is dynamic and is dependent of several factors. Various important components in EOs can be identified. Terpenes are hydrocarbons synthetized from the combination of several isoprene units and do not represent a group of constituents with high antimicrobial activity. The main terpenes are monoterpenes, which constitute up to 97% of citrus oils with different esters, alcohols and aldehydes, and sesquiterpenes. On the other hand, *terpenoids*, which are terpenes with biochemical modifications in oxygen molecules and methyl groups, show antimicrobial activity related to their functional groups, carvacrol being a good example. *Phenylpropenes* represent organic compounds synthesized from phenylalanine, an amino acid precursor, in plants, and constitute a small fraction in EOs, having representative compounds such as eugenol. Flavonoids are nonvolatile compounds that are categorized in six different classes: isoflavonoids, anthocyanins, flavans, flavonols, and flavanones. Fatty acids are present in low amounts oils, like citrus oils (Fisher and Phillips, 2008; Hyldgaard et al., 2012).

11.3.1 ANTIMICROBIAL ACTIVITY OF ESSENTIAL OILS

Throughout history, humans have used plants for different purposes including food preservation and different medicinal purposes, such as pharmaceuticals and natural therapies. As science has progressed, new techniques have made it possible to study, isolate, and characterize the bioactive secondary metabolites of plants. EOs show antimicrobial activity against a broad spectrum of microorganisms: Gram-positive and Gram-negative bacteria, viruses, fungi, and protozoa. Tables 11.1–11.4 summarize relevant insights about antimicrobial activity of EOs against various microorganisms.

As previously mentioned, EOs are composed of a wide variety of components which qualitative and quantitative composition determines their properties. The antimicrobial activity of the majority of the EOs is the result of the presence of different phenolic compounds, such as carvacrol and thymol. Importantly, these oils and their components do not appear to be specific to a certain target inside cells. In general, because of their hydrophobicity, EOs and their components are able to interact with cellular membranes, affecting the lipid ordering and the bilayer stability, decreasing the membrane integrity and increasing the proton passive flux across the membrane. As a consequence, membranes become more permeable and there can be leaking of different molecules and ions that, eventually, will result in cell death (Fisher and Phillips, 2008; Ben Arfa et al., 2006; Carson et al., 2002) (Fig. 11.1).

TABLE 11.1 Antibacterial Activity of Essential Oils and Compounds.

EO/compound	Bacteria	Concentration/method	References
Pinus densiflora	S. typhimurium	Filter paper disks	Hong et al. (2004)
Pinus koraiensis	L. monocytogenesis	Filter paper disks	
Chamaecyparis obtuse	E. coli K. pneumoniae S. aureus	Filter paper disks	
Cinnamomum zeylanicum	S. aureus B. subtilis K. pneumonia P. vulgaris P. aeruginosa E. coli	MIC: 3.2 mg/mL MIC: >1.6 mg/mL MIC: 3.2 mg/mL MIC: >1.6 mg/mL MIC: >0.8 mg/mL MIC: >1.6 mg/mL	Prabuseenivasan et al. (2006)
Satureja hortensis L.	A. actinomycetem-comitans F. nucleatum P. micra P. gingivalis P. intermedia P. nigrescens T. forsythia	MIC: <0.125 µL/mL	Gursoy et al. (2009)
Tamarix boveana	S. aureus S. epidermidis E. coli P. aeruginosa	1–0.3 mg/mL 4–0.5 mg/filter paper disk MIC: 0.3–0.8 mg/mL	Saïdana et al. (2007)
Eucalyptus camaldulensis	S. aureus L. monocytogenes E. durans E. coli P. aeruginosa	MIC: >1–0.5 (% v/v)	Akin et al. (2010)
Eucalyptus globulus	Methicillin-resistant S. aureus (MRSA) E. coli S. pyogenes S. agalactiae S. aureus S. pneumonia S. maltrophila	MIC: 85.6 µg/mL Disk diffusion method 50 µL/mL 25 µL/mL 1.25 µL/mL	Tohidpour et al. (2010), Prabuseenivasan et al. (2006), Cermelli et al. (2008)

TABLE 11.1 *(Continued)*

EO/compound	Bacteria	Concentration/method	References
Myrtus communis	*S. aureus*	MIC: >1–0.5 (% v/v)	Akin et al. (2010),
	L. monocytogenes	MIC 1.4–11.2 mg/MI	Rosato et al.
	E. durans		(2007)
	S. typhi		
	E. coli		
	P. aeruginosa		
	B. cereus		
Cinnamomum cassia	*E. coli* O157:H7	MIC: 0.05 (vol/vol)	Oussalah et al.
	S. typhimurium	MIC 0.025 (vol/vol)	(2006)
	S. aureus	MIC: 0.05 (vol/vol)	
	L. monocytogenes		
Dracocephalum foetidum	*B. subtilis*	MIC: 26–2592 µg/mL	Lee et al. (2007)
	S. aureus		
	M. lutens		
	E. hierae		
	S. mutans		
	E. coli		
Origanum vulgare	*B. subtilis*	MIC: 0.35–0.70 mg/mL	Rosato et al. (2007)
Carvacrol	*S. aureus*	MIC: 0.35–2.80 mg/mL	Rosato et al.
	S. epidermidis		(2007), Solór-zano-Santos and Miranda-Novales (2012)
Thymol	*S. aureus*	MIC: 0.7–1.40 mg/mL	Rosato et al. (2007)
Actinidia macrosperma	*S. aureus*	MIC: 0.78–25.50 µL/mL	Lu et al. (2007)
	B. subtilis		
	E. coli		
Salvia rosifolia Sm. (*Lamiaceae*)	MRSA	MIC: 125 µg/mL	Özek et al. (2010)
Thymus vulgaris L.	MRSA	MIC: 18.5 µg/mL	Tohidpour et al. (2010)
Eugenia caryophyllus	*E. coli* O157:H7	MIC: 0.1 (vol/vol)	Oussalah et al.
	S. typhimurium	MIC: 0.05 (vol/vol)	(2007), Prabu-seenivasan et al.
	S. aureus	MIC: 0.2 (vol/vol)	(2006)
	L. monocytogenes	MIC: >3.2 mg/mL	
	B. subtilis	MIC: >6.4 mg/mL	
	K. pneumonia	MIC: >3.2 mg/mL	
	P. vulgaris	MIC: >1.6 mg/mL	
	P. aeruginosa		

TABLE 11.1   *(Continued)*

EO/compound	Bacteria	Concentration/method	References
C. operculatus	*E. aerogenes* *E. coli* *S. enteritidis* *L. monocytogenes* *S. typhimurium*	MIC: 1–4 µL/mL	Dung et al. (2008)
Laurus nobilis	*E. faecalis* *L. monocytogenes* *S. aureus* *B. cereus* *Y. enterocolitica*	MIC: 0.02–0.2 % (v/v) MIC: 1% (v/v)	Erkmen and Ozcan (2008)
Mentha pulegium L.	*S. aureus* *V. cholerae* *B. cereus* *E. coli* *L. monocytogenes* *S. typhimurium*	MIC: 0.5–4 µL/mL	Mahboubi et al. (2008)
Mentha longifolia L.	*S. mutans* *S. pyrogenes* *K. pneumoniae* *L. monocytogenes*	Disk diffusion	Al-Bayati (2009), Mkaddem et al. (2009)
Mentha viridis L.	*K. pneumoniae* *L. monocytogenes*	Disk diffusion	Mkaddem et al. (2009)
Origanum compactum	*E. coli* O157:H7 *S. typhimurium* *S. aureus* *L. monocytogenes*	MIC: 0.025 (vol/vol) MIC: 0.05 (vol/vol) MIC: 0.013 (vol/vol) MIC: 0.1 (vol/vol)	Oussalah et al. (2007)
Pimenta dioica	*E. coli* O157:H7 *S. typhimurium* *S. aureus* *L. monocytogenes*	MIC: 0.1 (vol/vol) MIC: 0.2 (vol/vol)	Oussalah et al. (2007)
Lantana trifolia L.	*M. tuberculosis*	MIC: 80 µg/mL	Juliao et al. (2009)
Rosmarinus officinalis	*S. aureus* *B. subtilis* *K. pneumonia* *P. vulgaris* *P. aeruginosa* *E. coli*	MIC: >12.8 mg/mL MIC: >6.4 mg/mL MIC: >12.8 mg/mL MIC: >6.4 mg/mL	Prabuseenivasan et al. (2006)

MIC, minimum inhibitory concentration.

TABLE 11.2 Antifungal Activity of Essential Oils.

EO/compound	Fungi	Concentration/method	Reference
Chamaecyparis obtusa	*C. albicans*	Dilutions on filter paper disks	Hong et al. (2004)
Melaleuca alternifolia	*C. albicans* *C. glabrata* *S. cerevisiae*	0.25–1% (v/v)	Hammer et al. (2004)
Melaleuca alternifolia	Filamentous fungi Dermatophytes	MFC: 0.03–8% MIC: 0.004–0.25%	Hammer et al. (2004)
Citrus macroptera	*T. mentagrophytes*	MIC: 12.5 µg/mL	Waikedre et al. (2010)
Melissa officinalis	*Trichophyton*	MIC: 15–30 µL/mL	Mimica-Dukic et al. (2003)
Pimpinella anisum	*C. albicans* *C. parapsilosis* *C. tropicalis* *C. pseudotropicalis* *C. krusei* *C. glabrata* *T. rubrum* *T. mentagrophytes* *Microsporum canis* *Microsporum gypseum*	MIC: 0.1–1.56% v/v	Kosalec et al. (2005)
Artemisia absinthium	*Microsporum canis* *M. gypseum* *T. rubrum* *F. pedrosi*	Agar diffusion test	Lopez-Lutz et al. (2008)
Artemisia ludoviciana	*Microsporum canis* *M. gypseum* *T. rubrum* *F. pedrosi*	Agar diffusion test	Lopez-Lutz et al. (2008)
Calocedrus macrolepis var. *formosana*	*F. oxysporum* *R. solani* *P. funereal* *C. gloesporioides* *G. austral* *F. solani*	Antifungal index: 15% Antifungal index: 33.1% Antifungal index 65% Antifungal index: 16.7% Antifungal index: 22.5% Antifungal index: 52.1%	Chang et al. (2008)
Artemisia biennis	*M. canis* *M. gypseum* *T. rubrum* *F. pedrosoi*	Agar diffusion test	Lopez-Lutz et al. (2008)
Dracocephalum foetidum	*C. albicans* *S. cerevisiae*	MIC: 26–2592 µg/mL	Lee et al. (2007)

TABLE 11.2   *(Continued)*

EO/compound	Fungi	Concentration/method	Reference
Chenopodium ambrosioides	*A. niger* *A. fumigatus* *B. theobromae* *F. oxysporum* *S. rolfsii* *M. phaseolina* *C. cladosporioides* *P. debaryanum*	100 µg/mL	Kumar et al. (2007)
Laurus nobilis L.	*A. niger* *R. oryzae*	MIC: 0.02% (v/v)	Erkmen and Ozcan (2008)
Mentha longifolia	*A. ochraceus* *M. ramamnianus*	Diffusion test	Mkaddem et al. (2009)
Mentha viridis	*A. ochraceus* *M. ramamnianus*	Diffusion test	Mkaddem et al. (2009)
P. dulcis var. *amara*	*M. canis* *E. floccosum* *T. rubrum* *T. mentagrophytes*	Mycelial growth inhibition: 2–4 µL/mL	Ibrahim and El-Salam (2015)
Thymol	*F. oxysporum*	500 µL/L	Manganyi et al. (2015)
Eugenol	*F. oxysporum*	500 µL/L	Manganyi et al. (2015)
Salvia officinalis	*A. gossypii* *A. niger* *R. oryzae* *T. reesei*	MIC: 0.03–0.25 µL/mL	Bouaziz et al. (2009)
P. armeniace	*M. canis* *E. floccosum* *T. rubrum* *T. mentagrophytes*	Mycelial growth inhibition: 2–4 µL/mL	Ibrahim and El-Salam (2015)
Thymus vulgaris	*F. oxysporum*	500 µL/L	Manganyi et al. (2015)
O. europaea	*M. canis* *E. floccosum* *T. rubrum* *T. mentagrophytes*	Mycelial growth inhibition: 2–4 µL/mL	Ibrahim and El-Salam (2015)
M. piperita	*M. canis* *E. floccosum* *T. rubrum* *T. mentagrophytes*	Mycelial growth inhibition: 2–4 µL/mL	Ibrahim and El-Salam (2015)

MFC, minimum fungicidal concentration; MIC, minimum inhibitory concentration.
Antifungal index = $(1 - Da/Db) \times 100$, where Da is the diameter of growth zone in the test plate and Db is the diameter of growth zone in the control plate.

TABLE 11.3 Antiviral Activity of Essential Oils.

EO/compound	Virus	Concentration	References
Zingiber officinale	HSV-1 strain KOS	MIC: $\leq$0.1%	Schnitzler et al. (2007), Koch et al. (2008)
	HSV-2	IC_{50} (%): 0.0001	
Houttuynia cordata	HSV-1	ED_{50}: 0.00038–0.0091% (w/v)	Hayashi et al. (1995)
Cymbopogon spp.	MNV	Infectivity reduction at 2–4% (vol/vol)	Gilling et al. (2014a)
Zanthoxylum schinifolium	Calicivirus-F9	EC_{50}: 0.0007% (pretreatment)	Oh and Chung (2014)
L. nobilis	HSV-1	IC_{50} (μg/mL): 60	Loizzo et al. (2008)
	SARS-CoV	IC_{50} (μg/mL):120	
Hyssopus officinalis	HSV-1 strain KOS	MIC: $\leq$0.1%	Schnitzler et al. (2007), Koch et al. (2008)
	HSV-2	IC_{50} (%): 0.0006	
Pimienta sp.	MNV	Infectivity reduction at 4% (vol/vol)	Gilling et al. (2014b)
b-Triketone (*Leptospermum scoparium*)	HSV-1, HSV-2	IC_{50}: 0.58, 0.96 μg/mL	Reichling et al. (2005)
Santalum album	HSV-1 strain KOS	MIC: $\leq$0.1%	Schnitzler et al. (2007), Koch et al. (2008)
	HSV-2	IC_{50} (%): 0.0005	
T. orientalis	HSV-1	IC_{50} (μg/mL): >1000	Loizzo et al. (2008)
	SARS-CoV	IC_{50} (μg/mL): 130	
Eucalyptus globulus	*H. influenza*	MIC: 1.25 μL/mL	Cermelli et al. (2008)
	H. parainfluenzae MV	0.25 μL/mL (Mild antiviral activity)	
Zataria multiflora Boiss.	Hepatitis A virus	Reduced titers at 1% (vol/vol)	Sánchez and Aznar (2015), Elizaquível et al. (2013)
	MNV		
	FCV		
Thymus vulgaris	HSV-1 strain KOS	MIC: $\leq$0.1%	Schnitzler et al. (2007), Koch et al. (2008), El Moussaoui et al. (2013)
	HSV-2s	IC_{50} (%): 0.0007	
	MNV	2% (vol/vol)	
Thymol	FCV	Reduced titers:	Sánchez and Aznar (2015)
	MNV	0.5–2% (vol/vol)	
Melaleuca alternifolia	HSV-1	0.125% (vol/vol)	Garozzo et al. (2009)
	HSV-2		
Oregano sp.	MNV	4% (vol/vol)	Gilling et al. (2014a)

TABLE 11.3 *(Continued)*

EO/compound	Virus	Concentration	References
Origanum elongatum	MNV	2% (vol/vol)	El Moussaoui et al. (2013)
Origanum compactum	FCV	2% (vol/vol)	Elizaquível et al. (2013)
	MNV		
Carvacrol	MNV	0.25–1% (vol/vol)	Gilling et al. (2014b), Sánchez and Aznar (2015)
	FCV		
S. thymbra	HSV-1	IC_{50} (µg/mL): 220	Loizzo et al. (2008)
Santolina insularis	HSV-1	80% inhibition at 1.87 mg/mL	De Logu et al. (2000)
	HSV-2	80% inhibition at 1.25 mg/mL	
Mentha piperita	HSV-1	IC_{50}: 0.002%	Schuhmacher et al. (2003)
	HSV-2	IC_{50}: 0.0008%	

EC_{50}, effective concentration to reduce the 50% plaque number; ED_{50}, the 50% effective dose; FCV, feline calicivirus; IC_{50}, 50% inhibitory concentration; MIC, minimum inhibitory concentration; MNV, murine norovirus; MV, mumps virus.

TABLE 11.4   Antiprotozoa Activity of Essential Oils.

EO/Compound	Protozoa	Concentration	References
Annona vepretorum	*T. cruzi* *Plasmodium* spp.	IC_{50}: <20 µg/mL	Meira et al. (2014)
Annona squamosa	*T. cruzi* *Plasmodium* spp.	IC_{50}: <20 µg/mL	Meira et al. (2014)
Syzygium aromaticum	*Giardia lamblia* *T. cruzi* (epimastigote)	IC_{50}: 134 µg/mL IC_{50}/24 h: 99.5 µg/mL	Machado et al. (2011), Santoro et al. (2007a)
Syzygium cumini (L.)	*L. amazonensis*: Axenic amastigote Intracellular amastigote	IC_{50}: 43.9 µg/mL IC_{50}: 38.1 µg/mL	Rodrigues et al. (2015)
α -Pinene	*L. amazonensis*: Promastigote Axenic amastigote Intracellular amastigote	IC_{50}: 19.7 µg/mL IC_{50}: 16.1 µg/mL IC_{50}: 15.6 µg/mL	Rodrigues et al. (2015)
Calamintha officinalis	*Botrytis cinerea*	10–250 ppm	Bouchra et al. (2003)

TABLE 11.4 *(Continued)*

EO/Compound	Protozoa	Concentration	References
Origanum vulgare L.	*T. cruzi* (epimastigote) *T. cruzi* (trypomastigote)	IC_{50}/24 h: 175 μg/mL IC_{50}/24 h: 115 μg/mL	Santoro et al. (2007b)
Ocimum basilicum	*Giardia lamblia* *T. cruzi (epimastigote)*	2 mg/mL IC_{50}/24 h: 102 μg/mL	de Almeida et al. (2007), Santoro et al. (2007b)
Eugenol (*S. aromaticum*)	*Giardia lamblia*	IC_{50}: 101 μg/mL	Machado et al. (2011)
Thymus vulgaris L.	*T. cruzi* (epimastigote) *T. cruzi* (trypomastigote)	IC_{50}/24 h: 77 μg/mL IC_{50}/24 h: 38 μg/mL	Santoro et al. (2007b)
Lavandula angustifolia	*Giardia duodenalis* *Trichomonas vaginalis* *Hexamita inflata*	≤1% (vol/vol)	Moon et al. (2006)
Achillea millefolium L.	*T. cruzi* (epimastigote)	IC_{50}/24 h: 145.5 μg/mL	Santoro et al. (2007a)

IC_{50}, concentration that inhibits 50% parasite growth.

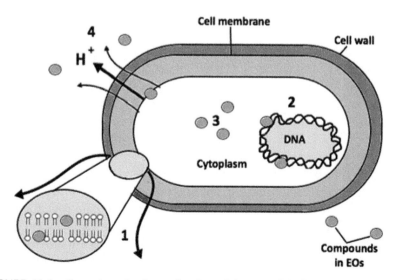

FIGURE 11.1 General mechanisms of action of the essential oils and their compounds against bacteria. Disruption of cell membrane and leakage of cell contents (1), interaction with DNA (2), disruption of cellular components, such as proteins (3), and depletion of the proton motive force (4).

In eukaryotic cells, EOs can act in mitochondrial membranes and decrease the membrane potential, causing depolarization of the mitochondrial membranes. This results in altered ionic Ca^{2+} cycling, reduction of the pH gradient, and disruption of the proton motive force, as in bacteria. Also, the fluidity of membranes is disrupted, facilitating the leakage of ions, radicals, cytochrome C, and proteins. Finally, the permeation of both outer and inner mitochondrial membranes leads to cell death (Bakkali et al., 2008).

When an EO damages the cellular and organelle membranes, it provokes *cytotoxicity*. When this occurs, EOs can act as prooxidants on different molecules, such as proteins and nucleic acids, such as DNA, with posterior production of reactive oxygen species. In other cases, however, these oils may be able to pass through cellular membranes without resulting in any sort of damage to the organisms until they are exposed to activating light. When this occurs, different molecules undergo radical reactions, as they are excited, producing oxygen singlet. This phenomenon is known as *phototoxicity* and, as a consequence, the cell may suffer damage in different molecules including DNA, lipids, and proteins. The fact that an EO may be cytotoxic or phototoxic is primarily due to the composition of molecules of the oil (Bakkali et al., 2008).

The chemical composition, proportions, and interactions among the components of an EO are found to reflect quite well the biophysical and biological features of the EOs from which they were isolated. Thus, there are different tendencies according to the behavior of the combination of EO or components. When a combined effect equals to the sum of other individual's activity, an *additive* effect is obtained. However, sometimes the activity of one or more compounds is less than the combination individually; this effect is known as *antagonism*. In the case of *synergism*, the activity of combined components or substances is greater than sum of individual effects (Burt, 2004; Ipek et al., 2005). There are different mechanisms in which these effects can be established. For example, in a previous work (Pei et al., 2009), it was hypothesized that the synergism shown by the combination of eugenol/carvacrol and thymol/eugenol against *E. coli* might result from the ability of carvacrol and thymol to disintegrate the bacterial outer membrane, enabling eugenol to penetrate into the cytoplasm and react with proteins. Also, other studies have shown that the synergistic effect of the combinations of EOs of oregano/basil against *E. coli* and oregano/perilla against *Saccharomyces cerevisiae* was able to disrupt the integrity of the cell membrane (Lv et al., 2011). Thus, the combination of different components with synergistic activity will diminish the concentration needed to induce the same effect compared with the sum of the components alone (Bassolé and Juliani, 2012).

Abroad screening of the antimicrobial activity of EOs, and their general characteristics, against different microorganisms has been presented. Now, a discussion about the general mechanism of action of the EOs and their components in the various microbes follows.

Because of the variability in structure and functional groups, the antibacterial activity of EOs is not the result of one single specific mechanism, but several acting to target different cellular components. The hydrophobicity shown by the EOs enables them to act in different locations through the bacterial wall, having consequences such as degradation of the cell wall, depletion of the proton motive force, damage to membrane proteins, leakage of cell contents, damage to cytoplasmic membrane, coagulation of cytoplasm, and disruption of DNA transcription and affection on gene expression (Burt, 2004; Bakkali et al., 2008; Oussalah et al., 2006).

Terpenes have shown the ability to pass through the cell wall in bacteria, affecting the lipid structure and leading to denaturalization of molecules, such as proteins, and the destruction of cell membrane. Consequently, the leakage of cytoplasmic content leads to cell death (Fisher and Philips, 2008). Also, it has been observed that the decrease in pH that occurs in bacteria is due to this cell membrane disruption and, thus, affecting the overall control of cellular metabolic pathways such as DNA transcription, protein synthesis, and enzyme activity (Oussalah et al., 2006). For example, an important compound that has shown to have important antibacterial activity and is present in a wide variety of plants is carvacrol, which is a lipophilic monoterpenoid phenol and, because of the free hydroxyl group in the chemical skeleton as well as delocalized electron system, affects the cytoplasmic membrane of bacteria to exert its antibacterial effects (Nabavi et al., 2015; De Sousa et al., 2012).

Important efforts have been made in elucidating if EOs have more antimicrobial activity against Gram-positive or Gram-negative bacteria. Prabuseenivasan et al. (2006) tested the antibacterial activity of 19 oils against Gram-negative (*E. coli*, *Klebsiella pneumonia*, *Pseudomonas aeruginosa*, and *Proteus vulgaris*) and Gram-positive bacteria (*Bacillus subtilis* and *S. aureus*) and concluded that, overall, Gram-positive bacteria were more resistant to the EOs than Gram-negative bacteria. However, some oils appeared more active exerting a greater inhibitory activity against Gram-positive bacteria. Oussalah et al. (2007) concluded that Gram-positive bacteria are more susceptible to the antimicrobial activity of EO compared with

Gram-negative bacteria with some exceptions, such as *Listeria monocytogenes*, as the cell wall of Gram-negative bacteria is composed of lipopolysaccharides, which prevents accumulation of the oils on the cell membrane, blocking the penetration of hydrophobic EOs into target cell membrane (Al-Bayati, 2008).

In some cases, EOs alone are not enough to display greater antimicrobial activity and the combination of two or more results in additive or synergistic effects. For example, a previous study demonstrated that the combination of EOs, *Thymus vulgaris* and *Pimpinella anisum* enhanced antimicrobial activity, having additive effects against *P. aeruginosa*, which showed prior resistance to the individual oils (Al-Bayati, 2008).

The components of EOs also show antibacterial activity. Eugenol, a major component of clove and oregano oil, has shown to be effective against methicillin-resistant and methicillin-sensitive *S. aureus*, being able to eradicate pre-established biofilms and decrease the expression of genes related to biofilm and enterotoxin. Eugenol can accomplish this effect because of its lipophilic characteristics, which enable the compound to pass through the cell wall and cytoplasmic membrane and permeate the cell membrane, eventually resulting in cell lysis (Yadav et al., 2015); also, other reports proved that leakage of intracellular components also occurs in Gram-negative bacteria, such as *Salmonella typhi*, as a result of the possible formation of pores in the membrane (Devi et al., 2010). Citral, an acyclic unsaturated monoterpene aldehyde, is also present in the EOs of many plants (lime, lemons, and lemongrass to cite some) and has been shown to be able to inhibit and cause sublethal damage in membranes against *E. coli*, requiring important amounts of energy to repair such damages (Somolinos et al., 2010).

Although the EOs and their compounds show potent antimicrobial activity, bacteria can have resistance to the antibacterial effect of these compounds. *P. aeruginosa* has intrinsic resistance to several antimicrobials, including the EO of *Melaleuca alternifolia*. This attribute has been associated with its outer membrane, which protects the bacteria from accumulation of compounds at toxic levels in cytoplasmic membranes; also, there might be other mechanisms conferring resistance to these compounds, such as active efflux of tea tree oil components, inhibition of porin production, and modifications to membrane structure (Mann et al., 2000). In addition, the compounds of EOs can influence bacteria to respond and adapt to its presence. According to Ultee et al. (2000), *Bacillus cereus* in the presence of sublethal concentrations of carvacrol in growth media can adapt to its presence, gaining resistance by changing the fatty acids ratio and composition of the bacterial membrane. Thus, EOs not applied properly may, in fact, promote the generation of resistant bacteria.

EOs also show potential antimicrobial activity in microbial eukaryotic organisms. The antifungal activity of EOs has also been studied and reported (Bansod and Rai, 2008; Moghtader, 2012). The mechanisms underlying this activity are dependent on the nature of the oil components. For example, thymol and eugenol show lipophilic nature, which enables them to diffuse between fatty acyl chains of the cell membrane, affecting its permeability and fluidity and, as a consequence, having impairment of ergosterol, which is the main sterol in fungi (Hyldgaard et al., 2012; Dupont et al., 2012). This mode of action is present in different EOs. For example, it was demonstrated that the EO of *Ocimum sanctum* and two of its main components, methyl chavicol and linalool, alter the integrity of the cytoplasmic membrane in *Candida*, disrupting structural and functional capacity of the lipid bilayer (Khan et al., 2010). Indeed, in previous studies, Manganyi et al. (2015) have shown that the antimicrobial activity of the EO of thyme and clove, capable of inhibiting the growth and biofilm formation of *Fusarium* isolates, could be associated with their major constituents, which are thymol and eugenol, respectively. Interestingly, an isolate was more resistant to the EO than the others, showing that some strains of the same species may have higher resistance or susceptibility to certain EOs or compounds than others. The mechanisms responsible for resistance against EOs are variable and complex such as altering the lipid layer of the cell membrane, resulting in impermeability to lipophilic compounds. The effect of EOs on the DNA of fungi has not been completely understood. Previous studies showed that the EO of coriander (*Coriandrum sativum* L.) does not interfere with DNA synthesis; however, when cultures were incubated with the oil, there was an increase in the DNA synthesis in response to the cell damage to repair the affected functions (Silva et al., 2011).

A fact that is not completely considered in this field is that the methodologies for obtaining the EOs and their application influence their antimicrobial effect. For example, larger phenolic compounds, such as thymol and eugenol (thyme, cinnamon, and clove), had best effect when applied directly to medium, whereas smaller compounds, such as allyl isothiocyanate and citral (mustard and lemongrass), showed to be more efficient when added as volatiles (Suhr and Nielsen, 2003). Thus, the antifungal activity of the EO and their compounds may vary depending of these circumstances.

In some cases, EOs might not show high antifungal activity; instead, a combination with other compounds can result in synergistic effects to enhance the antimicrobial activity of these oils. For example, ketoconazole can show higher antifungal effects in combination with the EO of *Pelargonium graveolens* or its main components against *Trichophyton* spp. (Shin and Lim, 2004). Another example is that anfotericin B shows synergism with the coriander EO (Silva et al., 2011).

EOs display antiviral activity for enveloped DNA and RNA viruses. Interestingly, the EOs do not significantly affect the nonenveloped viruses. Optimally, the most desirable characteristic in antiviral drugs is the capacity of a substance to act on specific steps of viral biosynthesis, such as the replication cycle, and should act at low concentrations without influencing the host cell machinery. Thus, important efforts have been made to elucidate if essential compounds and their components can provide such valuable ideal drugs (Reichling et al., 2009).

Past studies have concluded that the main mechanism of antiviral action against enveloped virus is the direct interaction of essential compounds and their components with the viral envelope, diminishing their adsorption in the host cell (Fig. 11.2). For example, Koch et al. (2008) found that EO from hyssop, anise, thyme, ginger, sandalwood, and chamomile affects HSV-2 before adsorption, suggesting that the oils interfere with the envelope structures of the virus or might mask viral compounds, which are necessary for adsorption or entry into host cells; Oh and Chung (2014) also suggested that

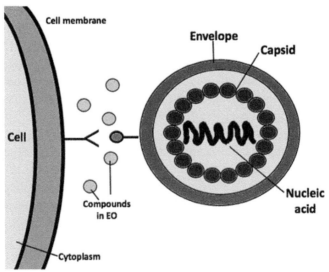

FIGURE 11.2 General mechanism of action of the essential oils and their compounds against enveloped virus. EOs and their compounds interact with the viral envelope, masking the critical sites to achieve adsorption in cells.

the EO of *Zanthoxylum schinifolium* affects FCV-F9 and murine norovirus (MNV)-1 by blocking virus attachment to the host cell. In other studies, De Logu et al. (2000) concluded that EOs from *Santolina insularis* inactivate the adsorption of HSV-1 and HSV-2, being ineffective after their attachment; Astani et al. (2010) and Schuhmacher et al. (2003) also support this mechanism of action.

In some instances, lacking the viral envelope may prevent the inactivation of EOs as reported by Cermelli et al. (2008), who observed that adenovirus, a nonenveloped virus, was not affected by the EO of eucalyptus due to lack of a viral envelope. The capsid in such nonenveloped viruses serves to protect the integrity of the viral nucleic acid and to initiate infection by adsorbing to the host cell (Cliver, 2009). The lemongrass oil and one of its major compounds, citral, showed the ability to leave the capsid and genome intact of the MNV but instead display their antiviral activity by coating the capsid and preventing specific adsorption, lowering the efficacy of infection (Gilling et al., 2014a). In addition, the authors reported that the allspice oil appeared to cause the viral capsid to lose its integrity in the MNV, resulting in exposure of the viral genome. The EO was shown to act upon the viral RNA, reducing the successful rate of infection due to the affected integrity of the viral genome. The EO of oregano and the compound carvacrol also share similarities in their mode of action compared with the EO of allspice (Gilling et al., 2014b) (Fig. 11.3).

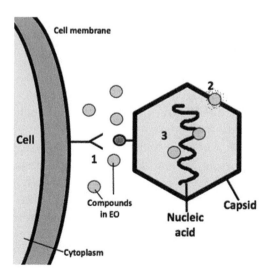

FIGURE 11.3   General mechanisms of action of the essential oils and their compounds are against nonenveloped virus; masking the critical sites to achieve adsorption in cells (1), loss of the capsids integrity (2), and interactions with the nucleic acids (3).

There have been efforts to prove that EOs can also target eukaryotic parasitic microbes other than fungi. These organisms are complex as they have, generally, different stages depending on their life cycle. In *Leishmania*, the IC_{50} (concentration that inhibits 50% of parasite growth) of different compounds was shown to be similar against promastigotes and amastigotes, suggesting that such molecules may target common molecules or pathways (Pastor et al., 2015); however, other reports have shown that intracellular amastigotes of *Trypanosoma cruzi* are as susceptible as bloodstream trypomastigotes, while epimastigote forms are more resistant (Santoro et al., 2007a). Also, the resistance and susceptibility of parasites may vary depending on the organisms. An example of this is the fact that *Trypanosoma* parasites are more resistant to the oil action than *Leishmania* (Santoro et al., 2007b). Nevertheless, the mode of action of the EOs against these organisms, as in bacteria, fungi, and virus, is dependent on the oil's characteristics. Because of the lipophilic manner, the compounds can pass through cytoplasmic membranes, affecting their composition and causing a deformation in its structure and functionality, rendering changes in permeability and promoting cell lysis. In addition, these components can disturb different metabolic pathways and organelles by interfering, for example, with the potential of the mitochondrial membranes, leading to a release of free radicals that can act on the DNA and lead to apoptosis or necrosis of the cells (de Almeida et al., 2007; Rodrigues et al., 2015).

Despite the activity of the EOs related to its major compounds, in some cases, the presence of other constituents renders an increase in the oil's activity. For example, it has been reported that the EO of *Syzygium aromaticum* induces changes in the viability and adherence of *Giardia lamblia* trophozoites; in contrast, eugenol inhibits only adhesion (Machado et al., 2011). The mode of action that results in inhibition of adherence involves the loss of cell polarity with consequently cytoskeleton alterations. Also, there are other reports that show synergistic activity of compounds, such as combinations of ascaridol–carvacrol against *Leishmania* (Pastor et al., 2015); Santoro et al. (2007b) also found better activity of the EO of clove compared with eugenol against *T. cruzi*. However, in some cases, the activity of a major component alone may show greater antiparasitic activity, such as α-pinene, which was shown to be more effective than the EO *Syzygium cumini* against *Leishmania* promastigotes (Rodrigues et al., 2015).

11.3.2 APPLICATION OF ESSENTIAL OILS IN INDUSTRY

The increasing research in EOs has highlighted multiple benefits of their commercial use to address current global trends. Several isolated compounds have proven to be GRAS and approved for use for different purposes. Eugenol, a phenolic compound present in several plants, has been approved by the European Union to use as a flavoring ingredient in foodstuffs (Yadav et al., 2015). Another well-characterized compound used as a food additive is carvacrol, which can prevent the bacterial growth and contamination in food (Nabavi et al., 2015). However, because of their volatility, EOs demand special storage conditions, such as avoiding exposure to light and the use of airtight reservoirs (Lang and Buchbauer, 2012).

To date, there are alternatives to EOs and their major compounds are available commercially as sanitizers or disinfectants. These products are derived from and based on the EOs or compounds of several plants and have gained acceptance due to their "natural" origin rather than synthetic. In addition, their standardized qualitative and quantitative characteristics with less problematic storage needs and suitability for different purposes, such as additives for food, disinfectants of surfaces, washing materials, and many other applications, make them attractive. Furthermore, the efficacy of some disinfectants, sanitizers, and hydrosols, based on EOs and their compounds, has been tested with promising results against different microorganisms, including the bacteria *Salmonella*, *E. coli* O157:H7, and *L. monocytogenes* (Valeriano et al., 2012; Upadhyay et al., 2012; Tornuk et al., 2011; Oliveira et al., 2010; García-Heredia et al., 2013). The EOs have also shown potential as pesticides (Isman et al., 2011), bioherbicide (Fisher and Phillips, 2008), larvadicidal, molluscicidal, and nematicidal (Sousa et al., 2015).

As EOs show important antimicrobial activity, pharmaceutical companies have also considered development of commercial products based on EOs and their components. Different studies have shown the potential of EOs for therapeutics to treat diabetes, as analgesics and anti-inflammatory remedies, as potent antitumors, and for other applications (Chung et al., 2010; Silva et al., 2003; Menichini et al., 2009). However, toxicity studies are required because, when overdosed, essential oils may be toxic and can show adverse effects on human cells (Nabavi et al., 2015).

There is increased research interest for different strategies to apply EOs and their compounds in the food industry in the near future. One of these technologies is the use of edible films. Edible films are physical, selective barriers and can serve as carriers to release antimicrobials onto food surface and thus controlling microbial growth; furthermore, these films can regulate the oxygen uptake, moisture transfer, loss of volatile aromas, and have shown to improve shelf-life and food quality (Avila-Sosa et al., 2012; Rojas-Graü et al., 2007). Different studies show that these edible films, with incorporation of different EOs and compounds, succeed in preserving food and act against microorganisms such as *Pseudomonas fluorescens, E. coli* O157:H7, *Aspergillus niger, Penicillium digitatum, Serratia marcescens, Shewanella putrefaciens, Achromobacter denitrificans, Photobacterium phosphoreum, Alcaligenes faecalis, Aeromonas hydrophila, Listeria innocua*, and *Lactobacillus acidophilus* (Avila-Sosa et al., 2012; Rojas-Graü et al., 2007; Gómez-Estaca et al., 2010; Ruiz-Navajas et al., 2013). Another strategy studied nanoencapsulation of bioactive compounds. This comprises an approach to increase the stability of active substances and protect them from interactions with other components and increase bioactivity (Donsì et al., 2011). In food areas, encapsulation can increase the concentration of the bioactive compounds where microorganisms are preferably located, such as water-rich phases or liquid–solid interfaces. Different studies have demonstrated the efficacy of nanocapsules to carry EO or compounds and act effectively against *E. coli, Lactobacillus delbrueckii*, and *S. cerevisiae* (Donsì et al., 2011; Lv et al., 2014). As noted, more studies are required to standardize the proper use of this technology in future.

11.4 CHALLENGES

As mentioned previously, these natural compounds are secondary metabolites produced with varying presence, concentration, and chemical composition across growth conditions, age of or source from the plant, etc., not to mention the variability that comes during extraction and collection assays (Prakash et al., 2015; Calo et al., 2015). Therefore, these plant sources of antimicrobials are often active due to a complex mixture of compounds. While this does have some social and environmental benefits, in practical applications, this provides a great challenge. For large-scale production of food products, chemical composition of the EOs or plant extracts should be ascertained via gas chromatography prior to their use and a consistent chemical composition is required to ensure consistent antimicrobial activity. Two options for addressing this include using pure compounds (less economical) or selecting an agent with a known main antimicrobial compound (Regnier et al., 2012).

Furthermore, while there have been few in vivo studies, it has been reported that EOs and plant extracts are more effective in culture medium, meaning higher concentrations are required to achieve the same antimicrobial effect in food formulations and matrices compared to those of in vitro. Reported differences include 2-fold in semiskim milk, 10-fold in pork sausage, 50-fold in soup, and 25–100-fold in soft cheese (Gyawali et al., 2015). This decreased efficacy within food applications is likely due to interactions with food components, such as lipids, protein, carbohydrates, water activity, pH, and enzymes (Prakash et al., 2015; Calo et al., 2015). For application to fruits and vegetables, the main components of concern for interaction with antimicrobial agents would be carbohydrates and pH. This effect of the complex food matrix on antimicrobial activity is not unique to EOs and plant extracts, but their poor water solubility does contribute to their poor incorporation into food products. In general, high water, salt, and simple sugar content, as well as low pH and low temperatures, will all favor antimicrobial activity of EOs. On the other hand, high fat, protein, and starch content will either encapsulate bacteria or absorb the EO, thus hindering its antimicrobial activity (Perricone et al., 2015). In addition, their high volatility complicates the delivery method, which is why they are normally suggested for use as fumigants in the vapor phase (Prakash et al., 2015).

These higher required concentrations would likely alter sensorial and organoleptic properties of the food such that flavor, odor, and texture would no longer be desirable to consumers. Furthermore, these levels could mask consumer perception of spoilage and leave them with an undesirable eating experience. Alternatively, these higher concentrations would no longer be cost-effective for producers. For example, it has been reported that the average cost of EO is six times that of a chemical fungicide necessary to treat the same quantity of citrus (Sivakumar and Bautista-Banos, 2014). For these reasons, current uses in food products are at inhibitory levels instead of microbicidal levels (Calo et al., 2015). While economy-of-scale production and utilization of EOs with synergistic effects can bring this cost down, it is a main consideration for producers and distributors. Existing GRAS status and proven nontoxicity means that legislative and regulatory issues are not major challenges for plant sources of antimicrobials; rather, the lack of collaboration between research and industry has perpetuated this lag time for significant industrial adoption in fruits and vegetables.

Given the limitations and challenges mentioned above, the future direction of this field of postharvest antimicrobial treatment of fruits and vegetables is toward improving the delivery method with in vivo experimentation. There is still much work in the area of synergistic combinations.

Some other proposed application methods involve combination with hurdle approaches, such as with organic acids or incorporation in edible films and waxes (Calo et al., 2015); with slow-release gels or encapsulation; smart packaging designed with antimicrobials incorporated (Regnier et al., 2012; Gyawali et al., 2015; Vergis et al., 2015); and even application as crop-protecting products for preharvest treatment (Regnier et al., 2012). As modern processing and ingredient technologies continue to advance, the efficacy of these antimicrobials may find more practical applications in fruits and vegetables.

Lastly, there have been many in vitro studies that test compounds and combinations of compounds against an array of microorganisms; however, prediction modeling may allow for better guidance on in vivo experimental design or understanding of the antimicrobial efficacy. In a study by Evrendilek, in vitro datasets were fit with multiple linear and nonlinear models to predict the most consistent high inhibitory and low inhibitory compounds for different bacteria (Perricone et al., 2015). All such efforts to better characterize the compounds conferring consistent antimicrobial effects will help with development of more directed commercial usage for preservation of the products and their sensory appeal.

11.5 CONCLUSIONS

Overall, research on plant sources of antimicrobial agents has been beneficial to many sectors of society, including produce industry. EOs and plant extracts have demonstrated broad in vitro antimicrobial activity against common postharvest pests in agriculture. Characterization of the main active compounds provides "structural motifs" for development of synthetic versions with reduced complexity and improved function for commercialization. Due to recent environmental and consumer concerns in the produce industry, the desire for alternatives to these natural compounds has led to experimentation with EOs and plant extracts on fruits and vegetables. The main advantages of using some natural antimicrobials could include biodegradation, lack of residues on food products, consumer acceptance, and activity retention in vapor phase. However, some limitations to overcome include compositional complexity, in vivo inconsistency, physicochemical interactions, cost, sensory, and organoleptic degradation. Therefore, the future direction of research will be aimed at improving their mode of delivery and consistent activity so as to minimize required effective concentrations and cost.

KEYWORDS

- antimicrobial
- essential oil
- plant extract
- mycotoxin
- preservation
- postharvest
- natural

REFERENCES

Abreu, A. C.; McBain, A. J.; Simoes, M. Plants as Sources of New Antimicrobials and Resistance-Modifying Agents. *Nat. Prod. Rep.* **2012**, *29* (9), 1007–1021.

Akin, M.; Aktumsek, A.; Nostro, A. Antibacterial Activity and Composition of the Essential Oils of Eucalyptus camaldulensis Dehn. and Myrtus communis L. Growing in Northern Cyprus. African J. Biotechnol. **2010**, *9*, 531–535.

Al-Bayati, F. A. Isolation and Identification of Antimicrobial Compound from *Mentha longifolia* L. Leaves Grown Wild in Iraq. *Ann. Clin. Microb. Anti.* **2009**, *8*, 20.

Al-Bayati, F. A. Synergistic Antibacterial Activity between *Thymus vulgaris* and *Pimpinella anisum* Essential Oils and Methanol Extracts. *J. Ethnopharmacol.* **2008**, *116* (3), 403–406.

Alghazeer, R.; Saeed, S.; Howell, N. K. Aldehyde Formation in Frozen Mackerel (*Scomber scombrus*) in the Presence and Absence of Instant Green Tea. *Food Chem.* **2008**, *108* (3), 801–810.

Alvarez, M. V.; Moreira, Roura, M. R.; S. I.; Ayala-Zavala, J. F.; González-Aguilar, G. A. Using Natural Antimicrobials to Enhance the Safety and Quality of Fresh and Processed Fruits and Vegetables: Types of Antimicrobials. In *Handbook of Natural Antimicrobials for Food Safety and Quality*; Taylor, M., Ed.; Woodhead Publishing: Cambridge, UK, 2015; pp 287–233.

Ananda Baskaran, S.; Kazmer, G. W.; Hinckley, L.; Andrew, S. M.; Venkitanarayanan, K. Antibacterial Effect of Plant-Derived Antimicrobials on Major Bacterial Mastitis Pathogens In Vitro. *J. Dairy Sci.* **2009**, *92* (4), 1423–1429.

Astani, A.; Reichling, J.; Schnitzler, P. Comparative Study on the Antiviral Activity of Selected Monoterpenes Derived from Essential Oils. *Phytother. Res.* **2010**, *24* (5), 673–679.

Avila-Sosa, R.; Palou, E.; Munguía, M. T. J.; Nevárez-Moorillón, G. V.; Cruz, A. R. N.; López-Malo, A. Antifungal Activity by Vapor Contact of Essential Oils Added to Amaranth, Chitosan, or Starch Edible Films. *Int. J. Food Microbiol.* **2012**, *153* (1), 66–72.

Bag, A.; Chattopadhyay, R. R. Evaluation of Synergistic Antibacterial and Antioxidant Efficacy of Essential Oils of Spices and Herbs in Combination. *PLoS ONE* **2015,** *10* (7), 131–321.

Bakkali, F.; Averbeck, S.; Averbeck, D.; Idaomar, M. Biological Effects of Essential Oils—A Review. *Food Chem. Toxicol.* **2008,** *46* (2), 446–475.

Bansod, S.; Rai, M. Antifungal Activity of Essential Oils from Indian Medicinal Plants against Human Pathogenic *Aspergillus fumigatus* and *A. niger. World J. Med. Sci.* **2008,** *3* (2), 81–88.

Bassolé, I. H. N.; Juliani, H. R. Essential Oils in Combination and their Antimicrobial Properties. *Molecules* **2012,** *17* (4), 3989–4006.

Belguith, H.; Kthiri, F.; Ben Ammar, A.; Jaafoura, H.; Ben Hamida, J.; Landoulsi, A. Morphological and Biochemical Changes of *Salmonella hadar* Exposed to Aqueous Garlic Extract. *Int. J. Morphol.* **2009,** *27* (3), 705–713.

Ben Arfa, A.; Combes, S.; Preziosi-Belloy, L.; Gontard, N.; Chalier, P. Antimicrobial Activity of Carvacrol Related to Its Chemical Structure. *Lett. Appl. Microbiol.* **2006,** *43* (2), 149–154.

Bisha, B.; Weinsetel, N.; Brehm-Stecher, B. F.; Mendonca, A. Antilisterial Effects of Gravinol-S Grape Seed Extract at Low Levels in Aqueous Media and Its Potential Application as a Produce Wash. *J. Food Protect.* **2010,** *73* (2), 266–273.

Bouaziz, M.; Yangui, T.; Sayadi, S.; Dhouib, A. Disinfectant Properties of Essential Oils from *Salvia officinalis* L. Cultivated in Tunisia. *Food Chem. Toxicol.* **2009,** *47*, 2755–2760.

Bouchra, C.; Achouri, M.; Idrissi Hassani, L. M.; Hmamouchi, M. Chemical Composition and Antifungal Activity of Essential Oils of Seven Moroccan *Labiatae* against *Botrytis cinerea* Pers.: Fr. *J. Ethnopharmacol.* **2003,** *89* (1), 165–169.

Braga, L. C.; Leite, A. A. M.; Xavier, K. G. S.; Takahashi, J. A.; Bemquerer, M. P.; Chartone-Souza, E.; Nascimento, A. M. A. Synergic Interaction between Pomegranate Extract and Antibiotics against *Staphylococcus aureus. Can. J. Microbiol.* **2005,** *51* (7), 541–547.

Brannan, R. G. Effect of Grape Seed Extract on Descriptive Sensory Analysis of Ground Chicken during Refrigerated Storage. *Meat Sci.* **2009,** *81* (4), 589–595.

Burt, S. Essential Oils: Their Antibacterial Properties and Potential Applications in Foods—A Review. *Int. J. Food Microbiol.* **2004,** *94* (3), 223–253.

Burt, S. A.; van der Zee, R.; Koets, A. P.; De Graaff, A. M.; van Knapen, F.; Gaastra, W.; Haagsman, H. P.; Veldhuizen, E. J. A. Carvacrol Induces Heat Shock Protein 60 and Inhibits Synthesis of Flagellin in *Escherichia coli* O157:H7. *Appl. Environ. Microb.* **2007,** *73* (14), 4484–4490.

Calo, J. R.; Crandall, P. G.; O'Bryan, C. A.; Ricke, S. C. Essential Oils as Antimicrobials in Food Systems–A Review. *Food Control* **2015,** *54*, 111–119.

Careaga, M.; Fernández, E.; Dorantes, L.; Mota, L.; Jaramillo, M. E.; Hernandez-Sanchez, H. Antibacterial Activity of Capsicum Extract against *Salmonella typhimurium* and *Pseudomonas aeruginosa* Inoculated in Raw Beef Meat. *Int. J. Food Microbiol.* **2003,** *83* (3), 331–335.

Carson, C. F.; Mee, B. J.; Riley, T. V. Mechanism of Action of *Melaleuca alternifolia* (Tea Tree) Oil on *Staphylococcus aureus* Determined by Time-Kill, Lysis, Leakage, and Salt Tolerance Assays and Electron Microscopy. *Antimicrob. Agents Ch.* **2002,** *46* (6), 1914–1920.

Castillo, S.; Heredia, N.; Arechiga-Carvajal, E.; García, S. Citrus Extracts as Inhibitors of Quorum Sensing, Biofilm Formation and Motility of *Campylobacter jejuni. Food Biotechnol.* **2014,** *28* (2), 106–122.

Castillo, S.; Heredia, N.; García, S. 2(5H)-Furanone, Epigallocatechin Gallate, and a Citric-Based Disinfectant Disturb Quorum-Sensing Activity and Reduce Motility and Biofilm Formation of *Campylobacter jejuni. Fol. Microbiol.* **2015,** *60* (1), 89–95.

Celestino, A.; Jaime, B.; Luévano, R.; Solís, L.; García, S.; Heredia, N. Reduction of Food-borne Pathogens in Parsley by an Improved Formulation Containing Lime and Oregano Extracts. *J. Food Agric. Environ.* **2014,** *12* (3 & 4), 6–11.

Cermelli, C.; Fabio, A.; Fabio, G.; Quaglio, P. Effect of Eucalyptus Essential Oil on Respiratory Bacteria and Viruses. *Curr. Microbiol.* **2008,** *56* (1), 89–92.

Chang, H. T.; Cheng, Y. H.; Wu, C. L.; Chang, S. T.; Chang, T. T.; Su, Y. C. Antifungal Activity of Essential Oil and Its Constituents from *Calocedrus macrolepis* var. Formosana Florin Leaf Against Plant Pathogenic Fungi. *Biores. Technol.* **2008,** *99*, 6266–6270.

Chinsembu, K. C. Plants and Other Natural Products Used in the Management of Oral Infections and Improvement of Oral Health. *Acta Trop.* **2016,** *154*, 6–18.

Chung, M. J.; Cho, S. Y.; Bhuiyan, M. J. H.; Kim, K. H.; Lee, S. J. Anti-diabetic Effects of Lemon Balm (*Melissa officinalis*) Essential Oil on Glucose- and Lipid-Regulating Enzymes in Type 2 Diabetic Mice. *Br. J. Nutr.* **2010,** *104* (02), 180–188.

Clatworthy, A. E.; Pierson, E.; Hung, D. T. Targeting Virulence: A New Paradigm for Antimicrobial Therapy. *Nat. Chem. Biol.* **2007,** *3* (9), 541–548.

Cliver, D. O. Capsid and Infectivity in Virus Detection. *Food Environ. Virol.* **2009,** *1* (3–4), 123–128.

Davidson, P. M. Chemical Preservatives and Natural Antimicrobial Compounds in Food Microbiology: Fundamentals and Frontiers; Doyle, M. P., Beuchat, L. R., Montville, T. J., Eds.; American Society for Microbiology: Washington, DC, 1997; pp 520–556.

de Almeida, I.; Alviano, D. S.; Vieira, D. P.; et al. Antigiardial Activity of *Ocimum basilicum* Essential Oil. *Parasitol. Res.* **2007,** *101* (2), 443–452.

De Logu, A.; Loy, G.; Pellerano, M. L.; Bonsignore, L.; Schivo, M. L. Inactivation of HSV-1 and HSV-2 and Prevention of Cell-to-Cell Virus Spread by *Santolina insularis* Essential Oil. *Antivir. Res.* **2000,** *48* (3), 177–185.

De Sousa, J. P.; De Araújo Torres, R.; De Azerêdo, G. A.; Figueiredo, R. C. B. Q.; Da Silva Vasconcelos, M. A.; De Souza, E. L. Carvacrol and 1,8-Cineole Alone or in Combination at Sublethal Concentrations Induce Changes in the Cell Morphology and Membrane Permeability of *Pseudomonas fluorescens* in a Vegetable-Based Broth. *Int. J. Food Microbiol.* **2012,** *158* (1), 9–13.

Deans, S. G.; Noble, R. C.; Hiltunen, R.; Wuryani, W.; Pénzes, L. G. Antimicrobial and antioxidant Properties of *Syzygium aromaticum* (L.) Merr. & Perry: Impact upon Bacteria, Fungi and Fatty Acid Levels in Ageing Mice. *Flavour Frag. J.* **1995,** *10* (5), 323–328.

Devi, K. P.; Nisha, S. A.; Sakthivel, R.; Pandian, S. K. Eugenol (an Essential Oil of Clove) Acts as an Antibacterial Agent against *Salmonella typhi* by Disrupting the Cellular Membrane. *J. Ethnopharmacol.* **2010,** *130* (1), 107–115.

Donsì, F.; Annunziata, M.; Sessa, M.; Ferrari, G. Nanoencapsulation of Essential Oils to Enhance their Antimicrobial Activity in Foods. *LWT—Food Sci. Technol.* **2011,** *44* (9), 1908–1914.

Dung, N. T.; Kim, J. M.; Kang, S. C. Chemical Composition, Antimicrobial and Antioxidant Activities of the Essential Oil and the Ethanol Extract of *Cleistocalyx operculatus* (roxb.) Merr and Perry Buds. *Food Chem. Toxicol.* **2008,** *46*, 3632–3639.

Dupont, S.; Lemetais, G.; Ferreira, T.; Cayot, P.; Gervais, P.; Beney, L. Ergosterol Biosynthesis: A Fungal Pathway for Life on Land? *Evolution* **2012,** *66* (9), 2961–2968.

El Moussaoui, N.; Sanchez, G.; El Ouardy, K.; Idaomar, M.; Mansour, A. I.; Abrini, J.; Aznar, R. Antibacterial and Antiviral Activities of Essential Oils of Northern Moroccan Plants. *Br. Biotechnol. J.* **2013**, *3* (3), 318–331.

Elizaquível, P.; Azizkhani, M.; Aznar, R.; Sanchez, G. The Effect of Essential Oils on Norovirus Surrogates. *Food Control* **2013**, *32*, 275–278.

Erkmen, O; Özcan, M. M. Antimicrobial Effects of Turkish Propolis, Pollen, and Laurel on Spoilage and Pathogenic Food-Related Microorganisms. *J. Med. Food* **2008**, *11*, 587–592.

Fisher, K.; Phillips, C. Potential Antimicrobial Uses of Essential Oils in Food: Is Citrus the Answer? *Trends Food Sci. Tech.* **2008**, *19* (3), 156–164.

Friedman, M.; Henika, P. R.; Levin, C. E.; Mandrell, R. E. Antibacterial Activities of Plant Essential Oils and their Components against *Escherichia coli* O157:H7 and *Salmonella enterica* in Apple Juice. *J. Agric. Food Chem.* **2004**, *52* (19), 6042–6048.

García, S.; Alarcón, G.; Gómez, M.; Heredia, N. *Haematoxylon brasiletto* Extracts Inhibit Growth, Enterotoxin Production, and Adhesion of *Vibrio cholerae. Food Biotechnol.* **2005**, *19* (1), 15–26.

García, S.; Alarcon, G.; Rodríguez, C.; Heredia, N. Extracts of *Acacia farnesiana* and *Artemisia ludoviciana* Inhibit Growth, Enterotoxin Production and Adhesion of *Vibrio cholerae. World J. Microb. Biot.* **2006**, *22*, 669–674.

García-Heredia, A.; Orue, N.; Heredia, N.; García, S. Efficacy of Citrus-Based Disinfectants to Inhibit Growth, Swarming, and Biofilm Formation of *Salmonella* and Decontaminate Parsley. *J. Food Agric. Environ.* **2013**, *11* (2), 1295–1299.

Garozzo, A.; Timpanaro, R.; Bisignano, B.; Furneri, P. M.; Bisignano, G.; Castro. A. *In vitro* Antiviral Activity of *Melaleuca alternifolia* Essential Oil. *Lett. Appl. Microbiol.* **2009**, *49*, 806–808.

Gilling, D.; Kitajima, M.; Torrey, J.; Bright, K. Antiviral Efficacy and Mechanisms of Action of Oregano Essential Oil and Its Primary Component Carvacrol against Murine Norovirus. *J. Appl. Microbiol.* **2014a**, *116* (5), 1149–1163.

Gilling, D. H.; Kitajima, M.; Torrey, J. R.; Bright, K. R. Mechanisms of Antiviral Action of Plant Antimicrobials against Murine Norovirus. *Appl. Environ. Microb.* **2014b**, *80* (16), 4898–4910.

Gómez-Estaca, J.; de Lacey, A. L.; López-Caballero, M.; Gómez-Guillén, M.; Montero, P. Biodegradable Gelatin–Chitosan Films Incorporated with Essential Oils as Antimicrobial Agents for Fish Preservation. *Food Microbiol.* **2010**, *27* (7), 889–896.

Gursoy, U. K.; Gursoy, M.; Gursoy, O. V.; Cakmakci, L.; Könönen, E.; Uitto, V. J. Antibiofilm Properties of *Satureja hortensis* L. Essential Oil Against Periodontal Pathogens. *Anaerobe* **2009**, *15*, 164–167.

Gyawali, R.; Hayek, S. A.; Ibrahim, S. A. Plant Extracts as Antimicrobials in Food Products: Mechanisms of Action, Extraction Methods, and Applications. In *Handbook of Natural Antimicrobials for Food Safety and Quality*; Taylor, M., Ed.; Woodhead Publishing: Cambridge, UK, 2015; pp 49–68.

Hammer, K. A.; Carson, C. F.; Riley, T. V. Antifungal Effects of *Melaleuca alternifolia* (Tea Tree) Oil and Its Components on *Candida Albicans, Candida Glabrata* and *Saccharomyces cerevisiae. J. Antimicrob. Chem.* **2004**, *53*, 1081–1085.

Hayashi, K.; Kamiya, M.; Hayashi, T. Virucidal Effects of the Steam Distillate from *Houttuynia cordata* and Its Components on HSV-1, Influenza Virus, and HIV. *Planta Med.* **1995**, *61*, 237–241.

Hayek, S. A.; Gyawali, R.; Ibrahim, S. A. Antimicrobial Natural Products. In *Microbial Pathogens and Strategies for Combating Them: Science, Technology and Education*; Méndez-Vilas, A., Ed.; Formatex Research Center: Badajoz, Spain, 2013; pp 910–921.

Heredia, N.; Escobar, M.; Rodríguez-Padilla, C.; García, S. Extracts of *Haematoxylon brasiletto* Inhibit Growth, Verotoxin Production, and Adhesion of Enterohemorrhagic *Escherichia coli* O157:H7 to HeLa Cells. *J. Food Protect.* **2005**, *68* (7), 1346–1351.

Hong, E. J.; Na, K. J.; Choi, I. G.; Choi, K. C.; Jeung, E. B. Antibacterial and Antifungal Effects of Essential Oils from Coniferous Trees. *Biol. Pharm. Bull.* **2004**, *27*, 863–866.

Hussain, A. I.; Anwar, F.; Nigam, P. S.; Ashraf, M.; Gilani, A. H. Seasonal Variation in Content, Chemical Composition and Antimicrobial and Cytotoxic Activities of Essential Oils from Four Mentha Species. *J. Sci. Food Agric.* **2010**, *90* (11), 1827–1836.

Hyldgaard, M.; Mygind, T.; Meyer, R. L. Essential Oils in Food Preservation: Mode of Action, Synergies, and Interactions with Food Matrix Components. *Front Microbiol.* **2012**, *3* (12), 1–24.

Ibrahim, S. Y.; El-Salam, M. M. A. Anti-Dermatophyte Efficacy and Environmental Safety of Some Essential Oils Commercial and *In Vitro* Extracted Pure and Combined Against Four Keratinophilic Pathogenic Fungi. *Environ. Health Prev. Med.* **2015**, *20*, 462.

Ibrahim, S.; Tse, T.; Yang, H.; Fraser, A. Antibacterial Activity of a Crude Chive Extract against *Salmonella* in Culture Medium, Beef Broth and Chicken Broth. *Food Prot. Trends* **2009**, *29* (3), 155–160.

Ipek, E.; Zeytinoglu, H.; Okay, S.; Tuylu, B. A.; Kurkcuoglu, M.; Baser, K. H. C. Genotoxicity and Antigenotoxicity of *Origanum* Oil and Carvacrol Evaluated by Ames *Salmonella*/ Microsomal Test. *Food Chem.* **2005**, *93* (3), 551–556.

Isman, M. B.; Miresmailli, S.; Machial, C. Commercial Opportunities for Pesticides Based on Plant Essential Oils in Agriculture, Industry and Consumer Products. *Phytochem. Rev.* **2011**, *10* (2), 197–204.

Juliao, L. S.; Bizzo, H. R.; Souza, A. M.; Lourenço, M. C.; Silva, P. E. A.; Tavares, E. S.; Rastrelli, L.; Leitão, S. G. Essential Oil from Two *Lantana* Species With Antimycobacterial Activity. *Nat. Prod. Commun.* **2009**, *4*, 1733–1736.

Khan, A.; Ahmad, A.; Akhtar, F.; Yousuf, S.; Xess, I.; Khan, L. A.; Manzoor, N. *Ocimum sanctum* Essential Oil and Its Active Principles Exert their Antifungal Activity by Disrupting Ergosterol Biosynthesis and Membrane Integrity. *Res. Microbiol.* **2010**, *161* (10), 816–823.

Koch, C.; Reichling, J.; Schneele, J.; Schnitzler, P. Inhibitory Effect of Essential Oils against Herpes Simplex Virus Type 2. *Phytomedicine* **2008**, *15* (1), 71–78.

Kosalec, I.; Pepeljnjak, S.; Kuštrak, D. Antifungal Activity of Fluid Extract and Essential Oil from Anise Fruits (*Pimpinella anisum* L., Apiaceae). *Acta Pharma.* **2005**, *55*, 377–385.

Kumar, R.; Kumar, A.; Mishra, A.; Dubey, N. K.; Tripathi, Y. B. Evaluation of *Chenopodium ambrosioides* Oil as a Potential Source of Antifungal, Antiaflatoxigenic and Antioxidant Activity. *Int. J. Food Microbiol.* **2007**, *115*, 159–164.

Lang, G.; Buchbauer, G. A Review on Recent Research Results (2008–2010) on Essential Oils as Antimicrobials and Antifungals. A Review. *Flavour Frag. J.* **2012**, *27* (1), 13–39.

Lavelli, V.; Vantaggi, C.; Corey, M.; Kerr, W. Formulation of a Dry Green Tea-Apple Product: Study on Antioxidant and Color Stability. *J. Food Sci.* **2010**, *75* (2), C184-C190.

Lee, S. B.; Cha, K. H.; Kim, S. N.; Altantsetseg, S.; Shatar, S.; Sarangerel, O.; Nho, C. W. The Antimicrobial Activity of Essential Oil from *Dracocephalum foetidum* against Pathogenic Microorganisms. *The J. Microbiol.* **2007**, *45*, 53–57.

Loizzo, M.; Saab, A. M.; Tundis, R.; Statti, G. A.; Menichini, F.; Lampronti, I.; Gambari, R.; Cinatl, J.; Doerr, H. W. Phytochemical Analysis and *In Vitro* Antiviral Activities of the Essential Oils of Seven Lebanon Species. *Chem. Biodivers.* **2008**, *5*, 461–470.

Lopes-Lutz, D.; Alviano, D. S.; Alviano, C. S.; Kolodziejczyk, P. P. Screening of Chemical Composition, Antimicrobial and Antioxidant Activities of *Artemisia* Essential Oils. *Phytochem.* **2008**, *69*, 1732–1738.

Lu, Y.; Zhao, Y. P.; Wang, Z. C.; Chen, S. Y.; Fu, C. X. Composition and Antimicrobial Activity of The Essential Oil of *Actinidia macrosperma* from China. *Nat. Prod. Res.* **2007**, *21*, 227–233.

Lucera, A.; Costa, C.; Conte, A.; Del Nobile, M. A. Food Applications of Natural Antimicrobial Compounds. *Front. Microbiol.* **2012**, *3*, 287.

Lv, F.; Liang, H.; Yuan, Q.; Li, C. *In vitro* Antimicrobial Effects and Mechanism of Action of Selected Plant Essential Oil Combinations against Four Food-Related Microorganisms. *Food Res. Int.* **2011**, *44* (9), 3057–3064.

Lv, Y.; Yang, F.; Li, X.; Zhang, X.; Abbas, S. Formation of Heat-Resistant Nanocapsules of Jasmine Essential Oil via Gelatin/Gum Arabic Based Complex Coacervation. *Food Hydrocolloid* **2014**, *35*, 305–314.

Machado, M.; Dinis, A.; Salgueiro, L.; Custódio, J. B.; Cavaleiro, C.; Sousa, M. Anti-Giardia Activity of *Syzygium aromaticum* Essential Oil and Eugenol: Effects on Growth, Viability, Adherence and Ultrastructure. *Exp. Parasitol.* **2011**, *127* (4), 732–739.

Mahboubi, M.; Haghi, G. Antimicrobial Activity and Chemical Composition of *Mentha pulegium* L. Essential Oil. *J. Ethnopharmacol.* **2008**, *119*, 325–327.

Manganyi, M.; Regnier, T.; Olivier, E. Antimicrobial Activities of Selected Essential Oils against *Fusarium oxysporum* Isolates and their Biofilms. *S. Afr. J. Bot.* **2015**, *99*, 115–121.

Mann, C.; Cox, S.; Markham, J. The Outer Membrane of *Pseudomonas aeruginosa* NCTC 6749 Contributes to Its Tolerance to the Essential Oil of *Melaleuca alternifolia* (tea tree oil). *Lett. Appl. Microbiol.* **2000**, *30* (4), 294–297.

Meira, C. S.; Guimarães, E. T.; Macedo, T. S.; da Silva, T. B.; Menezes, L. R. A.; Costa, E. V.; Soares, M. B. P. Chemical Composition of Essential Oils from *Annona vepretorum* Mart. and *Annona squamosa* L. (*Annonaceae*) Leaves and their Antimalarial and Trypanocidal Activities. *J. Essent. Oil Res.* **2014**, *27* (2), 160–168.

Menichini, F.; Conforti, F.; Rigano, D.; Formisano, C.; Piozzi, F.; Senatore, F. Phytochemical Composition, Anti-inflammatory and Antitumour Activities of Four *Teucrium* Essential Oils from Greece. *Food Chem.* **2009**, *115* (2), 679–686.

Mildner-Szkudlarz, S.; Zawirska-Wojtasiak, R.; Obuchowski, W.; Gośliński, M. Evaluation of Antioxidant Activity of Green Tea Extract and Its Effect on the Biscuits Lipid Fraction Oxidative Stability. *J. Food Sci.* **2009**, *74* (8), S362–S370.

Mkaddem, M.; Bouajila, J.; Ennajar, M.; Lebrihi, A.; Mathieu, F.; Romdhane, M. Chemical Composition and Antimicrobial and Antioxidant Activities of *Mentha* (*longifolia* L. and *viridis*) Essential Oils. *J. Food Sci.* **2009**, *74*, 358–363.

Moghtader, M. Antifungal Effects of the Essential Oil from *Thymus vulgaris* L. and Comparison with Synthetic Thymol on *Aspergillus niger*. *J. Yeast Fung. Res.* **2012**, *3* (6), 83–88.

Moon, T.; Wilkinson, J. M.; Cavanagh, H. M. A. Antiparasitic Activity of Two *Lavandula* Essential Oils against *Giardia duodenalis*, *Trichomonas vaginalis* and *Hexamita inflata*. *Parasitol. Res.* **2006**, *99*, 722–728.

Morcia, C.; Spini, M.; Malnati, M.; Stanca, A. M.; Terzi, V. Essential Oils and Their Components for the Control of Phytopathogenic Fungi that Affect Plant Health and Agri-food

Quality and Safety. In *Natural Antimicrobials in Food Safety and Quality*; Rai, M., Chikindas, M., Eds.; CAB International: Oxfordshire, UK, 2011; pp 224–241.

Nabavi, S. M.; Marchese, A.; Izadi, M.; Curti, V.; Daglia, M.; Nabavi, S. F. Plants Belonging to the Genus *Thymus* as Antibacterial Agents: From Farm to Pharmacy. *Food Chem.* **2015**, *173*, 339–347.

Nazer, A. I.; Kobilinsky, A.; Tholozan, J. L.; Dubois-Brissonnet, F. Combinations of Food Antimicrobials at Low Levels to Inhibit the Growth of *Salmonella* sv. *Typhimurium*: a Synergistic Effect? *Food Microbiol.* **2005**, *22* (5), 391–398.

Oh, M.; Chung, M. S. Effects of Oils and Essential Oils from Seeds of *Zanthoxylum schinifolium* against Foodborne Viral Surrogates. *J. Evidence-Based Complementary Altern. Med.* **2014**, *2014*, 6 p., Article ID 135797. DOI:10.1155/2014/135797.

Oliveira, M. M. M.; Brugnera, D. F.; Cardoso, M. G.; Alves, E.; Piccoli, R. H. Disinfectant Action of *Cymbopogon* sp. Essential Oils in Different Phases of Biofilm Formation by *Listeria monocytogenes* on Stainless Steel Surface. *Food Control* **2010**, *21* (4), 549–553.

Orue, N.; García, S.; Feng, P.; Heredia, N. Decontamination of *Salmonella*, *Shigella*, and *Escherichia coli* O157:H7 from Leafy Green Vegetables Using Edible Plant Extracts. *J. Food Sci.* **2013**, *78* (2), M290–M296.

Osman, K.; Evangelopoulos, D.; Basavannacharya, C.; Gupta, A.; Bhakta, S.; Gibbons, S. An Antibacterial from *Hypericum acmosepalum* Inhibits ATP-Dependent MurE Ligase from *Mycobacterium tuberculosis*. *Int. J. Antimicrob. Ag.* **2012**, *39* (2), 124–129.

Oussalah, M.; Caillet, S.; Lacroix, M. Mechanism of Action of Spanish Oregano, Chinese Cinnamon, and Savory Essential Oils against Cell Membranes and Walls of *Escherichia coli* O157:H7 and *Listeria monocytogenes*. *J. Food Protect.* **2006**, *69* (5), 1046–1055.

Oussalah, M.; Caillet, S.; Saucier, L.; Lacroix, M. Inhibitory Effects of Selected Plant Essential Oils on the Growth of Four Pathogenic Bacteria: *E. coli* O157:H7, *Salmonella typhimurium*, *Staphylococcus aureus* and *Listeria monocytogenes*. *Food Control* **2007**, *18* (5), 414–420.

Özek, G.; Demirci, F.; Özek, T.; Tabanca, N.; Wedge, D. E.; Khan, S. I.; Başer, K. H. C.; Duran, A.; Hamzaoglu, E. Gas Chromatographic–Mass Spectrometric Analysis of Volatiles Obtained by Four Different Techniques from *Salvia rosifolia* sm., and Evaluation for Biological Activity. *J. Chromat. A.* **2010**, *1217*, 741–748.

Paraschos, S.; Mitakou, S. L.; Skaltsounis, A. Chios Gum Mastic: A Review of Its Biological Activities. *Curr. Med. Chem.* **2012**, *19* (14), 2292–2302.

Pastor, J.; García, M.; Steinbauer, S.; Setzer, W. N.; Scull, R.; Gille, L.; Monzote, L. Combinations of Ascaridole, Carvacrol, and Caryophyllene Oxide against *Leishmania*. *Acta Trop.* **2015**, *145*, 31–38.

Pei, R. S.; Zhou, F.; Ji, B. P.; Xu, J. Evaluation of Combined Antibacterial Effects of Eugenol, Cinnamaldehyde, Thymol, and Carvacrol against *E. coli* with an Improved Method. *J. Food Sci.* **2009**, *74* (7), M379–M383.

Perez-Montaño, J. A.; Gonzalez-Aguilar, D.; Barba, J.; Pacheco-Gallardo, C.; Campos-Bravo, C.; Garcia, S.; Heredia, N. L.; Cabrera-Diaz, E. Frequency and Antimicrobial Resistance of *Salmonella* Serotypes on Beef Carcasses at Small Abattoirs in Jalisco State, Mexico. *J. Food Protect.* **2012**, *75* (5), 867–873.

Perricone, M.; Arace, E.; Corbo, M. R.; Sinigaglia, M.; Bevilacqua, A. Bioactivity of Essential Oils: A Review on their Interaction with Food Components. *Front. Microbiol.* **2015**, *6* (76), 1–7.

Perumalla, A. V. S.; Hettiarachchy, N. S. Green Tea and Grape Seed Extracts Potential Applications in Food Safety and Quality. *Food Res. Int.* **2011,** *44* (4), 827–839.

Pina-Pérez, M.; Martínez-López, A.; Rodrigo, D. Cocoa Powder as a Natural Ingredient Revealing an Enhancing Effect to Inactivate *Cronobacter sakazakii* Cells Treated by Pulsed Electric Fields in Infant Milk Formula. *Food Cont.* **2013,** *32* (1), 87–92.

Prabuseenivasan, S.; Jayakumar, M.; Ignacimuthu, S. *In Vitro* Antibacterial Activity of Some Plant Essential Oils. *BMC Complement. Altern. Med.* **2006,** *6* (1), 39.

Prakash, B.; Kedia, A.; Mishra, P. K.; Dubey, N. Plant Essential Oils as Food Preservatives to Control Moulds, Mycotoxin Contamination and Oxidative Deterioration of Agri-Food Commodities—Potentials and Challenges. *Food Cont.* **2015,** *47,* 381–391.

Regnier, T.; Combrinck, S.; Du Plooy, W. Essential Oils and Other Plant Extracts as Food Preservatives. In *Progress in Food Preservation*; Bhat, R., Alias, A. K., Paliyath, G., Eds.; Wiley-Blackwell: West Sussex, UK, 2012; pp 539–580.

Reichling, J.; Koch, C.; Stahl-Biskup, E.; Sojka, C.; Schnitzler, P. Virucidal Activity of a Beta-Triketone-Rich Essential Oil of *Leptospermum scoparium* (Manuka Oil) against HSV-1 and HSV-2 in Cell Culture. *Planta Med.* **2005,** *71* (12), 1123–1127.

Reichling, J.; Schnitzler, P.; Suschke, U.; Saller, R. Essential Oils of Aromatic Plants with Antibacterial, Antifungal, Antiviral, and Cytotoxic Properties—An Overview. *Forsch. Komplementärmed.* **2009,** *16* (2), 79–90.

Rodrigues, K. A. F.; Amorim, L. V.; Dias, C. N.; Moraes, D. F. C.; Carneiro, S. M. P.; de Amorim Carvalho, F. A. *Syzygium cumini* (L.) Skeels Essential Oil and Its Major Constituent α-Pinene Exhibit Anti-*Leishmania* Activity through Immunomodulation In Vitro. *J. Ethnopharmacol.* **2015,** *160,* 32–40.

Rojas-Graü, M. A.; Avena-Bustillos, R. J.; Olsen, C.; Friedman, M.; Henika, P. R.; Martín-Belloso, O.; Pan, Z.; McHugh, T. H. Effects of Plant Essential Oils and Oil Compounds on Mechanical, Barrier and Antimicrobial Properties of Alginate–Apple Puree Edible Films. *J. Food Eng.* **2007,** *81* (3), 634–641.

Roller, S.; Board, R. G. Naturally Occurring Antimicrobial Systems. In *Food Preservatives*; Russel, N. J., Gould, G. W., Eds.; Springer: Berlin-Heidelberg, 2003; pp 262–290.

Rosas-Taraco, A.; Sanchez, E.; García, S.; Heredia, N.; Bhatnagar, D. Extracts of *Agave americana* Inhibit Aflatoxin Production in *Aspergillus parasiticus. World Mycotoxin J.* **2011,** *4* (1), 37–42.

Rosato, A.; Vitali, C.; Laurentis, N. D.; Armenise, D.; Milillo, M. A. Antibacterial Effect of Some Essential Oils Administered Alone or In Combination with Norfloxacin. *Phytomed.* **2007,** *14,* 727–732.

Rosenblat, M.; Volkova, N; Coleman, R; Almagor, Y; Aviram, M. Antiatherogenicity of Extra Virgin Olive Oil and Its Enrichment with Green Tea Polyphenols in the Atherosclerotic Apolipoprotein-E-Deficient Mice: Enhanced Macrophage Cholesterol Efflux. *J. Nutr. Biochem.* **2008,** *19* (8), 514–523.

Ruiz-Navajas, Y.; Viuda-Martos, M.; Sendra, E.; Perez-Alvarez, J.; Fernández-López, J. *In Vitro* Antibacterial and Antioxidant Properties of Chitosan Edible Films Incorporated with *Thymus moroderi* or *Thymus piperella* Essential Oils. *Food Control* **2013,** *30* (2), 386–392.

Saïdana, D.; Mahjoub, M. A.; Boussaada, O.; Chriaa, J.; Cheraif, I.; Daami, M.; Mighri, Z.; Helal, A. N. Chemical Composition and Antimicrobial Activity of Volatile Compounds of *Tamarix boveana* (Tamaricaceae). *Microbiol. Res.* **2007,** *163,* 445–455.

Sánchez, E.; Dávila-Aviña, J.; Castillo, S. L.; Heredia, N.; Vázquez-Alvarado, R.; García, S. Antibacterial and Antioxidant Activities in Extracts of Fully Grown Cladodes of 8 Cultivars of Cactus Pear. *J. Food Sci.* **2014,** *79* (4), M659–M664.

Sánchez, E.; García, S.; Heredia, N. Extracts of Edible and Medicinal Plants Damage Membranes of *Vibrio cholerae*. *Appl. Environ. Microb.* **2010,** *76* (20), 6888–6894.

Sánchez, E.; Heredia, N.; Camacho-Corona, M. R.; García, S. Isolation, Characterization and Mode of Antimicrobial Action against *Vibrio cholerae* of Methyl Gallate Isolated from *Acacia farnesiana*. *J. Appl. Microbiol.* **2013,** *115* (6), 1307–1316.

Sánchez, G.; Aznar, R. Evaluation of Natural Compounds of Plant Origin for Inactivation of Enteric Viruses. *Food Environ. Virol.* **2015,** *7,* 183–187.

Santoro, G. F.; Cardoso, M. G.; Guimarães, L. G. L.; Mendonça, L. Z.; Soares, M. J. *Trypanosoma cruzi*: Activity of Essential Oils from *Achillea millefolium* L., *Syzygium aromaticum* L. and *Ocimum basilicum* L. on Epimastigotes and Trypomastigotes. *Exp. Parasitol.* **2007a,** *116* (3), 283–290.

Santoro, G. F.; das Graças Cardoso, M.; Guimarães, L. G. L.; Salgado, A. P. S.; Menna-Barreto, R. F.; Soares, M. J. Effect of Oregano (*Origanum vulgare* L.) and Thyme (*Thymus vulgaris* L.) Essential Oils on *Trypanosoma cruzi* (Protozoa: Kinetoplastida) Growth and Ultrastructure. *Parasitol. Res.* **2007b,** *100* (4), 783–790.

Schnitzler, P.; Koch, C.; Reichling, J. Susceptibility of Drug-Resistant Clinical *Herpes Simplex Virus* Type 1 Strains to Essential Oils of Ginger, Thyme, Hyssop, and Sandalwood. *Antimicrob. Agents Ch.* **2007,** *51* (5), 1859–1862.

Schuhmacher, A.; Reichling, J.; Schnitzler, P. Virucidal Effect of Peppermint Oil on the Enveloped Viruses Herpes Simplex Virus Type 1 and Type 2 In Vitro. *Phytomedicine* **2003,** *10* (6), 504–510.

Shan, B.; Cai, Y. Z.; Brooks, J. D.; Corke, H. The In Vitro Antibacterial Activity of Dietary Spice and Medicinal Herb Extracts. *Int. J. Food Microbiol.* **2007,** *117* (1), 112–119.

Shin, S.; Lim, S. Antifungal Effects of Herbal Essential Oils Alone and in Combination with Ketoconazole against *Trichophyton* spp. *J. Appl. Microbiol.* **2004,** *97* (6), 1289–1296.

Shiu, W. K. P.; Rahman, M. M.; Curry, J.; Stapleton, P.; Zloh, M.; Malkinson, J. P.; Gibbons, S. Antibacterial Acylphloroglucinols from *Hypericum olympicum*. *J. Nat. Prod.* **2011,** *75* (3), 336–343.

Silva, F.; Ferreira, S.; Duarte, A.; Mendonça, D. I.; Domingues, F. C. Antifungal Activity of *Coriandrum sativum* Essential Oil, Its Mode of Action against *Candida* Species and Potential Synergism with Amphotericin B. *Phytomedicine* **2011,** *19* (1), 42–47.

Silva, J.; Abebe, W.; Sousa, S.; Duarte, V.; Machado, M.; Matos, F. Analgesic and Anti-inflammatory Effects of Essential Oils of *Eucalyptus*. *J. Ethnopharmacol.* **2003,** *89* (2), 277–283.

Sivakumar, D.; Bautista-Banos, S. A Review on the Use of Essential Oils for Postharvest Decay Control and Maintenance of Fruit Quality During Storage. *Crop Prot.* **2014,** *64,* 27–37.

Smilanick, J.; Sorenson, D. Control of Postharvest Decay of Citrus Fruit with Calcium Polysulfide. *Postharvest Biol. Technol.* **2001,** *21* (2), 157–168.

Solórzano-Santos, F.; Miranda-Novales, M. G. Essential Oils from Aromatic Herbs as Antimicrobial Agents. *Curr. Opin. Biotechnol.* **2012,** *23* (2), 136–141.

Somolinos, M.; García, D.; Condón, S.; Mackey, B.; Pagán, R. Inactivation of *Escherichia coli* by Citral. *J. Appl. Microbiol.* **2010,** *108* (6), 1928–1939.

Sousa, R. M. O.; Rosa, J. S.; Silva, C. A.; Almeida, M. T. M.; Novo, M. T.; Cunha, A. C.; Fernandes-Ferreira, M. Larvicidal, Molluscicidal and Nematicidal Activities of Essential Oils and Compounds from *Foeniculum vulgare*. *J. Pest Sci.* **2015,** *88* (2), 413–426.

Suhr, K. I.; Nielsen, P. V. Antifungal Activity of Essential Oils Evaluated by Two Different Application Techniques against Rye Bread Spoilage Fungi. *J. Appl. Microbiol.* **2003,** *94* (4), 665–674.

Tajkarimi, M.; Ibrahim, S.; Cliver, D. Antimicrobial Herb and Spice Compounds in Food. *Food Control* **2010,** *21* (9), 1199–1218.

Teixeira, B.; Marques, A.; Ramos, C.; Neng, N. R.; Nogueira, J. M. F.; Saraiva, J. A.; Nunes, M. L. Chemical Composition and Antibacterial and Antioxidant Properties of Commercial Essential Oils. *Ind. Crop. Prod.* **2013,** *43*, 587–595.

Termentzi, A.; Fokialakis, N.; Leandros Skaltsounis, A. Natural Resins and Bioactive Natural Products Thereof as Potential Anitimicrobial Agents. *Curr. Pharm. Des.* **2011,** *17* (13), 1267–1290.

Tiwari, B. K.; Valdramidis, V. P.; O'Donnell, C. P.; Muthukumarappan, K.; Bourke, P.; Cullen, P. Application of Natural Antimicrobials for Food Preservation. *J. Agric. Food Chem.* **2009,** *57* (14), 5987–6000.

Tohidpour, A.; Sattari, M.; Omidbaigi, R.; Yadegar, A.; Nazemi, J. Antibacterial Effect of Essential Oils from Two Medicinal Plants Against Methicillin-Resistant *Staphylococcus aureus* (MRSA). *Phytomed.* **2010,** *17*, 142–145.

Tornuk, F.; Cankurt, H.; Ozturk, I.; Sagdic, O.; Bayram, O.; Yetim, H. Efficacy of Various Plant Hydrosols as Natural Food Sanitizers in Reducing *Escherichia coli* O157:H7 and *Salmonella typhimurium* on Fresh Cut Carrots and Apples. *Int. J. Food Microbiol.* **2011,** *148* (1), 30–35.

Ultee, A.; Kets, E. P.; Alberda, M.; Hoekstra, F. A.; Smid, E. J. Adaptation of the Food-Borne Pathogen *Bacillus cereus* to Carvacrol. *Arch. Microbiol.* **2000,** *174* (4), 233–238.

Upadhyay, A.; Upadhyaya, I.; Mooyottu, S.; Kollanoor-Johny, A.; Venkitanarayanan, K. Efficacy of Plant-Derived Compounds Combined with Hydrogen Peroxide as Antimicrobial Wash and Coating Treatment for Reducing *Listeria monocytogenes* on Cantaloupes. *Food Microbiol.* **2014,** *44*, 47–53.

Valeriano, C.; De Oliveira, T. L. C.; De Carvalho, S. M.; Cardoso, M. G.; Alves, E.; Piccoli, R. H. The Sanitizing Action of Essential Oil-Based Solutions against *Salmonella enterica* Serotype Enteritidis S64 Biofilm Formation on AISI 304 Stainless Steel. *Food Control* **2012,** *25* (2), 673–677.

Valtierra-Rodríguez, D.; Heredia, N. L.; García, S.; Sánchez, E. Reduction of *Campylobacter jejuni* and *Campylobacter coli* in Poultry Skin by Fruit Extracts. *J. Food Protect.* **2010,** *73* (3), 477–482.

Vergis, J.; Gokulakrishnan, P.; Agarwal, R.; Kumar, A. Essential Oils as Natural Food Antimicrobial Agents: A Review. *Crit. Rev. Food Sci.* **2015,** *5* (10), 1320–1323.

Waikedre, J.; Dugay, A.; Barrachina, I.; Herrenknecht, C.; Cabalion, P.; Fournet, A. Chemical Composition and Antimicrobial Activity of the Essential Oils from New Caledonian *Citrus macroptera* and *Citrus hystrix*. *Chem. Biodiv.* **2010,** *7*, 871–877.

Wu, Y.; Chen, Z.; Li, X.; Li, M. Effect of Tea Polyphenols on the Retrogradation of Rice Starch. *Food Res. Int.* **2009,** *42* (2), 221–225.

Yadav, M. K.; Chae, S. W.; Im, G. J.; Chung, J. W.; Song, J. J. Eugenol: A Phyto-Compound Effective against Methicillin-Resistant and Methicillin-Sensitive *Staphylococcus aureus* Clinical Strain Biofilms. *PLoS ONE* **2015,** *10* (3). DOI:10.1371/journal.pone.0119564.

Yoda, Y.; Hu, Z. Q.; Shimamura, T.; Zhao, W. H. Different Susceptibilities of *Staphylococcus* and Gram-Negative Rods to Epigallocatechin Gallate. *J. Infect. Chemother.* **2004,** *10* (1), 55–58.

CHAPTER 12

POLYAMINES FOR PRESERVING POSTHARVEST QUALITY

SWATI SHARMA[1], ANIL KUMAR SINGH[2],
SANJAY KUMAR SINGH[3], KALYAN BARMAN[2*],
SUNIL KUMAR[4], and VISHAL NATH[3]

[1]Division of Crop Production, ICAR-Indian Institute of Vegetable Research, Varanasi, Uttar Pradesh, India

[2]Department of Horticulture, Institute of Agricultural Sciences, Banaras Hindu University, Varanasi, Uttar Pradesh, India

[3]ICAR-National Research Centre on Litchi, Mushahari, Muzaffarpur, Bihar, India

[4]Department of Horticulture, North-Eastern Hill University, Tura, Meghalaya, India

*Corresponding author. E-mail: barman.kalyan@gmail.com

ABSTRACT

Polyamines (PAs) are polycationic aliphatic amines ubiquitously present in all living organisms. In plants, polyamines play vital role in myriad of physiological functions including, growth, development and biotic and abiotic stress responses. The most common polyamines present in the plants are putrescine, spermidine and spermine. Postharvest application of polyamines has been found effective in delaying ripening, senescence and extending shelf life of fruits and vegetables. Polyamine reduces the biosynthesis of ethylene by sharing the common precursor S-adenosyl methionine, required for production of ethylene. In addition, PAs also plays important role in maintaining firmness, reducing respiration rate, color changes, mechanical damage, chilling injury, etc. during postharvest storage of fresh fruits and vegetables. In this chapter, the influence of postharvest exogenous polyamines on fresh fruits and vegetables, safety and acceptance issues, and future line of work is discussed.

12.1   INTRODUCTION

Polyamines (PAs) have been referred to as "the Cinderellas of cell biology" by Dale Walters in 1987 (Edreva, 1996) indicating the need to elucidate understanding of their biological role and the underlying mechanisms. They are organic polycations with ubiquitous presence in all living organisms and vital role in a myriad of biological processes, namely, cell division, elongation, differentiation, flowering initiation, initiation of development of floral primordial, growth of floral buds, organ development, leaf senescence, fruit drop, fruit set, size, yield, quality, ripening, development of physiological disorders and abiotic stress responses to salt, and temperature (Tabor and Tabor, 1985; Crisosto et al., 1988a; Galston and Sawhney, 1990; Edreva, 1996; Kumar et al., 1997; Pandey et al., 2000; Tanguy, 2001; Malik and Singh, 2006; Takahashi and Kakehi, 2010). All plant organs vegetative as well as reproductive contain PAs. They are present in floral parts like pollen, stamens, pistil, leaves, stems, roots, and seeds. Xylem, phloem, and parenchyma tissues contain PAs (Vallee et al., 1983; Felix and Harr, 1987; Evans and Malmberg, 1989; Edreva, 1996). Endogenous PA levels and the flowering initiation and morphogenesis processes during developmental processes are associated with each other (Kakkar and Rai, 1993). However, specificity of the process has not been elucidated till now regarding which particular PAs conjugate with the flowering process or how the process is effectively controlled. PAs are aliphatic nitrogen containing polycations of low-molecular weight. The major plant PAs are putrescine (diamine), spermidine (triamine), and spermine (tetramine). They play a significant role in regulation of ripening and senescence. Homospermidine, 1,3-diaminopropane, cadaverine, and canavalmine are uncommon PAs which have been identified in plants, animals as well as algae and bacteria. The PA concentrations in plants are much higher than hormonal levels and de novo synthesis in plant cells indicate that PAs might be grouped with endogenous growth regulators rather than plant hormone. However, the status of PA as a hormone or endogenous growth regulator is still not established due to lack of clarity about the physiological significance of PA transport from synthesis to target locations (Edreva, 1996; Kumar et al., 1997).

Protective properties of PAs against senescence were first observed in oat protoplast cultures by Brenneman and Galston (1975). It reduces stress by scavenging the free radicals and reducing the reactive oxygen species (Drolet et al., 1986). PA treatment reduces mechanical injury symptoms and the rise in free PAs, spermidine, or spermine level may be a physiological marker of mechanical damage (Vicente et al., 2002). The ripening delay due to methyl jasmonate treatment in loquat, peach, and *Cucurbita pepo* fruits is concomitant with the upregulation of PAs (Wang and Buta, 1994; Ziosi et al., 2009; Cao et al., 2014). They suggested that the role of PAs is both as anti-stress and rejuvenating molecules. The antistress responses of PAs might be facilitated by nitric oxide production (Wimalasekera et al., 2011). Nonetheless, various studies have shown that the PAs have differential effect on fruit quality reliant on many factors (Bregoli et al., 2006). Keeping in mind the high postharvest losses in fresh horticultural produce, all around the world, several research studies are being conducted to identify potential technologies to augment their keeping quality. PAs being natural ubiquitous molecules can be considered as biologically safe and environmentally compatible technique for delaying senescence and enhancing keeping quality of fruits and vegetables. The polycationic nature of PAs is instrumental in stabilizing the membranes. The enhanced membrane stability, lower lipid peroxidation, inhibition of ethylene biosynthesis, enhanced antioxidant capacity, antistress response, and prevention of degrading enzymatic activities result in slowing down the ripening and deterioration process. This chapter tries to give an insight into the general introduction about PAs; how the preharvest and postharvest PAs treatments influence the size, yield, and keeping quality of fresh horticultural produce; what are the safety and acceptance issues in incorporating them into supply chain; and future line of study and concluding remarks.

12.2 BIOSYNTHESIS, ASSIMILATION, REGULATION, AND ROLE IN GROWTH, DEVELOPMENT, AND RIPENING

The simplified representation of pathways of PAs biosynthesis, the involved enzymes along with relationship with ethylene and *S*-adenosyl methionine (SAM), has been presented in Figure 12.1.

The diamine putrescine can be synthesized by ornithine decarboxylase (ODC), arginine decarboxylase, and putrescine synthase pathways (Slocum, 1991a; Palavan-Ünsal, 1995; Edreva, 1996). The principal pathway for PA biosynthesis in plants is by decarboxylation of arginine to agmatine (enzyme–arginine decarboxylase), intermediate *N*-carbomyl putrescine (enzyme-agmatine imino hydrolase), and then formation of putrescine (enzyme-*N*-carbamoyl putrescine amido hydrolase) (Edreva, 1996). ODC enzyme catalyzes the decarboxylation of ornithine to putrescine (Edreva, 1996). It has been reported that the ODC pathway plays a major role in cell division while arginine decarboxylase in maturation and response to stresses. The enzyme ODC is compartmentalized in cytoplasm and arginine decarboxylase in chloroplast (Serrano et al., 2016). Difluoromethylornithine and difluoromethylarginine are specific inhibitors of ODC and arginine decarboxylase enzymes. Spermidine and spermine are synthesized by addition of one aminopropyl group to putrescine and spermidine, respectively, catalyzed by enzymes spermidine synthase and spermine synthase. This aminopropyl group is provided by decarboxylated SAM (Slocum, 1991a; Edreva, 1996). Remarkably, this is the point which may have a critical role in balancing the levels of PAs and ethylene in plants thus either promoting or retarding ripening and senescence since SAM is precursor for the synthesis of ethylene hormone (Kushad and Dumbroff, 1991).

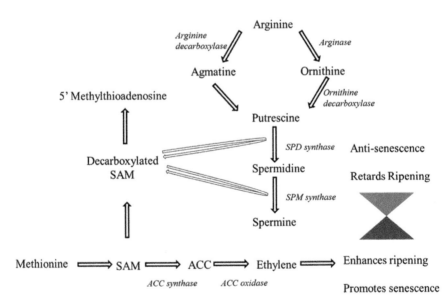

FIGURE 12.1   (See color insert.) Simplified schematic pathway of polyamines biosynthesis (Adapted from Pandey et al., 2000).

PAs are contained primarily in vacuoles in plant cells. They are also bound to nucleus, mitochondria, chloroplasts, ribosomes, cell walls, and membranes (Bagni, 1989; Bors et al., 1989; Evans and Malmberg, 1989; Bagni and Pistocchi, 1991; Slocum, 1991b; Edreva, 1996). Several experiments were conducted to understand the uptake and transport mechanism of PAs in plants (Slocum et al., 1984; Edreva, 1996). It was reported that

PA movement is bidirectional, studies confirming influx as well as efflux. It was noted that the influx movement requires energy and Ca^{2+} ions. It is also dependent on pH and hormonal status of the cells (Bagni and Pistocchi, 1991; Edreva, 1996). There are contrasting reports regarding the long-distance transport. Some workers like Young and Galston (1983) deny any such transport while Bagni and Pistocchi (1991) have obtained supportive results. It is noted that the translocation primarily occurs through the xylem and is transpiration dependent. However, it was also observed that translocation of PAs is rather a slow process and the quantity involved is not high (Bagni and Pistocchi, 1991; Edreva, 1996).

PAs interact with nucleic acids, proteins, and cell substructures, playing a vital role in several physiological functions related to plant growth and development (Kumar et al., 1997). PAs levels are low in dormancy while exogenous treatment of PAs results in breaking of dormancy (Serafini-Fracassini, 1991). Exogenous application of PAs has been reported to promote cell division (Biasi et al., 1988; Egea-Cortines and Mizrahi, 1991). Growth, ensuing from cell division and cell differentiation, has been correlated to PAs level (Bagni, 1989; Edreva, 1996). It plays a critical role in flower initiation, development, fertilization, fruit set, fruit growth, and ripening. Critical role of PAs associated with flower initiation, sex expression, pollen germination, pollen tube growth, ovule fertility, and proper morphology of reproductive organs has been stated in literature. Kaur-Sawhney et al. (1988) reported that PAs particularly spermidine induced flowering in vegetative buds while spermidine synthase inhibitor changed flowering phase to vegetative in tobacco PA mutants. Fall in PAs and their biosynthetic enzymes levels were recorded by Galston and Kaur-Sawhney (1987) during senescence progression which might be due to the senescence protective properties of PAs. Exogenous treatments of PAs retard senescence and the associated metabolic processes. The mechanisms fundamental to the antisenescence activities of PAs are reported to be principally based on the stabilization of membranes, reducing activity of cell-wall-degrading enzymes, inhibition of lipid peroxidation, reducing ethylene biosynthesis, slowing respiration, preventing chilling injury, and interactions with proteins and nucleic acids (Fig. 12.2). Romero et al. (2014) have reported that the perturbation of PA catabolism mechanism affects ripening of grape berries. The underlying mechanisms for PA action may be studied by employing genetic analysis of mutants with changed PA levels or different sensitivity to PAs. This will help in regulation of plant growth, development, ripening, and senescence in future.

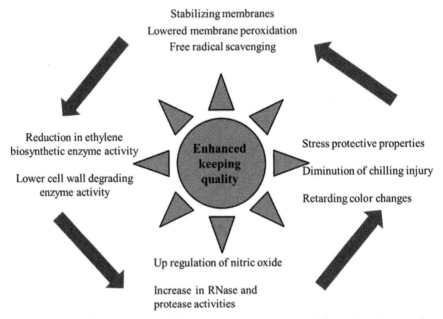

FIGURE 12.2   (See color insert.) Schematic diagram of mechanism of action of polyamines for enhanced keeping quality of horticultural produce.

12.3 MODULATION OF QUALITY BY PREHARVEST POLYAMINE SPRAYS

PAs are known to act as stress-protecting agents and increase cell division and morphogenesis (Liu et al., 2007). Their treatment has aided delaying maturing and ripening in grape (Ponce et al., 2002). PAs and aliphatic polycations have been reflected as antiethylene substances (Apelbaum et al., 1981). Exogenous preharvest application of PAs has been reported to improve fruit retention and yield in apple (Costa et al., 1986), olive (Rugini and Mencuccini, 1985), litchi (Stern and Gazit, 2000), and mango (Singh and Singh, 1995; Singh and Janes, 2000). The PAs treatments during preharvest stages have been reported to significantly influence the growth, development, retention, quality, and postharvest attributes of various fruits. Some of these have been elaborated in the following text.

PAs have been reported to be associated with the development of creasing, a peel-related physiological disorder in sweet orange (Hussain and Singh, 2015). Putrescine spray at fruit set or golf ball stage was reported to be more effective than other treatments by enhancing the rind thickness while the treatment with PA inhibitor methylglyoxal bis(guanylhydrazone) (MGBG) enhanced injury symptoms. Mora et al. (2005) studied the effects of putrescine treatments on fruit set in pears. They concluded that 1 mM putrescine enhanced fruit set in Japanese pear by stimulating pollen germination. Putrescine was employed as a treatment by Farag et al. (2007) to ascertain the effects on delaying peach fruit maturity and ripening. They observed that 10 mM putrescine delayed fruit maturity by 7 days, thus extending the harvesting and marketing period of peaches. Pear trees are shy bearers due to heavy flower and fruit drop. Crisosto et al. (1988a,b) applied putrescine at anthesis to know its effects on yield parameters of pear. They noted increase in fruit set, fruit size, weight, seeds, and yield. The increase in fruit set and yield by PA treatment has also been observed in apple (Costa and Bagni, 1983). They suggested that the enhanced cell division, ovule longevity, pollen germination and pollen tube growth rate might have increased fruit set. PA sprays after blooming in apple and olive enhanced fruit size during early growth period (Costa and Bagni, 1983; Rugini and Mencuccini, 1985). Increase in ovule longevity and extended effective pollination period was reported by Crisosto et al. (1986, 1992). PA treatments influenced ovule senescence, fertilization time, and fruit set in pear (Crisosto et al., 1988b). The ovule longevity and effective pollination period were extended significantly. The pollen tube growth rate was increased and they reached the micropyle earlier than in untreated flowers. They concluded that putrescine treatment may be useful in increasing pear yield (Crisosto et al., 1988b). Kassem et al. (2011) probed the effects of preharvest foliar sprays of PAs on jujube trees after fruit set. It was noted that putrescine spray delayed initial harvest date, increased fruit firmness, and maintained its quality. The PAs enhance the sink capacity of the fruit, stimulating photo-assimilates translocation and sugars accumulation, consequently reducing flower and fruit drop and increasing yield. Enhanced firmness in treated fruits was explained by ethylene biosynthesis inhibition, reducing cell-wall-degrading enzyme activity and higher cell-wall stability (Kassem et al., 2011). Marzouk and Kassem (2011) sprayed grape vines with putrescine at two berry development stages and veraison stage. Putrescine enhanced cluster and berry-quality characters like firmness, size, berry weight, berry adherence, and keeping quality. Putrescine was reported to delay initial harvest date than control. Berry shattering and percentage of unmarketable berries was reduced by putrescine treatment. Application of PA enhances cell-wall stability (Messiaen et al., 1997) and reduces polygalacturonase and pectin methyl esterase enzyme activities increasing rigidity and firmness (Valero, 1998; Romero et al., 2002; Vicente et al., 2002). Malik and Singh (2006) investigated the effects of PAs on fruit retention, yield, and fruit quality of mango, lemon, and other fruits. They found that spermine was more effective in increasing fruit retention and yield. The PAs application at final fruit stage retarded fruit skin color development, sugars, and total soluble solids (TSS) and increased fruit acidity. They concluded that the type and dose of PAs and stage of application influence their effects. The reduction in flower and fruit drop by exogenous application of PAs might be due to inhibition of ethylene biosynthesis, a well-known trigger in abscission (Brown, 1997) and enhanced levels of endogenous PAs. Putrescine sprays are reported to increase fruit retention in litchi and apple (Costa and Bagni, 1983; Mitra and Sanyal, 1990). Further, it improves development of floral organs, pollination, fertilization, embryo, and initial fruit development. PAs slowed color development of peach, lemon, and papaya fruits (Valero, 1998; Novita and Purvoko, 2004; Kaur et al., 2013). The effect of putrescine on keeping quality of tomato was studied (Law et al., 1991; Babu et al., 2014). The putrescine treatments reduced decay and ethylene production rates and enhanced fruit firmness. In strawberry, apricot, and sweet cherry fruits, PA treatment lowered ethylene production and prevented softening during storage (Khosroshahi and Ashari, 2008). Changes in fruit ripening patterns of nectarines by PA treatments were noted and enhancement in firmness, TSS, lessened ethylene production, and delayed abscission was observed (Torrigiani et al., 2004). They concluded that PAs can be used for extension of postharvest quality. Ali et al. (2014) planned studies on foliar application of putrescine on peach tree. They noted an increase in fruit weight, diameter, pulp, stone ratio, fruit firmness, and yield over control. El-Migeed et al. (2013) studied effect of PA sprays on date palm at bloom and recommended combination of potassium citrate and putrescine for improvement in fruit quantity and quality by increasing fruit set, retention, and fruit weight. It improved flesh/seed ratio, reducing sugars, and total sugars.

12.4 MODULATION OF QUALITY BY POSTHARVEST POLYAMINE TREATMENTS

PAs are vital for innumerable vital cell growth and development functions. As ripening hormone, ethylene, and endogenous growth regulators, PAs share a common precursor SAM for their biosynthesis. The tug for SAM, a common intermediate, and the inhibition of ethylene biosynthetic enzymes by PAs result in a balance between PAs and ethylene levels which determines the developmental, ripening, and senescence stage of fruit or vegetables (Li et al., 1992; Hong and Lee, 1996; Lee et al., 1997). PAs have been reported to inhibit ethylene biosynthetic enzymes, enhance firmness by stabilizing membranes, increase antioxidant capacity, diminution of chilling injury, inhibition of membrane lipid peroxidation, as antistress agents, and delay senescence. PAs actively scavenge the free radicals (Drolet et al., 1986). In several fruits and vegetables, authors have reported decline in PAs content with advancement in ripening (Hong and Lee, 1996). Among the PAs and other known senescence retardants, spermine has been reported to be more effective in delaying senescence and maintaining fruit quality (Apelbaum et al., 1981; Kaur-Sawhney and Galston, 1991). The effect of PAs on the quality of horticultural produce varies with the species, cultivar, ripening stage, dose of PAs, method of application, and duration of treatment among others. It has been reported to influence ethylene production rates, fruit quality, keeping quality, appearance of chilling injury symptoms, and content of bioactive compounds. PAs have been reported to enhance the keeping quality of both climacteric as well as nonclimacteric fruits, vegetables, and flowers (Mirdehghan et al., 2007; Cherian et al., 2014; Champa et al., 2015). Several studies have been conducted to ascertain and understand the effects of exogenous PAs treatment on quality and postharvest storage life of fresh horticultural produce. Fruit reaches its maximum edible quality at ripening after which deterioration and senescence processes deteriorate the quality. Prestorage exogenous PA treatment aids in diminution of the progression in senescence process with advancement in storage period and thus extends the keeping quality of the horticultural produce. Table 12.1 summarizes the findings of studies conducted on modulation of quality and extension of postharvest storage life of fresh horticultural produce by exogenous PAs treatment.

TABLE 12.1 Summarization of Research Inferences of Postharvest Polyamine Treatments on Quality of Various Horticultural Commodities.

PA treatment	Effects	Conclusion	Reference
Lemon			
Vacuum infiltration 1 mM PUT	Reduced mechanical deformation Higher fruit firmness	Increase in endogenous levels of SPM and ABA can be considered as physiological markers in lemons exposed to external mechanical stress	Romero et al. (1999)
Apricot			
PUT (1 mM)	Increased fruit firmness Reduced bruising zones Reduced RR and EE rates	Endogenous SPD levels can be considered as a physiological marker of mechanical stress	Romero et al. (2002)
PUT and SPD (1 mM)	Reduced EE, maintained fruit color and firmness	Exogenous PUT and SPD treatments delay ripening and alleviate chilling injury during low temperature storage	Koushesh-Saba et al. (2012)
Blueberry			
10 mM PUT and 1 mM SPM	Better fruit quality	Combination of PA with heat treatment did not improve quality	Basiouny (1996)
Papaya			
Immersion treatment 1, 3, and 10 mM PUT; SPD and SPM at 0.3, 1, and 3 mM	Delayed softening Delayed rise in TSS Delayed color change	Extended shelf-life was observed	Purwoko et al. (1998)
Mango			
Pressure infiltration PUT, SPD, and SPM (0.3 and 3 mM)	Delayed fruit softening Consistently lower PLW not recorded	Pressure infiltration of 0.72 kg/cm^2 for PA treatment inferred to be high	Purwoko et al. (1998)

TABLE 12.1 *(Continued)*

PA treatment	Effects	Conclusion	Reference
0.5, 1, 1.5 and 2 mM PUT	Reduced RR and EE Rates Higher fruit firmness, phenolics, antioxidants, and antioxidant enzyme activities Inhibited enzyme activities	2 mM PUT was reported best treatment for maintaining quality in mango	Razzaq et al. (2014)
PUT, SPD, and SPM at 0.25 and 0.5 mM	Reduced chilling injury	SPM (0.5 mM) was most effective in alleviating chilling injury	Nair and Singh (2004)
0.01, 0.5, and 1 mM PUT, SPD, and SPM	Delayed softening and color change Reduced PLW and ascorbic acid loss	0.5 mM SPM was highly effective in extending shelf-life without affecting fruit quality	Malik and Singh (2005)
Grape			
PUT and SPD (0.5, 1, and 1.5 mM)	Maintained firmness, phenolics content, TSS, TA, and berry color suppressed enzymatic activity Reduced electrolyte leakage and decay	Lowest dose was found to be most effective while highest dose (1.5 mM) of both PAs showed damaging effects	Champa et al. (2014)
PUT (1 and 2 mM)	Maintained higher total phenolics, catechin, total quercetin, and antioxidant activity	Postharvest 1 mM PUT + 1% chitosan treatment is effective in extending shelf-life and maintaining overall quality for up to 60 days of storage	Shiri et al. (2013)
Kiwifruit			
Immersion treatment SPD (1, 1.5, 2 mM) and SPM (0.5, 1, 1.5 mM)	Higher ascorbic acid content Low PLW and EE rates	SPM (1.5 mM) was adjudged the best treatment which increased the shelf-life up to 15 days	Jhalegar et al. (2012a)

TABLE 12.1 *(Continued)*

PA treatment	Effects	Conclusion	Reference
SPM (0.5, 1, and 1.5 mM) and SPD (1, 1.5, and 2 mM)	Low RR Low polygalacturonase and lipoxygenase enzyme activities	Under ambient storage conditions, SPM (1.5 mM) and SPD (2 mM) maintained best fruit quality	Jhalegar et al. (2012b)
SPM (0.5, 1, and 1.5 mM), SPD (1, 1.5, and 2 mM)	Suppressed RR and EE rates Lowered polygalacturonase and lipoxygenase enzyme activities	1.5 mM SPM and 2 mM SPD was highly effective in extending shelf-life of kiwifruit	Jhalegar et al. (2011)
Plum			
0.1, 1, and 2 mM PUT	Delayed fruit softening Reduced EE rates Lower ethylene biosynthesis enzyme activities Reduced fruit softening enzyme activities	2 mM PUT treatment was most effective	Khan et al. (2007)
1 mM PUT	Reduction in EE rates Maintained higher fruit firmness Lower PLW Delayed color changes	Exogenous PUT is effective for enhancing shelf-life of plums. Opposing trend between ethylene and polyamine pathways was inferred	Serrano et al. (2003)
Pressure infiltration 1 mM PUT	Reduced mechanical damage Increased firmness Reduced RR AND EE rates	Free endogenous SPD might be a physiological marker of mechanical stress	Vicente et al. (2002)
Pomegranate			
PUT 2 mM	Reduced CI Better fruit quality Lower fruit softening, RR and EE rates	PUT + carnauba wax treatment effectively alleviates CI and exhibits extended postharvest life for up to 60 days	Barman et al. (2011)
PUT 2 mM	Retained higher anthocyanin, ascorbic acid, antioxidant activity, and better sensory quality	PUT + carnauba wax pretreatment was most effective in preserving functional and sensory quality up to 60 days of storage	Barman et al. (2014)

TABLE 12.1 *(Continued)*

PA treatment	Effects	Conclusion	Reference
PUT, SPD 1 mM Pressure infiltration and immersion	Alleviation of CI Reduced softening and PLW	Pressure infiltration was suggested to be better for beneficial effects in pomegranate	Mirdehghan et al. (2007)
Litchi			
1 mM/L PUT, SPD, and SPM	Lowered browning, peroxide level, EE rate, and cell leakage	Every PA treatment increased the levels of other two PA levels SPM was reported as most effective Application of SPM or SPD (1 mM/L) can slow deterioration of litchi fruit	Jiang and Chen (1995)
Zucchini			
1 mM PUT, SPD, and SPM	Better cold tolerance Lowered PLW and CI	PUT initiated stress defense mechanisms which resulted in improved postharvest quality	Palma et al. (2015)
PUT, SPD, and SPM (0.1, 0.25, 0.5, 2, and 4 mM)	Alleviated CI Reduced ion leakage and polygalacturonase activity	0.25 and 0.5 mM SPD was most effective treatment in alleviating chilling injury and extending shelf-life	Martínez-Téllez et al. (2002)
Strawberry			
(0.3, 0.5, 1, and 2 mM) PUT	Increase in fruit firmness No significant PLW Lowered EE rates	Better fruit quality in 2-mM-treated fruit	Khosroshahi et al. (2007)
Pistachio shoots			
PUT (0.1 and 1 mM), SPD (0.1 and 1 mM), and SPM (0.1 and 1 mM)	SPM but not PUT decreased the physiological disorders and increased the yield	SPM dominantly played an important role in the growth and development of pistachio nuts	Khezri et al. (2010)
Carnation			
PUT and SPM	Did not delay senescence Higher EE production rates	PA treatment does not always increase the flower longevity and may even result in an accelerated senescence	Downs and Lovell (1986)

TABLE 12.1   *(Continued)*

PA treatment	Effects	Conclusion	Reference
SPM	Reduced EE rates	Delayed senescence of cut carnation flowers	Lee et al. (1997)
	Lower ACC content		
	Lower ACC oxidase and ACC synthase activities		
Apple			
SPD (0.25 mM), SPM (1 mM)	Increased fruit firmness	Delayed softening	Kramer et al. (1991)
		Inhibited chilling injury development	
Peach			
PUT 1 mM	Increased fruit firmness	Delayed ripening	Martinez-Romero et al. (2000)
	Reduced RR and EE rates	Reduced mechanical stress	
	Susceptible to mechanical injury		
Muskmelon			
SPD and SPM (0.25 and 0.5 M)	Delayed color change	Extended marketability	Lester (2000)
	Lower membrane peroxidation and LOX activity		

CI, chilling injury; EE, ethylene evolution; RR, respiration rate; SPD, spermidine; SPM, spermine; PA, polyamine; PUT, putrescine.

12.5   ACCEPTANCE AND SAFETY ISSUES

PAs are endogenous growth regulators involved in innumerable growth and developmental processes in biological world. They are ubiquitously present in all living organisms, namely, bacteria, plants, and mammals (Kalac, 2014). It is a naturally occurring molecule, which is synthesized endogenously and metabolized by cells for various vital cell functions. Keeping this in mind, the exogenous use of PAs for maintaining the postharvest quality of fresh horticultural produce can be considered a biologically and environmentally safe and acceptable alternative to chemical means. The PAs such as putrescine, spermidine, and spermine are commonly found in food, both plant and animal sources. Their postharvest application does increase the endogenous concentration; however, the levels generally remain far lesser than the damaging harmful or toxic levels (Bardocz, 1995; Kalac, 2014; Serrano et al., 2016). The oral acute toxicity of putrescine, spermidine, and spermine was determined in rats as 2000, 600, and 600 mg/kg, respectively. The no-observed adverse-effect concentrations were 180, 83, and 19 mg/kg body weight, respectively (Kalac, 2014). These are extremely high levels of PAs and dietary intakes cannot be supposed. PAs are vital for cellular growth, maintenance, and function. They are endogenously synthesized in human body, produced by intestinal bacteria as well as obtained from dietary sources. However, the daily recommended dietary intake for individual PAs has not been suggested till date and only limited information is available on PA contents in different foods (Kalac et al., 2005). Kalac (2014) reported that foods with high PA contents should be avoided by patients with tumors (Table 12.2) while higher intake of dietary PAs is beneficial during wound healing, postoperational recovery (except for tumors), liver regeneration, etc. (Bardocz, 1995). They have antioxidant property and also help in building up immunity and preventing allergies. Breast milk is quite rich in PAs, which might be the reason for protecting infants from various allergies. In general, dietary PAs at levels normally present in food are nontoxic (Bardocz, 1995). Nevertheless, there have also been some apprehensions regarding the source of commercial extraction of PAs for use in postharvest management of fresh horticultural produce. Some scholars are of the view that if any animal source is used for PAs extraction, then it might not be accepted by the people having vegetarian diet preference. However, PAs are ubiquitously present in all living organisms and have been commercially extracted from plant sources like leaves and stems of corn (*Zea mays* L.), cucumber (*Cucumis sativus* L.), oat (*Avena sativa* L.), and radish (*Raphanus sativus* L.) (Asrey et al., 2008). Further, the application of PAs for maintenance of postharvest quality can easily be integrated into supply chain since the treatment application methods are as simple as immersion treatments.

TABLE 12.2 Fruit and Vegetables Rich in Polyamines Which Should Be Avoided in Patients with Tumors (Source: Kalac, 2014).

Horticultural commodity	Rich source of polyamine
Orange and related fruits and their juices	PUT
Passion fruit	SPM
Pear	SPD
Cauliflower	SPD
Broccoli	SPD

SPD, spermidine; SPM, spermine; PUT, putrescine.

12.6 SCOPE OF COMMERCIALIZATION

Presently, PAs are not being used at a commercial scale for postharvest management of horticultural produce as far as our knowledge goes. Plausible scope in future depends on many factors like commercial source of extraction, degree of beneficial influences on fruits and vegetables, their levels in food, application methods, regulatory laws, and consumer acceptance. A patent was filed in United States proposing use of PAs for enhancing shelf-life and maintaining fruit quality (Law et al., 1988; Serrano et al., 2016). However, any further significant development has since not taken place including PA treatment in the supply chain. Europe does not have any particular regulations regarding postharvest use of PAs for postharvest management. Nevertheless, PAs is a biologically acceptable, eco-friendly technique of enhancing fresh horticultural produce fruit quality and thus it may find its place as an effective biological molecule in postharvest management supply chain and food industry.

12.7 FUTURE PROSPECTS

Further studies are required to decisively establish the status of PAs in plant biology whether they are secondary messengers, factors correlated to growth and development or hormone like factors. Several mechanism(s) of action of PAs, their perception, signal transduction, and regulatory pathways are still in the dark, which need to be elucidated. In the future, steps may be taken for genetic manipulation of PA biosynthetic genes to have new varieties with extended postharvest life and improved fruit quality. There is need to clarify the relationship and mode of action of auxin, abscisic acid, ethylene, and nitric oxide in with relation to PAs. The mechanism and effects of preharvest treatments of PAs on postharvest fruit quality needs to be elucidated with clarity. There is need to document the changes in endogenous levels of PAs content during processing and storage.

12.8 CONCLUSION

PAs play a key role in a myriad of growth and developmental processes in plants including cell division, elongation, organ development, flowering, fertilization, fruit development, stress tolerance, ripening, and senescence. The ubiquitous presence of PAs in all the living organisms is an indicator of their significance for the biological world. PAs have a very significant role for postharvest management of horticultural crops since they act as antisenescence agents. By modulating the biosynthesis of ethylene, reducing the activities of cell-wall-degrading and ethylene biosynthetic enzymes, upregulating the antioxidant enzymes, and maintaining the firmness and texture, they help in delaying ripening and senescence. Yet, we should keep in mind that PAs effects vary depending upon several factors like species, cultivar, stage, dose, and mode of application. Even so, it is accepted as an endogenous growth regulators and therefore we must understand that for all dependent physiological functions, it is not that high level of PAs are required but rather a balanced level tilted in favor of delay in ripening and senescence processes of fresh horticultural produce to serve our aim of reducing postharvest losses and maintaining quality.

KEYWORDS

- polyamines
- postharvest
- ethylene
- senescence
- shelf life

REFERENCES

Ali, I.; Abbasi, N. A.; Hafiz, I. A. Physiological Response and Quality Attributes of Peach Fruit cv. Florida King as Affected by Different Treatments of Calcium Chloride, Putrescine and Salicylic Acid. *Pak. J. Agric. Sci.* **2014,** *51* (1), 33–39.

Apelbaum, A.; Burgoon, A. C.; Anderson, J. D.; Lieberman, M. Polyamines Inhibit Biosynthesis of Ethylene in Higher Plant Tissue and Fruit Protoplasts. *Plant Physiol.* **1981,** *68,* 239–247.

Asrey, R.; Sasikala, C.; Barman, K.; Koley, T. K. Advances in Postharvest Treatments of Fruits—A Review. *Ann. Hortic.* **2008,** *1* (1), 1–10.

Babu, R.; Singh, K.; Jawandha, S. K.; Alam, M. S.; Jindal, S. K.; Khurana, D. S.; Narsaiah, K. Effect of Pre-harvest Spray of Putrescine on Shelf Life and Quality of Tomato During Storage. *Int. J. Adv. Res.* **2014,** *2* (10), 861–865.

Bagni, N. Polyamines in Plant Growth and Development. In *The Physiology and Biochemistry of Polyamines*; Bachrach, U., Heimer, Y., Eds.; CRC Press: Boca Raton, FL, 1989; Vol. II, pp 107–120.

Bagni, N.; Pistocchi, R. Uptake and Transport of Polyamines and Inhibitors of Polyamine Metabolism in Plants. In *Biochemistry and Physiology of Polyamines in Plants*; Slocum, R., Flores, H., Eds.; CRC Press: Boca Raton, FL, 1991; pp 105–120.

Bardocz, S. Polyamines in Food and their Consequences for Food Quality and Human Health. *Trends Food Sci. Technol.* **1995**, *6*, 341–346.

Barman, K.; Asrey, R.; Pal, R. K. Putrescine and Carnauba Wax Pretreatments Alleviate Chilling Injury, Enhance Shelf Life and Preserve Pomegranate Fruit Quality During Cold Storage. *Sci. Hortic.* **2011**, *130*, 795–800.

Barman, K.; Asrey, R.; Pal, R. K.; Kaur, C.; Jha, S. K. Influence of Putrescine and Carnauba Wax on Functional and Sensory Quality of Pomegranate (*Punica granatum* L.) Fruits During Storage. *J. Food Sci. Technol.* **2014**, *51* (1), 111–117.

Basiouny, F. M. Blueberry Fruit Quality and Storability Influenced by Postharvest Application of Polyamines and Heat Treatments. *Proc. Fla. State Hortic. Soc.* **1996**, *109*, 269–272.

Biasi, R.; Bagni, N.; Costa, G. Endogenous Polyamines in Apple and their Relationship to Fruit Set and Fruit Growth. *Physiol. Plant* **1988**, *73*, 201–205.

Bors, W.; Langebartels, C.; Michel, C.; Sanderman, J. H. Polyamines as Radical Scavengers and Protectants against Ozone Damage. *Phytochemistry* **1989**, *28*, 1589–1596.

Bregoli, A. M.; Ziosi, V.; Biondi, S.; Claudio, B; Costa, G.; Torrigiani, P. A Comparison between Intact Fruit and Fruit Explants to Study the Effect of Polyamines and Aminoethoxyvinylglycine (AVG) on Fruit Ripening in Peach and Nectarine (*Prunus persica* L. Batch). *Postharvest Biol. Technol.* **2006**, *42*, 31–40.

Brenneman, F.; Galston, A. Experiments on the Cultivation of Protoplasts and Calli of Agriculturally Important Plants. I. Oat (*Avena sativa* L.). *Biochem. Physiol. Pflanzen* **1975**, *168*, 453–471.

Brown, K. M. Ethylene and Abscission. *Physiol. Plant* **1997**, *100*, 567–576.

Cao, S.; Cai, Y.; Yang, Z.; Joyce, D. C.; Zheng, Y. Effect of MeJA Treatment on Polyamine, Energy Status and Anthracnose Rot of Loquat Fruit. *Food Chem.* **2014**, *145*, 86–89.

Champa, W. A. H.; Gill, M. I. S.; Mahajan, B. V. C.; Arora, N. K. Postharvest Treatment of Polyamines Maintains Quality and Extends Shelf-Life of Table Grapes (*Vitis vinifera* L.) cv. Flame Seedless. *Postharvest Biol. Technol.* **2014**, *91*, 57–63.

Champa, W. A. H.; Gill, M. I. S.; Mahajan, B. V. C.; Bedi, S. Exogenous Treatment of Spermine to Maintain Quality and Extend Postharvest Life of Table Grapes (*Vitis vinifera* L.) cv. Flame Seedless under Low Temperature Storage. *LWT—Food Sci. Technol.* **2015**, *60*, 412–419.

Cherian, S.; Figueroa, C. R.; Nair, H. 'Movers and Shakers' in the Regulation of Fruit Ripening: A Cross-Dissection of Climacteric versus Non-climacteric Fruit. *J. Exp. Bot.* **2014**, *65*, 4705–4722.

Costa, G.; Bagni, N. Effect of Polyamines on Fruit Set of Apple. *Hortic. Sci.* **1983**, *18*, 59–61.

Costa, G.; Biasi, R.; Bagni, N. Effect of Putrescine on Fruiting Performance on Apple (cv. Hi Early). *Acta Hortic.* **1986**, *149*, 189–195.

Crisosto, C. H.; Lombard, P. B.; Richardson, D. G.; Tetley, R. Putrescine extends Effective Pollination Period in 'Comice' Pear (*Pyrus communis* L.) Irrespective of Post-anthesis Ethylene Levels. *Sci. Hortic.* **1992**, *49* (3–4), 211–221.

Crisosto, C. H.; Lombard, P. B.; Sugal, D.; Polito, V. S. Putrescine Influences Ovule Senescence, Fertilization Time and Fruit Set in 'Comice' Pear. *J. Am. Soc. Hortic. Sci.* **1988a**, *113* (5), 708–712.

Crisosto, C. H.; Sugar, D.; Lombard, P. B. Effect of Putrescine Sprays at Anthesis on 'Comice' Pear Yield Components. *Adv. Hortic. Sci.* **1988b**, *2*, 27–29.

Crisosto, C. H.; Vasilakakis, M.; Lombard, P. B.; Richardson, D. G.; Tetley, R. Effect of Ethylene Inhibitors on Fruit Set, Ovule Longevity, and Polyamine Levels in 'Comice' Pear. *Acta Hortic.* **1986**, *179*, 229–236.

Downs, G. C.; Lovell, P. H. The Effect of Spermidine and Putrescine on the Senescence of Cut Carnations. *Physiol. Plant* **1986**, *66*, 679–684.

Drolet, G.; Dumbroff, E. B.; Legge, R. L.; Thompson, J. E. Radical Scavenging Properties of Polyamines. *Phytochemistry* **1986**, *25* (2), 367–371.

Edreva, A. Polyamines in Plants. *J. Plant Physiol.* **1996**, *22* (1–2), 73–101.

Egea-Cortines, M.; Mizrahi, Y. Polyamines in Cell Division, Fruit Set and Development, and Seed Germination. In *Biochemistry and Physiology of Polyamines in Plants*; Slocum, R., Flores, H., Eds.; CRC Press: Boca Raton, FL, 1991; pp 143–158.

El-Migeed, M. M. M. A.; Mostafa, E. A.; Ashour, N. E.; Hassan, H. A. S.; Mohamed, D. M.; Saleh, M. M. S. Effect of Potassium and Polyamine Sprays on Fruit Set, Fruit Retention, Yield and Fruit Quality of Amhat Date Palm. *Int. J. Agric. Res.* **2013**, *8* (2), 77–86.

Evans, P.; Malmberg, R. Do Polyamines Have Roles in Plant Development? *Ann. Rev. Plant Physiol. Plant Mol. Biol.* **1989**, *40*, 235–269.

Farag, K. M.; Ismail, A. A.; Essa, A. A.; El-Sabagh, A. S. Effect of Putrescine, Gibberellic Acid and Calcium on Quality Characteristics and Maturity Delay of "Desert Red" Peach Fruit Cultivar. B: Chemical Properties of the Fruit. *J. Agric. Environ. Sci. Alex Univ., Egypt* **2007**, *6* (1), 35–66.

Felix, H.; Harr, J. Association of Polyamines to Different Parts of Various Plant Species. *Physiol. Plant* **1987**, *71*, 245–250.

Galston, A.; Kaur-Sawhney, R. Polyamines and Senescence in Plants. In *Plant Senescence: Its Biochemistry and Physiology*; Thompson, W., Nothnagel, E., Huffaker, R., Eds.; Amer. Soc. Plant Physiol.: Rockville, MD, 1987; pp 167–181.

Galston, A. W.; Sawhney, R. K. Polyamines in Plant Physiology. *Plant Physiol.* **1990**, *94*, 406–410.

Hong, S. J.; Lee, S. K. Changes in Endogenous Putrescine and the Relationship to the Ripening of Tomato Fruits. *J. Korean Soc. Hortic. Sci.* **1996**, *37*, 369–373.

Hussain, Z.; Singh, Z. Involvement of Polyamines Increasing of Sweet Orange [*Citrus sinensis* (L.) Osbeck] Fruit. *Sci. Hortic.* **2015**, *190*, 203–210.

Jhalegar, M. J.; Sharma, R. R.; Pal, R. K.; Arora, A.; Dahuja, A. Analysis of Physiological and Biochemical Changes in Kiwifruit (*Actinidia deliciosa* cv. Allison) after the Postharvest Treatment with 1-Methylcyclopropene. *J. Plant Biochem. Biotechnol.* **2011**, *20* (2), 205–210.

Jhalegar, M. J.; Sharma, R. R.; Pal, R. K. Post-harvest Treatments of Polyamines Influence Shelf-Life and Quality of Kiwifruit (*Actinidia deliciosa*). *Indian J. Agric. Sci.* **2012a**, *82* (1), 81–84.

Jhalegar, M. J.; Sharma, R. R.; Pal, R. K.; Rana, V. Effect of Postharvest Treatments with Polyamines on Physiological and Biochemical Attributes of Kiwifruit (*Actinidia deliciosa*) cv. Allison. *Fruits* **2012b**, *67*, 13–22.

Jiang, Y. M.; Chen, F. A. Study on Polyamine Change and Browning of Fruit during Cold Storage of Litchi (*Litchi chinensis* Sonn.). *Postharvest Biol. Technol.* **1995**, *5*, 245–250.

Kakkar, R. K.; Rai, V. K. Plant Polyamines in Flowering and Fruit Ripening. *Phytochemistry* **1993**, *33* (6), 1281–1288.

Kalac, P. Health Effects and Occurrence of Dietary Polyamines: A review for the Period 2005–Mid-2013. *Food Chem.* **2014**, *161*, 27–39.

Kalac, P.; Krizek, M.; Pelikanov, T.; Langov, M.; Veskrn, O. Contents of Polyamines in Selected Foods. *Food Chem.* **2005**, *90*, 561–564.

Kassem, H. A.; Al-Obeed, R. S.; Ahmed, M. A.; Omar, A. K. H. Productivity, Fruit Quality and Profitability of Jujube Trees Improvement by Preharvest Application of Agro-chemicals. *Middle-East J. Sci. Res.* **2011**, *9* (5), 628–637.

Kaur, B.; Jawandha, S. K.; Singh, H.; Thakur, A. Effect of Putresine and Calcium on Colour Changes of Stored Peach Fruits. *Int. J. Ag. Environ. Biotechnol.* **2013**, *6* (2), 301–304.

Kaur-Sawhney, R.; Galston, A. Physiological and Biochemical Studies on the Antisenescence Properties of Polyamines in Plants. In *Biochemistry and Physiology of Polyamines in Plants*; Slocum, R., Flores, H., Eds.; CRC Press: Boca Raton, FL, 1991; pp 201–211.

Kaur-Sawhney, R.; Tiburcio, A.; Galston, A. Spermidine and Flower Bud Differentiation in Thin-Layer Explants of Tobacco. *Planta* **1988**, *173*, 282–284.

Khan, A. S.; Singh, Z.; Abbasi, N. A. Pre-storage Putresine Application Suppresses Ethylene Biosynthesis and Retards Fruit Softening during Low Temperature Storage in 'Angelino' Plum. *Postharvest Biol. Technol.* **2007**, *46* (1), 36–46.

Khezri, M.; Talaie, A.; Javanshah, A.; Hadavi, F. Effect of Exogenous Application of Free Polyamines on Physiological Disorders and Yield of 'Kaleh-Ghoochi' Pistachio Shoots (*Pistacia vera* L.). *Sci. Hortic.* **2010**, *125*, 270–276.

Khosroshahi, M. R. Z.; Ashari, E. M. Effect of Putrescine Application on Post-harvest Life and Physiology of Strawberry, Apricot, Peach and Sweet Cherry Fruits. *J. Sci. Technol. Agric. Nat. Res.* **2008**, *45*, 219–230.

Khosroshahi, M. R. Z.; Ashari, M. E.; Ershadi, A. Effect of Exogenous Putrescine on Post-harvest Life of Strawberry (*Fragaria ananassa* Duch.) Fruit, Cultivar Selva. *Sci. Hortic.* **2007**, *114*, 27–32.

Koushesh-Saba, M.; Arzani, K.; Barzegar, M. Postharvest Polyamine Application Alleviated Chilling Injury and Affects Apricot Storage Ability. *J. Agric. Food Chem.* **2012**, *60* (36), 8947–8953.

Kramer, G. F.; Wang, C. Y.; Conway, W. S. Inhibition of Softening by Polyamines Application in "Golden Delicious" and "McIntosh" Apples. *J. Am. Soc. Hortic. Sci.* **1991**, *116*, 813–817.

Kumar, A.; Altabella, T.; Taylor, M. A.; Tiburcio, A. F. Recent Advances in Polyamine Research. *Trends Plant Sci.* **1997**, *2* (4), 124–130.

Kushad, M.; Dumbroff, E. Metabolic and Physiological Relationships between the Polyamine and Ethylene Biosynthetic Pathways. In *Biochemistry and Physiology of Polyamines in Plants*; Slocum, R., Flores, H., Eds.; CRC Press: Boca Raton, FL, 1991; pp 77–92.

Law, D. M.; Davies, P. J.; Mutschler, M. A. Method of Extending Shelf Life and Enhancing Keeping Quality of Fruits. *US Patent 4,957,757*, 1988.

Law, D. M.; Davies, P. J.; Mutschler, M. A. Polyamines-Induced Prolongation of Storage in Tomato Fruits. *Plant Growth Reg.* **1991**, *10*, 283–290.

Lee, M. M.; Lee, S. H.; Park, K. Y. Effects of Spermine on Ethylene Biosynthesis in Cut Carnation (*Dianthus caryophyllus* L.) Flowers During Senescence. *J. Plant Physiol.* **1997**, *151*, 68–73.

Lester, G. E. Polyamines and their Cellular Anti-senescence Properties in Honey Dew Muskmelon Fruit. *Plant Sci.* **2000**, *160*, 105–112.

Li, N.; Parsons, B. L.; Liu, D.; Mattoo, A. K. Accumulation of Wound-Inducible ACC Synthase Transcript in Tomato Fruits is Inhibited by Salicylic Acid and Polyamines. *Plant Mol. Biol.* **1992**, *18*, 477–487.

Liu, J. H.; Kitashiba, H.; Wang, J.; Ban, Y.; Moriguchi, T. Polyamines and their Ability to Provide Environmental Stress Tolerance to Plants. *Plant Biotechnol.* **2007**, *24*, 117–126.

Malik, A. U.; Singh, Z. Improved Fruit Retention, Yield and Fruit Quality in Mango with Exogenous Application of Polyamines. *Sci. Hortic.* **2006**, *110*, 167–174.

Malik, A. U.; Singh, Z.; Khan, A. S. Role of Polyamines in Fruit Development, Ripening, Chilling Injury, Storage and Quality of Mango and Other Fruits: A Review. In *Proc. Int. Conf. Mango Date Palm: Cult. Export*, 2005; pp 182–197.

Martinez-Romero, D.; Valero, D.; Serrano, M.; Burlo, F.; Carbonell, A.; Burgos, L.; Riquelme, F. Exogenous Polyamines and Gibberellic Acid Effects on Peach (*Prunus persica* L.) Storability Improvement. *J. Food Sci.* **2000**, *65*, 288–294.

Martínez-Téllez, M. A; Ramos-Clamont, M. G.; Gardea, A. A.; Vargas-Arispuro, I. Effect of Infiltrated Polyamines on Polygalacturonase Activity and Chilling Injury Responses in Zucchini Squash (*Cucurbita pepo* L.). *Biochem. Biophys. Res. Commun.* **2002**, *295* (1), 98–101.

Marzouk, H. A.; Kassem, H. A. Improving Yield, Quality and Shelf Life of Thompson Seedless Grapevine by Preharvest Foliar Applications. *Sci. Hortic.* **2011**, *130*, 425–430.

Messiaen, J.; Cambier, P.; Cutsem, P. V. Polyamines and Pectins. *Plant Physiol.* **1997**, *113*, 387–395.

Mirdehghan, S. H.; Rahemid, M.; Castillo, S.; Romero, D. M.; Serrano, M.; Valero, D. Pre-storage Application of Polyamines by Pressure or Immersion Improves Shelf-Life of Pomegranate Stored at Chilling Temperature by Increasing Endogenous Polyamine Levels. *Postharvest Biol. Technol.* **2007**, *44* (1), 26–33.

Mitra, S. K.; Sanyal, D. Effect of Putrescine on Fruit Set and Fruit Quality of Litchi. *Gartenbauwissenschaft* **1990**, *55*, 83–84.

Mora, O. F.; Tanabe, K.; Itai, A. F. T. Effects of Putrescine Application on Fruit Set in 'Housui' Japanese Pear (*Pyrus pyrifolia* Nakai). *Sci. Hortic.* **2005**, *104*, 265–273.

Novita, T.; Purvoko, B. S. Role of Polyamine in Ripening of Solo Papaya Fruits (*Carica papaya* L.). *J. Stigma* **2004**, *11*, 78–81.

Palavan-Ünsal, N. Stress and Polyamine Metabolism. *Bulg. J. Plant Physiol.* **1995**, *21* (2–3), 3–14.

Palma, F.; Carvajala, F.; Ramosa, J. M.; Jamilena, M.; Garrido, D. Effect of Putrescine Application on Maintenance of Zucchini Fruit Quality during Cold Storage: Contribution of GABA Shunt and Other Related Nitrogen Metabolites. *Postharvest Biol. Technol.* **2015**, *99*, 131–140.

Pandey, S.; Ranade, S. A.; Nagar, P. K.; Kumar, N. Role of Polyamines and Ethylene as Modulators of Plant Senescence. *J. Biosci.* **2000**, *25* (3), 291–299.

Ponce, M. T.; Guinazú, M.; Tizio, R. Effect of Putrescine on Embryo Development in the Stenospermocarpic Grape cvs. Emperatriz and Fantasy. *Vitis* **2002**, *41*, 53–54.

Purwoko, B. S.; Kesmayanti, N.; Susanto, S.; Nasution, M. Z. Effect of Polyamines on Quality Changes in Papaya and Mango Fruits. *Acta Hortic.* **1998**, *464*.

Razzaq, K.; Khan, A. S.; Malik, A. U.; Shahid, M.; Ullah, S. Role of Putrescine in Regulating Fruit Softening and Antioxidative Enzyme Systems in 'Samar Bahisht Chaunsa' Mango. *Postharvest Biol. Technol.* **2014**, *96*, 23–32.

Romero, D. M.; Serrano, M.; Carbonell, A.; Burgos, L.; Riquelme, F.; Valero, D. Effects of Postharvest Putrescine Treatment on Extending Shelf Life and Reducing Mechanical Damage in Apricot. *J. Food Sci.* **2002**, *67* (5), 1706–1712.

Romero, D. M.; Valero, D.; Serrano, M.; Sanchez, F. M.; Riquelme, F. Effects of Post-harvest putrescine and Calcium Treatments on Reducing Mechanical Damage and Polyamines and Abscisic Acid Levels during Lemon Storage. *J. Sci. Food Agric.* **1999**, *79*, 1589–1595.

Nair, S.; Singh, Z. Chilling Injury in Mango Fruit in Relation to Biosynthesis of Free Poly-amines. *J. Hort. Sci. Biotech.* **2004**, *79*, 515–522.

Romero, P. A.; Ali, K.; Choi, Y. H.; Sousa, L.; Verpoorte, R.; Tiburcio, A. F.; Fortes, A. M. Perturbation of Polyamine Catabolism Affects Grape Ripening of *Vitis vinifera* cv. Trinca-deira. *Plant Physiol. Biochem.* **2014**, *74*, 141–155.

Rugini, E.; Mencuccini, M. Increased Yield in Olives with Putrescine Treatment. *Hortic. Sci.* **1985**, *20*, 102–103.

Serafini-Fracassini, D. Cell Cycle-Dependent Changes in Plant Polyamine Metabolism. In *Biochemistry and Physiology of Polyamines in Plants*; Slocum, R., Flores, H., Eds.; CRC Press: Boca Raton, FL, 1991; pp 159–173.

Serrano, M.; Romero, D. M.; Guillen, F.; Valero, D. Effects of Exogenous Putrescine on Improving Shelf Life of Four Plum Cultivars. *Postharvest Biol. Technol.* **2003**, *30*, 259–271.

Serrano, M.; Zapata, P. J.; Romero, D. M.; Diaz-Mula, H. M.; Valero, D. Polyamines as an Eco-Friendly Postharvest Tool to Maintain Fruit Quality. In *Eco-Friendly Technology for Postharvest Produce Quality*; Siddiqui, M. W., Ed.; Elsevier: Amsterdam, 2016; pp 219–242.

Shiri, M. A.; Ghasemnezhad, M.; Bakhshi, D.; Sarikhani, H. Effect of Posthar-vest Putrescine Application and Chitosan Coating on Maintaining Quality of Table Grape cv. "Shahroudi" during Long-Term Storage. *J. Food Proc. Preserv.* **2012**. DOI:10.1111/j.1745-4549.2012.00735.x.

Singh, Z.; Singh, L. Increased Fruit Set and Retention in Mango with Exogenous Application of Polyamines. *J. Hortic. Sci.* **1995**, *70*, 271–277.

Singh, Z.; Janes, J. Regulation of Fruit Set and Retention in Mango with Exogenous Appli-cation of Polyamines and their Biosynthesis Inhibitors. *Acta Hortic.* **2000**, *509*, 675–680.

Slocum, R. Polyamine Biosynthesis in Plants. In *Biochemistry and Physiology of Polyamines in Plants*; Slocum R., Flores, H., Eds.; CRC Press: Boca Raton, FL, 1991a; pp 23–40.

Slocum, R. Tissue and Subcellular Localization of Polyamines and Enzymes of Polyamine Metabolism. In *Biochemistry and Physiology of Polyamines in Plants*; Slocum, R., Flores, H., Eds.; CRC Press: Boca Raton, FL, 1991b, pp 93–103.

Slocum, R. D.; Sawhney, R. K.; Galston, A. W. The Physiology and Biochemistry of Poly-amines in Plants. *Arch. Biochem. Biophys.* **1984**, *235* (2), 283–303.

Stern, R. A.; Gazit, S. Application of the Polyamine Putrescine Increased Yield of 'Mauritius' Litchi (*Litchi chinensis* Sonn.). *J. Hortic. Sci. Biotechnol.* **2000**, *75*, 612–614.

Tabor, C.; Tabor, H. Polyamines. *Microbiol. Rev.* **1985**, *49*, 81–99.

Takahashi, T.; Kakehi, J. Polyamines: Ubiquitous Polycations with Unique Roles in Growth and Stress Responses. *Ann. Bot.* **2010**, *105*, 1–6.

Tanguy, J. M. Metabolism and Function of Polyamines in Plants: Recent Development (New Approaches). *Plant Growth Regul.* **2001**, *34*, 135–148.

Torrigiani, P.; Bregoli, A. M.; Ziosi, V.; Scaramagli, S.; Ciriaci, T.; Rasori, A.; Biondi, S.; Costa, G. Pre-harvest Polyamine and Aminoethoxyvinylglycine (AVG) Applications

23/10/2024

01777703-0009

Modulate Fruit Ripening in Stark Red Gold Nectarines (*Prunus persica* L. Batsch). *Postharvest Biol. Technol.* **2004,** *33*, 293–308.

Valero, D. Influence of Postharvest Treatment with Putrescine and Calcium on Endogenous Polyamines, Firmness, and Abscisic Acid in Lemon (*Citrus lemon* L. Burm cv. Verna). *J. Agric. Food Chem.* **1998,** *46*, 2102–2109.

Vallee, J. C.; Vansuyt, G.; Negrel, J.; Perdrizet, E.; Prevost, J. Mise en evidence d'amines liées a des structures cellulaires chez *Nicotiana tabacum* et *Lycopersicum esculentum*. *Physiol. Plant* **1983,** *57*, 143–148.

Vicente, A. P.; Romero, D. M.; Carbonell, A.; Serrano, M.; Riquelme, F.; Guillen, F.; Valero, D. Role of Polyamines in Extending Shelf Life and the Reduction of Mechanical Damage During Plum (*Prunus salicina* Lindl.) Storage. *Postharvest Biol. Technol.* **2002,** *25*, 25–32.

Wang, C. Y.; Buta, J. G. Methyl Jasmonate Reduces Chilling Injury in *Cucurbita pepo* through Its Regulation of Abscisic Acid and Polyamine Levels. *Environ. Exp. Bot.* **1994,** *34* (4), 427–432.

Wimalasekera, R.; Tebartz, F.; Scherer, G. F. E. Polyamines, Polyamine Oxidases and Nitric Oxide in Development, Abiotic and Biotic Stresses. *Plant Sci.* **2011,** *181*, 593–603.

Young, N.; Galston, A. Are Polyamines Transported in Etiolated Peas? *Physiol. Plant* **1983,** *73*, 912–914.

Ziosi, V.; Bregoli, A. M.; Fregola, F.; Costa, G.; Torrigiani, P. Jasmonate-Induced Ripening Delay Is Associated with Up-regulation of Polyamine Levels in Peach Fruit. *J. Plant Physiol.* **2009,** *166*, 938–946.

INDEX